Methods in Molecular Biology

Series Editor
John M. Walker
School of Life Sciences
University of Hertfordshire
Hatfield, Hertfordshire, AL10 9AB, UK

For further volumes:
http://www.springer.com/series/7651

Yeast Genetics

Methods and Protocols

Edited by

Jeffrey S. Smith

Department of Biochemistry and Molecular Genetics, University of Virginia School of Medicine, Charlottesville, VA, USA

Daniel J. Burke

Department of Biochemistry and Molecular Genetics, University of Virginia School of Medicine, Charlottesville, VA, USA

Humana Press

Editors
Jeffrey S. Smith
Department of Biochemistry
and Molecular Genetics
University of Virginia School of Medicine
Charlottesville, VA, USA

Daniel J. Burke
Department of Biochemistry
and Molecular Genetics
University of Virginia School of Medicine
Charlottesville, VA, USA

ISSN 1064-3745 ISSN 1940-6029 (electronic)
ISBN 978-1-4939-1362-6 ISBN 978-1-4939-1363-3 (eBook)
DOI 10.1007/978-1-4939-1363-3
Springer New York Heidelberg Dordrecht London

Library of Congress Control Number: 2014947342

Printed on acid-free paper

Humana Press is a brand of Springer
Springer is part of Springer Science+Business Media (www.springer.com)

Preface

Almost anyone working in a laboratory that utilizes *Saccharomyces cerevisiae* for their research projects has no doubt heard the phrase, "The awesome power of yeast genetics." This statement has long been a source of pride within the yeast research community. It refers to the simple nature of a single-cell eukaryotic organism, the relative ease of manipulating its genome, and the ability to interchangeably exist in both haploid and diploid states. Genes can be deleted, mutated, engineered, and tagged at will. Tetrad dissection of diploids is the workhorse of classical yeast genetics and remains a critical technique, complementing the widespread use of synthetic genetic array (SGA) technologies that can generate various mutant combinations in high throughput. Protocols for all of these methods are included in this book.

Saccharomyces cerevisiae has played a major role in the elucidation of multiple conserved cellular processes including MAP kinase signaling, splicing, transcription, and many others. With the advent of RNAi, the ability to reduce gene expression in higher eukaryotes has allowed some model organisms to "catch up" with the yeast system in terms of high-throughput analysis, and CRISPR technology is sure to push this even further in mammalian cells. However, the simplicity of the yeast system still makes it a highly attractive, and truly genetic powerhouse. This is especially true given the wide range of genome-wide resources that are readily available for analysis. Such collections simply do not exist for most other model organisms, at least not to such a degree. The general idea of this protocols book is to provide a balanced blend of classic and more modern genetic methods relevant to a wide range of research areas. It should be widely used as a reference in yeast labs.

Charlottesville, VA, USA

Jeffrey S. Smith
Daniel J. Burke

Contents

Contributors

DAVID C. AMBERG • *Department of Biochemistry and Molecular Biology, SUNY Upstate Medical University, Syracuse, NY, USA*

BRENDA J. ANDREWS • *The Department of Molecular Genetics, University of Toronto, Toronto, ON, Canada; Donnelly Center for Cellular and Biomedical Research, Toronto, ON, Canada; Banting and Best Department of Medical Research, Toronto, ON, Canada*

OSCAR M. APARICIO • *Molecular and Computational Biology Program, University of Southern California, Los Angeles, CA, USA*

LUIS ARAGÓN • *Cell Cycle Group, MRC Clinical Sciences Centre, Imperial College, London, UK*

JACQUELINE M. BARBER • *School of Biological Sciences, Victoria University of Wellington, Wellington, New Zealand*

ANASTASIA BARYSHNIKOVA • *The Department of Molecular Genetics, University of Toronto, Toronto, ON, Canada; Donnelly Center for Cellular and Biomedical Research, Toronto, ON, Canada*

DAVID S. BELLOWS • *School of Biological Sciences, Victoria University of Wellington, Wellington, New Zealand*

ANJALI BISARIA • *Lewis-Sigler Institute for Integrative Genomics, Princeton University, Princeton, NJ, USA*

CHARLES BOONE • *The Department of Molecular Genetics, University of Toronto, Toronto, ON, Canada; Donnelly Center for Cellular and Biomedical Research, Toronto, ON, Canada; Banting and Best Department of Medical Research, Toronto, ON, Canada*

SIMON D. BOURQUE • *Department of Biology, Concordia University, Montreal, QC, Canada*

DANIEL J. BURKE • *Department of Biochemistry and Molecular Genetics, University of Virginia School of Medicine, Charlottesville, VA, USA*

SANDRA CLAUDER-MÜNSTER • *Genome Biology Unit, EMBL, Heidelberg, Germany*

NAMAL V.C. COOREY • *School of Biological Sciences, Victoria University of Wellington, Wellington, New Zealand*

MICHAEL COSTANZO • *Donnelly Centre for Cellular and Biomolecular Research, Toronto, ON, Canada*

LIOR DAVID • *Department of Animal Sciences, R.H. Smith Faculty of Agriculture, Food and Environment, The Hebrew University of Jerusalem, Rehovot, Israel*

JOE R. DELANEY • *Department of Pathology, University of Washington, Seattle, WA, USA; The Molecular and Cellular Biology Program, University of Washington, Seattle, WA, USA*

JENNIFER E.G. GALLAGHER • *Department of Biology, West Virginia University, Morgantown, WV, USA*

JENNIFER M. GARDNER • *Stowers Institute for Medical Research, Kansas City, MO, USA*

R. DANIEL GIETZ • *Department of Biochemistry and Medical Genetics, University of Manitoba, Winnipeg, MB, Canada*

JINGYU GUO • *Department of Genetics, University of Alabama at Birmingham, Birmingham, AL, USA*

BRIAN HAARER • *Department of Biochemistry and Molecular Biology, SUNY Upstate Medical University, Syracuse, NY, USA*

MICHAEL HAMPSEY • *Department of Biochemistry and Molecular Biology, Robert Wood Johnson Medical School, Rutgers University, Piscataway, NJ, USA*
JOHN L. HARTMAN IV • *Department of Genetics, University of Alabama at Birmingham, Birmingham, AL, USA*
PASCAL HERSEN • *Laboratoire Matière et Systèmes Complexes, Unité Mixte de Recherche (UMR), Centre National de la Recherche Scientifique (CNRS), Université Paris Diderot, Paris, France*
SUE L. JASPERSEN • *Stowers Institute for Medical Research, Kansas City, MO, USA; Department of Molecular and Integrative Physiology, University of Kansas Medical Center, Kansas City, KS, USA*
JOSEPH M. JOHNSON • *Department of Biochemistry and Molecular Genetics, University of Virginia School of Medicine, Charlottesville, VA, USA*
MATT KAEBERLEIN • *Department of Pathology, University of Washington, Seattle, WA, USA; Institute of Aging Research, Guangdong Medical College, Dongguan, China*
PETER KAISER • *Department of Biological Chemistry, School of Medicine, University of California Irvine, Irvine, CA, USA*
JILL B. KEENEY • *Department of Biology, Juniata College, Huntingdon, PA, USA*
ANN L. KIRCHMAIER • *Department of Biochemistry and Purdue University Center for Cancer Research, Purdue University, West Lafayette, IN, USA*
SEPP D. KOHLWEIN • *Institute of Molecular Biosciences, BioTechMed Graz and BioImaging Graz, University of Graz, Graz, Austria*
ELENA KUZMIN • *The Department of Molecular Genetics, University of Toronto, Toronto, ON, Canada; Donnelly Center for Cellular and Biomedical Research, Toronto, ON, Canada*
PHILIPPE LEFRANÇOIS • *Department of Molecular, Cellular and Developmental Biology, Yale University, New Haven, CT, USA*
MICHAEL LICHTEN • *Laboratory of Biochemistry and Molecular Biology, Center for Cancer Research, National Cancer Institute, Bethesda, MD, USA*
RADHIKA MATHUR • *Department of Biological Chemistry, School of Medicine, University of California Irvine, Irvine, CA, USA*
MARIA D. MAYÁN • *Cell Cycle Group, MRC Clinical Sciences Centre, Imperial College, London, UK; Rheumatology Division, Cartilage Biology Research Group, INIBIC—Hospital Universitaro A. Coruña, Coruña, Spain*
MEGAN N. MCCLEAN • *Lewis-Sigler Institute for Integrative Genomics, Princeton University, Princeton, NJ, USA*
ANDREW MILLER • *Department of Biochemistry and Purdue University Center for Cancer Research, Purdue University, West Lafayette, IN, USA*
CHAD L. MYERS • *Department of Computer Science and Engineering, University of Minnesota-Twin Cities, Minneapolis, MN, USA*
LUCY F. PEMBERTON • *Department of Microbiology, Immunology and Cancer Biology, Center for Cell Signaling, University of Virginia, Charlottesville, VA, USA*
VINCENT R. RICHARD • *Department of Biology, Concordia University, Montreal, QC, Canada*
JOHN RODGERS • *Department of Genetics, University of Alabama at Birmingham, Birmingham, AL, USA*
LIAM D.P. SAMPSON • *School of Biological Sciences, Victoria University of Wellington, Wellington, New Zealand*
DAVID A. SCHNEIDER • *Department of Biochemistry and Molecular Genetics, University of Alabama at Birmingham, Birmingham, AL, USA*

SARA SHARIFPOOR • *Donnelly Centre for Cellular and Biomolecular Research, Toronto, ON, Canada*
BADRI NATH SINGH • *Department of Biochemistry and Molecular Biology, Robert Wood Johnson Medical School, Rutgers University, Piscataway, NJ, USA*
JEFFREY S. SMITH • *Department of Biochemistry and Molecular Genetics, University of Virginia School of Medicine, Charlottesville, VA, USA*
MICHAEL SNYDER • *Department of Genetics, Stanford University School of Medicine, Stanford, CA, USA*
LARS M. STEINMETZ • *Genome Biology Unit, EMBL, Heidelberg, Germany*
GEORGE L. SUTPHIN • *Department of Pathology, University of Washington, Seattle, WA, USA; The Molecular and Cellular Biology Program, University of Washington, Seattle, WA, USA*
VLADIMIR I. TITORENKO • *Department of Biology, Concordia University, Montreal, QC, Canada*
SANDRA K. VILLWOCK • *Molecular and Computational Biology Program, University of Southern California, Los Angeles, CA, USA*
MARK WEIR • *Department of Biology, Juniata College, Huntingdon, PA, USA*
HEIMO WOLINSKI • *Institute of Molecular Biosciences, University of Graz, Graz, Austria*

Chapter 1

Yeast Transformation by the LiAc/SS Carrier DNA/PEG Method

R. Daniel Gietz

Abstract

Efficient transformation of yeast has been the corner stone of molecular studies in yeast for numerous years. It is essential for techniques such as the yeast two-hybrid system as well as genome modification. Here I present protocols for a number of different transformation applications. A quick and easy protocol is for situations in which large numbers of transformants are not required. The high-efficiency protocol generates large numbers of transformants from small amounts of DNA that can be scaled up for the library transformation protocol. A transformation protocol for a microtiter plate format is also included. Finally, a protocol for the production of frozen competent yeast cells is included, which allows cells to be taken from the freezer at a moment's notice and transformed with high efficiency.

Key words *Saccharomyces cerevisiae*, Transformation, High efficiency, Microtiter plate, Frozen competent yeast cells

1 Introduction

The transformation of yeast has become a method used widely for genetic modification of various strains for many different purposes. The ways this can be accomplished has been reviewed [1]. The LiAc/ssDNA/PEG method of transformation will be discussed here. Here, I describe a number of transformation methods, from Quick and Easy to a High-Efficiency Transformation Protocol. In addition, I have included methods for transformation in microtiter plates, as well as a protocol for producing highly efficiently transforming frozen competent yeast cells.

The Quick and Easy Transformation Protocol can be used to produce a small number of transformants from a specific plasmid. The High-Efficiency Transformation Protocol can be employed to transform complex DNA mixtures into a specific strain. It is also useful for transforming a particular yeast strain with a plasmid or oligonucleotide for integration [2, 3], to simultaneously transform

Jeffrey S. Smith and Daniel J. Burke (eds.), *Yeast Genetics: Methods and Protocols*, Methods in Molecular Biology, vol. 1205, DOI 10.1007/978-1-4939-1363-3_1, © Springer Science+Business Media New York 2014

a yeast strain with multiple plasmids [4]. The Library Screen Transformation Protocol is used to generate the large numbers of transformants to efficiently cover complex DNA libraries. This protocol can be used for two-hybrid and similar screens [5, 6]. The Microtiter Plate Transformation can be used to transform plasmid libraries in a 96-well format into a single yeast strain, or used to transform a large number of strains in one operation. The production of frozen competent yeast cells gives the researcher to option to perform a high-efficiency transformation at a moment's notice.

2 Materials

2.1 General Equipment

1. A microtiter centrifuge is required for the Microtiter Plate Transformation Protocols.
2. A 96-prong replicator (Fisher Scientific, 112 Colonnade Road, Ottawa, ON, Canada) and an eight-channel micropipettor (Eppendorf) are required for the Microtiter Plate Transformation Protocols.
3. A rotary shaker for agitating microtiter plates.

2.2 Media

2.2.1 YPAD (Yeast Extract-Peptone-Adenine-Dextrose) Medium

This medium is used for routine growth of yeast strains; adenine is added to decrease the selective advantage of *ade2* to *ADE2* reversions. We use YPAD for initial grow and double strength YPAD broth, 2× YPAD, to regrow cultures to log phase before transformation. YPAD: Yeast extract 8 g, Peptone 16 g, Dextrose 16 g, Adenine hemisulfate 80 mg, Agar 12 g, distilled/deionized water (d/d water) 800 mL. Add Agar when making plates. 2× YPAD: Yeast Extract 16 g, Peptone 32 g, Dextrose 32 g, Adenine hemisulfate 80 mg, d/d water 800 mL. Dissolve ingredients and autoclave to sterility.

2.2.2 SC (Synthetic Complete) Selection Medium

SC selection medium is used to select for plasmids containing nutritional genes that complement mutations in the yeast strain required to take up the DNA. SC selection medium: Difco Yeast Nitrogen base w/o amino acids 5.4 g, Amino acid mix 1.6 g, Dextrose 16 g, Difco Bacto agar 12 g, d/d water 800 mL. Agar is omitted to make liquid SC selection medium. Dissolve ingredients in distilled water. Adjust the pH to 5.6 with 1.0 N NaOH. Autoclave to sterility, cool to 60 °C before pouring plates. The amino acid mixture [7] used contains: Adenine hemisulfate 0.5 g, L-Arginine HCl 2.0 g, Aspartic Acid 2.0 g, Glutamic Acid 2.0 g, Histidine HCl 2.0 g, Inositol 2.0 g, IsoLeucine 2.0 g, Leucine 4.0 g, Lysine HCl 2.0 g, Methionine 2.0 g, Phenylalanine 2.0 g, Serine 2.0 g, Threonine 2.0 g, Tryptophan 2.0 g, Tyrosine 2.0 g, Uracil 2.0 g, Valine 2.0 g, p-Aminobenzoic acid 0.2 g. All components are added to a plastic bottle containing two or three glass marbles and then mixed by shaking. Remember to omit the specific compound used in your selection.

2.3 Solutions

2.3.1 Lithium Acetate (1.0 M)

Add 10.2 g of lithium acetate dihydrate (Sigma) to 100 mL of d/d water, dissolve, autoclave for sterility, and store at room temperature.

2.3.2 PEG MW 3350 (50 % w/v)

Add 100 g of PEG 3350 (Sigma) to 60 mL of d/d water in a 400 mL beaker. Dissolve while stirring and add heat if necessary. Once dissolved, cool to room temperature and make the volume up to 200 mL. Transfer the solution to a glass bottle and autoclave for sterility. This can be stored at room temperature. The bottle must be securely capped to prevent evaporation, as increasing PEG concentrations will negatively affect transformation efficiency.

2.3.3 Single-Stranded Carrier DNA (2.0 mg/mL)

Dissolve 200 mg of salmon sperm DNA (Sigma) in 100 mL of TE: 10 mM Tris–HCl, 1 mM Na_2 EDTA pH 8.0, using a stir plate overnight at 4 °C. It may be necessary to draw the solution up and down in a 10 mL pipette to break up the large clumps prior to stirring overnight. Aliquot into microcentrifuge tubes and store at −20 °C. Denature the carrier DNA in a boiling water bath for 5 min and quench immediately in an ice/water bath before use. Denatured carrier DNA can be frozen and thawed 3 or 4 times, without any decline in its activity; however, it should be denatured again prior to use if its activity begins to decline.

2.3.4 Transformation (TRAFO) Mix

All transformation protocols use the same basic mix. The recipe below is for the transformation of 1×10^8 cells; the volumes can be amended as appropriate for larger and smaller numbers of cells. TRAFO Mix can be made up in bulk the same day and chilled until required. The recipe for TRAFO mix for one transformation reaction with 1×10^8 cells is listed in Table 1.

Table 1
TRAFO mix

Component	Volume
PEG 3350 (50 % [w/v])	240 mL
LiAc 1.0 M	36 μL
SS Carrier DNA (2.0 mg/mL)	50 μL
Plasmid DNA plus d/d sterile water	34 μL
Total volume	360 μL

3 Methods

3.1 Quick and Easy Transformation Protocol

Day 1

1. Inoculate a patch (2 cm^2) of your yeast strain onto YPAD agar plate or start a 2 mL liquid YPAD and incubate overnight at 30 °C (*see* **Note 1**).

Day 2

2. Denature a tube of carrier DNA in a boiling water bath for 5 min and chill immediately in ice/water.
3. Scrape an aliquot of yeast (the approximate size of a match head or 50 μL in volume) from the YPAD plate and suspend the cells in 1.0 mL of sterile d/d water in a 1.5 mL microcentrifuge tube or take 1 mL of YPAD culture and transfer over to a 1.5 mL microcentrifuge tube. This should give about $2–5 \times 10^8$ cells.
4. Pellet the cells at top speed in a microcentrifuge for 30 s and discard the supernatant.
5. Add 360 μL of TRAFO Mix to the cell pellet. Resuspend the cell pellet by vortex mixing briskly. For a single transformation the ingredients should be added in the order listed and mixed vigorously.
6. Incubate the tube in a water bath at 42 °C for 30–180 min (*see* **Notes 2** and **3**).
7. Microcentrifuge the transformation tube at 10,000 × *g* for 30 s and remove the TRAFO Mix with a micropipettor.
8. Pipette 1.0 mL of sterile d/d water into the transformation tube. Resuspend the cell pellet by vortex mixing.
9. Plate the cell suspension onto plates of appropriate selection medium. Transformants can be isolated after incubation at 30 °C for 3 or 4 days.

3.2 High-Efficiency Transformation Protocol

Day 1

1. Inoculate your yeast strain into 5 mL of 2× YPAD or 15 mL of appropriate SC selection medium (for maintaining a plasmid) and incubate overnight at 30 °C with shaking. In addition place some 2× YPAD and a culture flask in the incubator as well.

Day 2

2. Determine the titre of the yeast culture. This can be done using either of the methods below.
   (a) Dilute overnight culture 1/100 into sterile water and deliver into a spectrophotometer cuvette, mix thoroughly by inversion and measure the OD at 600 nm (a suspension containing 1×10^6 cells/mL will give an OD_{600} of 0.1).

(b) Dilute the overnight culture 1/10 in sterile water in a microcentrifuge tube and mix thoroughly. Deliver 10 μL of this dilution onto the counting grid of an improved Neubauer haemocytometer with cover slide, count the number of cells in the 25 large grid squares under a microscope. Multiply this number by 100,000 to obtain the titre of the original culture in cells/mL.

3. Set up the growth culture by adding 2.5×10^8 cells to a final volume of 50 mL of the prewarmed 2× YPAD in the prewarmed culture flask. The starting titre should be 5×10^6 cells/mL. This is sufficient for ten transformation reactions.
4. Incubate the flask in the shaking incubator at 30 °C and 200 rpm until the cell titre is at least 2×10^7 cells/mL. This should take up to 4 h or more depending on the strain.
5. Before the yeast cells have achieved the correct density, denature a sample of carrier DNA in a boiling water bath for 5 min and chill immediately in an ice/water bath.
6. Harvest the cells by centrifugation at 3,000 × *g* for 5 min, wash the pellet twice in 25 mL of sterile d/d water and resuspend the cells in 1.0 mL of sterile water.
7. Transfer the cell suspension to a 1.5 mL microcentrifuge tube, centrifuge for 30 s at maximum speed and discard the supernatant.
8. Resuspend the cells in a final volume of 0.5 mL of sterile d/d water and deliver 50 μL samples of 1×10^8 cells into 1.5 mL microfuge tubes, one for each transformation. Centrifuge at 10,000 × *g* for 30 s and remove the supernatant.
9. Make up sufficient TRAFO Mix (*see* Subheading 2) for the planned number of transformations plus one extra. TRAFO Mix can be kept on ice if made in advance.
10. Add 360 μL of TRAFO Mix to each transformation tube and resuspend the cells by vortex mixing vigorously. Alternatively, components of the TRAFO Mix can be added individually to each tube containing the cell pellet starting with the PEG.
11. Incubate the tube in a water bath at 42 °C for 40 min or more depending on the strain (*see* **Note 4**).
12. Harvest the cells after the incubation is complete by centrifugation at 10,000 × *g* in a microcentrifuge for 30 s and remove the TRAFO Mix with a micropipettor.
13. Resuspend the cell pellet in 1.0 mL of sterile d/d water. Stir the pellet with a sterile micropipette tip to resuspend the cells and then mix on a vortex mixer.
14. Plate appropriate amount of the cell suspension onto SC selection medium (*see* **Note 5**).
15. Incubate the plates at 30 °C for 3–4 days and harvest transformants.

3.3 The Library Screen Transformation Protocol

This protocol is used to generate the large numbers of transformants required to screen complex DNA libraries. It is advisable to use the high-efficiency protocol to test the effects of increasing plasmid DNA on transformation efficiency and transformation yield before using this protocol. Scale up the transformation reaction 30-, 60-, or 120-fold with an appropriate plasmid concentration to obtain the number of transformants required. This protocol is for a large-scale screen, any specific requirements for the two-hybrid and similar screens are listed in **Notes 6–11**.

Day 1

1. Inoculate your yeast strain into 50 mL of 2× YPAD in a 250 mL flask. Incubate at 30 °C overnight with shaking. Warm 200 mL (30×), 400 mL (60×), or 800 mL (120×) of 2× YPAD broth and a culture flask (500 mL—30×, 1,000 mL—60×, 2,000 mL—120×) overnight at 30 °C. (Two-hybrid screen, *see* **Note 10**).

Day 2

2. Determine the titre of the culture. Transfer the volume containing 6.25×10^8 cells (30× scale-up), 1.25×10^9 cells (60× scale-up), or 2.5×10^9 (120× scale-up) cells into 50 mL centrifuge the cells at 3,000 × *g* for 5 to pellet the cells. Resuspend the pellet(s) in warm 2× YPAD broth and transfer to the culture flask(s). Add sufficient 2× YPAD broth to bring the final cell titre to 5×10^6/mL (Two-hybrid screen, *see* **Note 10**).
3. Incubate the flask at 30 °C and 200 in a shaking incubator until the cells have undergone two divisions. This will take at least 4 h.
4. Denature the SS carrier DNA reagent (30×—2.0 mL, 60×—3.5 mL, 120×—7.0 mL) for 5 min in a boiling water bath and chill in ice/water.
5. Make up appropriate volumes of TRAFO Mix and keep in ice/water.

Ingredients	30×	60×	120×
PEG 50 % w/v	7.2 mL	14.4 mL	28.8 mL
LiAc 1.0 M	1.08 mL	2.16 mL	4.32 mL
SS carrier DNA (2 mg/mL)	1.5 mL	3.0 mL	6.0 mL
Plasmid DNA + d/d sterile water	1.02 mL	2.04 mL	4.08 mL
Total volume	10.8 mL	21.6 mL	43.2 mL

6. Harvest the cells by centrifugation at 3,000 × *g* for 5 and resuspend them in 1/5 volume with d/d sterile water. Centrifuge and wash the cells again in the same volume of water and discard the supernatant.

7. Add the TRAFO Mix onto the cell pellet and suspend the cells by vortexing the tube vigorously.
8. Incubate the cell suspension at 42 °C for 60 min. Mix the contents of the tube by inversion at 5 min intervals to ensure quick temperature equilibration.
9. Centrifuge at 3,000 × *g* for 5 min. Pour off the TRAFO Mix, centrifuge again, and remove the remainder of the TRAFO Mix with micropipettor.
10. Resuspend the cells in d/d sterile water (30×—20 mL; 60×—40 mL; 120×—40 mL) and spread 400 μL samples onto 150 mm plates of SC selection agar (30×—50 plates; 60×, 120×—100 plates). (Two-hybrid screen, *see* **Note 11**.)
11. Incubate the plates at 30 °C for 4–7 days and count and recover transformants.

3.4 Microtiter Plate Transformation Protocols

Transformation of yeast cells in 96-well microtiter plates can be tailored for many different purposes (*see* **Note 12**). For large numbers of transformations we use a 96-prong replicator and 150 mm petri dishes of medium. The TRAFO Mix for these protocols is prepared without PEG; it is less viscous than regular TRAFO Mix and makes is easier to resuspend the cell pellet. The PEG is added after the cell pellets have been resuspended.

3.4.1 Agar Plate Protocol

Day 1

1. Dip the prongs of a replicator into a dish of 70 % ethanol and sterilize them by passing them through a Bunsen flame.
2. Gently place the prongs onto the agar surface of a plate of YPAD and press so that all of them make an imprint but do not break the surface of the agar.
3. Use an inoculating loop or sterile flat toothpicks to patch the yeast strain(s) onto the imprints. Orient this "master" plate(s) by marking the bottom and incubate overnight at 30 °C.

Day 2

4. Pipette 150 μL samples of d/d sterile water into the wells of a microtiter plate.
5. Sterilize a replicator and cool.
6. Align the patches of yeast with the tips of the prongs and lower the cooled replicator onto the plate making sure all prongs make contact. Move the replicator very gently in small circles to transfer cells to the prongs.
7. Lower the replicator into the microtiter plate and move in a mixing motion to suspend the cells from the ends of the prongs. This should give about 1×10^7 cells per well. Repeating the transfer process will increase the number of cells. Mark the orientation of the microtiter plate.

8. Denature a tube of carrier DNA (2 mg/mL) in a boiling water bath for 5 min and chill in ice/water.
9. Pellet the cells at 1,380 × *g* in a microtiter plate centrifuge.
10. Remove the supernatant from the wells. This can be accomplished by aspiration with a sterile micropipette tip attached to a vacuum line. Or I find it easier to dump the water out of the wells into a sink; however, it is best to practice this operation before a critical experiment.
11. Prepare TRAFO Mix minus PEG. The volumes listed are for one transformation (one well). Make sufficient for 100 transformations if you intend to use all 96 wells. You can use more or less than the listed amount of plasmid DNA. Keep the TRAFO Mix minus PEG on ice.

Component	One well
LiAc 1.0 M	15.0 μL
Carrier DNA (2 mg/mL)	20.0 μL
Plasmid DNA (20–50 ng) + d/d sterile Water	15.0 μL
Total volume	50.0 μL

12. Pipette 50 μL TRAFO Mix minus PEG into each well. Clamp the plate on a rotary shaker and shake it for 2 min at 400 rpm to resuspend the cell pellets.
13. Pipette 100 μL PEG 3350 (50 % w/v) into each well. Shake the plate for 5 min at 400 rpm and check that the cell suspensions are homogeneous.
14. Seal the plate in a plastic bag and incubate it at 42 °C for 1–4 h (*see* **Note 13**).
15. Pellet the cells by centrifugation for 10 min at 1,380 × *g* and remove the TRAFO Mix by aspiration or dumping.

Quantitative Sampling

16. Pipette 100 μL of d/d sterile water into each well and shake the plate at 400 rpm for 5 min to resuspend the cells. Pipette 5 μL samples into 100 μL puddles of water on plates of SC selection medium.

Qualitative Sampling

17. Pipette 50 μL of d/d sterile water into each well and resuspend the cells as above. Sterilize a replicator and use it "prongs down" to print samples (approximately 10 μL) onto plates of SC selection medium. Additional samples can be overlaid if necessary.
18. Incubate the plates at 30 °C for 2–4 days and recover and/or count the transformants.

3.4.2 Liquid Culture Protocol

The yeast culture is grown overnight and regrown for two divisions as in the High-Efficiency Transformation Protocol. The cells of the regrown culture are harvested, washed, and resuspended in water and the cell titre determined as described in the High-Efficiency Transformation Protocol.

1. Adjust the titre of the cell suspension to 4×10^8 cells/mL and dispense 100 μL samples of the suspension into the wells of the microtiter plate.
2. Continue from **step 6** of the Agar Plate protocol but increase the amount of plasmid DNA to 100 ng per transformation.
3. Seal and incubate the plates at 42 °C for 60 min.
4. Sample the wells by plating or replica plating onto SC selection medium.
5. Incubate the plates at 30 °C for 2–4 days and recover and/or count the transformants.

3.5 Transformation-Competent Frozen Yeast Cells

This method can be used to produce frozen competent yeast cells when a single strain is used repeatedly. Yeast cultures are regrown for at least two divisions and used to produce transformation-competent cells that can be frozen and used at another time.

3.5.1 Preparation

1. Grow the yeast strain overnight and then regrow in 2× YPAD to a titer of 2×10^7 cells/mL as described in Subheading 3.2. One hundred samples of 1×10^8 frozen competent cells will require 500 mL of regrown culture (1×10^{10} cells).
2. Harvest the cells by centrifugation at 3,000 × *g* for 5 min, wash the cells in 0.5 volumes of sterile water and resuspend in 0.01 volumes of sterile water. Transfer to a suitable sterile centrifuge tube and pellet the cells at 3,000 × *g* for 5 min.
3. Resuspend the cell pellet in 0.01 volumes of Frozen Competent Cell (FCC) solution (5 % v/v glycerol, 10 % v/v DMSO) Use the best quality DMSO for best results.
4. Dispense 50 μL samples into an appropriate number of 1.5 mL microfuge tubes.
5. Place the microfuge tubes into a 100-tube Styrofoam rack with lid (Sarstedt) or similar type rack. It is best to place this container upright in a larger box (Styrofoam or cardboard) with additional insulation such as styrofoam chips or news paper to reduce the air space around the samples. This will result in the samples freezing slowly, which is essential for high survival rates.
6. Place the container at −80 °C overnight. The tubes can then be removed from the tube rack container and stored at −80 °C in bulk. These cells can be stored for up to 1 year with little loss of transformation efficiency.

3.5.2 Transformation of Frozen Competent Yeast Cells

These cells are transformed using Subheading 3.2 but with the differences listed below:

1. Thaw cells in a 42 °C water bath for 15 s.

2. Pellet the cells at 10,000 × *g* in a microfuge for 2 min and remove the supernatant.
3. Add 360 μL of FCC TRAFO mix (260 μL 50 % PEG, 36 μL 1.0 M LiOAC, 50 μL denatured carrier DNA, and 14 μL of DNA and d/d sterile water) and mix vigorously with a vortex mixer to resuspend the cell pellet. Note the difference in PEG volume.
4. Incubate in a 42 °C water bath for 20–60 min depending on the strain. Centrifuge as above to remove the supernatant and resuspend the cell pellet in 1 mL of sterile water.
5. Plate appropriate dilutions onto SC selection medium (*see* **Note 5**).

4 Notes

1. This protocol can be used to transform freshly grown yeast cells as well as those that have been stored in a refrigerator or at room temperature. The yield will be reduced but there will generally be sufficient transformants of the desired genotype.
2. Incubation at 42 °C for 30 min will usually result in several thousand transformants per transformation. Extending the duration of the heat shock can increase the yield of transformants significantly with many yeast strains. Up to 1×10^5 transformants/μg plasmid have been obtained after 60 min incubation and $>1 \times 10^6$/μg plasmid DNA after 180 min.
3. The addition of 5 % (v/v) DMSO to the TRAFO Mix increases the yield of transformants with some strains.
4. The average heat shock time is 40 min but each strain can be different. The optimum heat shock duration should be previously determined for each distinct strain utilized.
5. Typically when screening for rare transformant events and plating onto 85 mm petri plates the plating volume can range from 1 to 200 μL per plate depending on the transformation efficiency. Plate volumes less than 100 μL into a 100 μL puddle of sterile water.
6. Two-hybrid screens typically require the transformation of "bait" and "prey" plasmids into a specific yeast strain. The genotypes of suitable yeast strains and procedures for the construction and testing of fusion plasmids can be found in [6].
7. A two-hybrid screen involves the transformation of the "prey" plasmid library into yeast cells carrying the "bait" plasmid. Use the high-efficiency protocol to transform the yeast strain contain the "bait" plasmid with a range from 100 ng to 10 μg of the "prey" plasmid library to determine the appropriate scale up factor.
8. The "bait" plasmid and the "prey" plasmid library can be co-transformed into the yeast strain in a single operation. The high

transformation efficiencies obtained with these protocols can result in up to 40 % of the transformed yeast cells containing both plasmids. Co-transformation may be necessary if the "bait" plasmid affects the growth or viability of your yeast strain.

9. Inoculate the strain carrying "bait" plasmid into liquid SC selection medium. Use 50 mL medium in a 250 mL flask for a 30× scale up and 100 mL medium in a 500 mL flask for a 60× scale-up or double this for a 120× scale up.
10. The strain carrying the "bait" plasmid can be cultured in 2× YPAD for the two divisions prior to transformation without significant loss of the plasmid in most cases. However prior to the growth in 2× YPAD the strain must be grown on SC selection medium to maintain the plasmid.
11. In a two-hybrid screen the yeast strain contains a reporter gene that is activated by interaction of the protein products of the "bait" and "prey" plasmids. Details of the selection and detection of reporter gene activation are given in [6].
12. The microtiter plate protocols can be adapted for a number of purposes.
    (a) Many different yeast strains can be grown on a master plate, sampled with a replicator into the wells of a microtiter plate and tested for transformation efficiency with a single plasmid.
    (b) A single strain can be transformed with many different plasmids (e.g. a plasmid library in a 96-well format).
    (c) Many yeast strains can be grown on a master plate, transferred to wells containing 150 μL of 2× YPAD, regrown in sealed plates on a shaker at 200 rpm and then transformed in situ with a single plasmid.
    (d) One or more strains can be tested for response to variation in the composition of the TRAFO Mix.
    (e) One or more strains can be tested for response to variation in the duration of incubation at 42 °C.
13. After incubation at 42 °C for 60 min we have obtained an efficiency of 2×10^5 and a yield of 573 transformants per well; extending the incubation to 4 h resulted in an efficiency of 3.9×10^6 and 6,210 transformants per well.

References

1. Gietz RD, Woods RA (2001) Genetic transformation of yeast. BioTechniques 30:816–831
2. Linske-O'Connell LI, Sherman F, McLendon G (1995) Stabilizing amino acid replacements at position 52 in yeast iso-1-cytochrome c: in vivo and in vitro effects. Biochemistry 34:7094–7102
3. Yamamoto T, Moerschell LRP, Wakem P et al (1992) Parameters affecting the frequencies of transformation and co-transformation with synthetic oligonucleotides in yeast. Yeast 8: 935–948
4. Gietz RD, Schiestl RH (1991) Applications of high efficiency transformation of intact yeast

cells using single stranded nucleic acids as carrier. Yeast 7:253–263
5. Bartel PL, Fields S (1997) The yeast two-hybrid system. Oxford University Press, Oxford, UK
6. MacDonald PN (ed) (2001) Methods in molecular biology, two-hybrid systems: methods and protocols, vol 177. Humana, Totowa, NJ
7. Rose MD (1987) Isolation of genes by complementation in yeast. Methods Enzymol 152:481–504

Chapter 2

Tetrad, Random Spore, and Molecular Analysis of Meiotic Segregation and Recombination

Michael Lichten

Abstract

The power of *Saccharomyces cerevisiae* as an experimental organism derives from its genetic tractability. Mutant variants can be isolated or constructed and phenotypically characterized with relative ease. In addition, the ability to recover and characterize all four products of meiosis, as haploid spores in a tetrad ascus, greatly facilitates determining the allelic composition of variants, measuring linkage relationships between alleles, and constructing new allele combinations for the analysis of genetic interactions. *Saccharomyces cerevisiae* also is a preeminent model organism for the study of meiotic recombination, by analysis of tetrads, by analysis of populations of single spores (often called random spore analysis), and by direct monitoring of recombination at the DNA level. This chapter contains methods for tetrad dissection, for random spore preparation, and for preparing DNA for molecular analysis from liquid cultures undergoing synchronous meiosis.

Key words Tetrad dissection, Meiosis, Recombination, Budding yeast, DNA preparation, Electrophoresis, Southern blotting

1 Introduction

While molecular approaches such as defined deletions [1] and whole-genome mutation mapping [2] threaten to supplant traditional methods of genetic analysis in *Saccharomyces cerevisiae*, there is still considerable value in the use of meiotic segregation for the rapid construction of strains, characterization and mapping of mutations and of suppressors, and in the use of meiosis to study homologous recombination. When diploid yeast cells are starved for nutrients, in particular for nitrogen and for fermentable carbon sources, they undergo meiosis. During meiosis, the diploid genome is duplicated, and then segregated through two successive nuclear divisions to produce four haploid products, which in yeast are recovered as an unordered ascus with four haploid spores [3]. These haploid products of meiosis can be separated, germinated, and subjected to genetic analysis, thus allowing a comprehensive

Jeffrey S. Smith and Daniel J. Burke (eds.), *Yeast Genetics: Methods and Protocols*, Methods in Molecular Biology, vol. 1205, DOI 10.1007/978-1-4939-1363-3_2, © Springer Science+Business Media New York 2014

picture of marker segregation to be drawn. In addition, the ability to induce sporulation by starvation has led to the development of methods to perform meiosis in liquid mass cultures with relative synchrony, which in turn has made it possible to monitor meiotic recombination in real time, using DNA prepared from cells at different stages of meiosis. This chapter describes methods for preparing the spore products of meiosis for subsequent genetic analysis. It also describes a simple method for preparing DNA from meiotic cells, which allows detected of recombined DNA molecules.

1.1 Tetrad Dissection

This is the traditional method for isolating and analyzing meiotic segregants. Asci are first prepared for dissection by enzymatic treatment to remove the outer cell wall surrounding the four spores. Digested asci are placed on an agar surface, and are moved about and spores are separated using a glass-fiber needle attached to a micromanipulator. Spores are deposited in a grid pattern, facilitating their subsequent analysis by replica plating (Fig. 1a). Tetrad dissection requires a specialized microscope and micromanipulator,

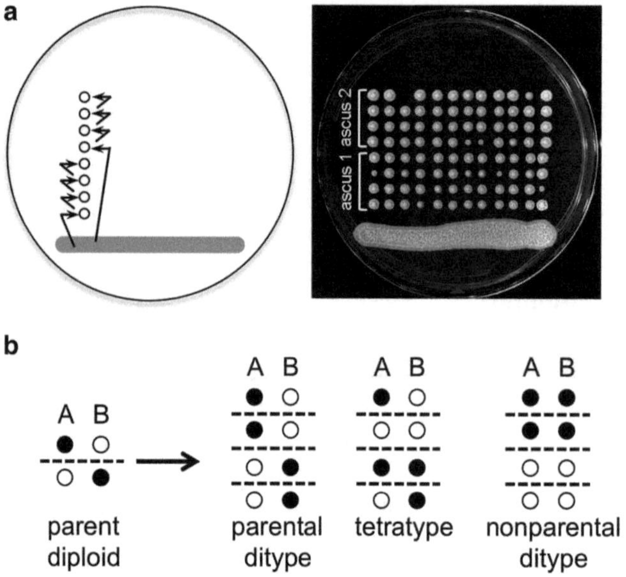

Fig. 1 Tetrads. (**a**) *Left*—strategy for dissecting tetrads. A streak of asci is laid down at the bottom of the plate. A tetrad is picked up, placed at the first position, and dissected, leaving a single spore behind. The remaining three spores are placed at the second position, and the process repeated until all four spores are deposited. A second tetrad is picked up, deposited at the fifth position, and dissected as before. *Right*—spore colonies grown on a full dissection plate. (**b**) Types of two-marker segregation. Tetrads from a diploid heterozygous at two loci can contain: two pairs of spores with parental marker configurations (parental ditype); four spores, each with a unique parental marker configuration (tetratype); or two pairs of spores, each with a nonparental marker configuration (nonparental ditype)

and also requires the development of some skill in the use of the needle, which is more than ten times the diameter of the tetrad being dissected. Most researchers become sufficiently adept with several hours of practice, and video demonstrations are available online [4, 5].

While a complete discussion of the advantages of tetrad analysis is beyond the scope of this chapter, one major advantage is the ability to monitor patterns of segregation for all of the alleles present in a single meiosis (Fig. 1b). In a cross where two genes are in a heterozygous allelic state, three segregation patterns are possible: parental ditype (PD), where alleles are recovered in spores in the two parental configurations; tetratype (T), where four different allele configurations are recovered; and nonparental ditype (NPD), where only two allele configurations, both nonparental, are recovered (Fig 1b). Linkage relationships between two genes can be calculated from the ratio of these tetrad types. If the two genes are completely unlinked, they will produce a PD:T:NPD ratio of 1:4:1. If parental ditypes are significantly greater than nonparental ditypes, the two genes are linked, and the map distance between them can be calculated by the formula of Perkins [6]. Tetrad analysis is also the primary genetic method, in studies of recombination, used to detect gene conversion events, where parental alleles are recovered in non-Mendelian ratios [7], and also to determine levels of crossover interference [8].

In the genomic age, where distances between genes are known to the nucleotide, tetrad analysis is rarely used to determine distances between loci. However, tetrad dissection is frequently used to combine markers during strain construction. In such cases, if the genetic map positions of two loci, available from the *Saccharomyces* Genome Database [9], indicate that they are linked, the Perkins formula can be used to estimate the number tetrads required to obtain the desired recombinant. Tetrad dissection is also used to identify synthetic genetic interactions, where double mutants display a growth advantage or disadvantage relative to the two single mutants. Tetrad dissection is also useful in determining if an observed mutant phenotype is due to mutation at a single locus, or if it is due to the combined effects of mutations that can be separated by recombination or chromosome segregation. Tetrad dissection is essential when constructing double-mutant strains where the two mutations of interest are marked by the same selectable marker; double mutants can be identified in nonparental ditype tetrads as the two spores that carry the selectable marker.

1.2 Random Spores

While tetrad dissection is the most commonly used method to obtain haploid spores for analysis, sonication of asci suspended in liquid can also be used to separate spores. A certain amount of segregation information is lost relative to tetrad dissection, but this is compensated by the relative ease with which a large population of single spores can be obtained. This is especially useful when low-frequency recombination events must be scored, but random spore isolation is also useful for simple strain constructions. Random spore isolation also has the advantage of not requiring a specialized tetrad dissecting microscope and micromanipulator, since spores are separated using a sonicator, a more common piece of laboratory equipment.

1.3 Analysis of Recombination Using DNA from Meiotic Cells

While much can be learned from the genetic analysis of meiotic segregants in spores, many yeast mutants defective in recombination either fail to sporulate or produce inviable spores as a consequence of massive chromosome mis-segregation, thus precluding conventional genetic analysis. In addition, because spores are an end-product of meiosis, they contain no information regarding the relative timing of different meiotic recombination events. To gain further insight into recombination mechanisms, restriction site or restriction fragment length polymorphisms are used as genetic markers, and recombination is scored by restriction enzyme digestion and Southern blotting [10], using DNA isolated from meiotic cells (Fig. 2). DNA isolation methods that allow branched

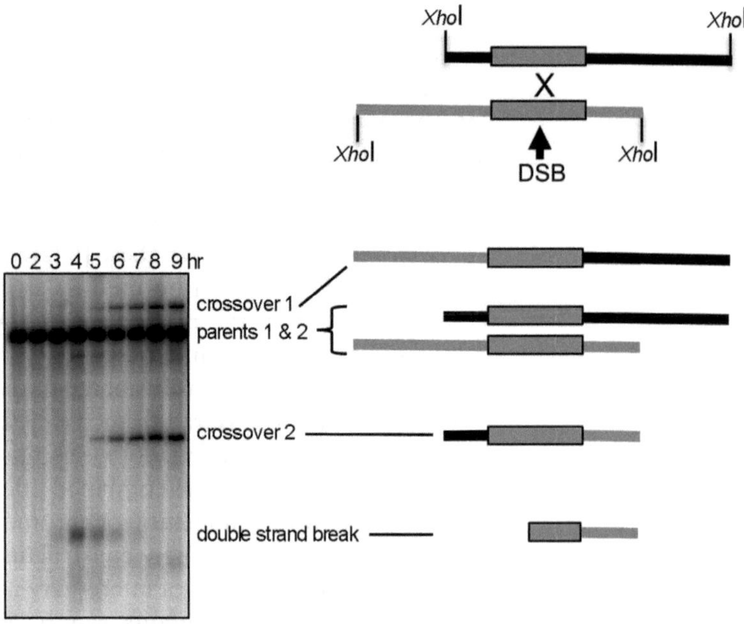

Fig. 2 Scoring recombination on Southern blots. Southern blot used to score crossing-over and meiotic double-strand break formation in a recombination reporter insert on chromosome III (for details, *see* ref. 13). *XhoI* restriction sites differently located around a region of interest that contains a strong meiotic double-strand break site (DSB) allow distinction of the two crossover products from parental fragments, and also allow detection of the DSB. DNA samples were taken at the indicated time after initiation of meiosis

recombination intermediates to be purified and characterized have been described in a previous volume [11]. This chapter contains a less-demanding protocol that does not preserve recombination intermediate structure, but that is well suited to simple analysis of meiotic double-strand DNA breaks and recombinant products.

2 Materials

The supplier is named only when deemed important; otherwise, any high-quality source will suffice.

2.1 Sporulation for Genetic Analysis

1. YPD agar: 2 % (w/v) Bacto™ peptone (BD, Franklin Lakes, NJ), 1 % (w/v) Bacto™ yeast extract, 2 % (w/v) Bacto™ agar, 2 % (w/v) D-glucose, 0.004 % (w/v) adenine. Adjust pH to 5.5 with 1 N HCl and autoclave. Pour into 100 mm diameter petri dishes: for general use, ~40 mL/dish; for tetrad dissection, ~30 mL/dish on a level surface (*see* **Note 1**). Plates used for tetrad dissection should be made in small batches and stored at room temperature (*see* **Note 2**).
2. YPD broth: 2 % (w/v) Bacto™ peptone (BD, Franklin Lakes, NJ), 1 % (w/v) Bacto™ yeast extract, 2 % (w/v) Bacto™ agar, 2 % (w/v) D-glucose, 0.004 % (w/v) adenine. Mix all ingredients except D-glucose in 0.8× final volume, adjust pH to 5.5 with 1 N HCl and autoclave. Before using, add D-glucose from a sterile 20 % (w/v) stock.
3. KAc agar: 2 % (w/v) potassium acetate, 0.22 % (w/v) Bacto™ yeast extract, 0.05 % (w/v) D-glucose, 0.087 % (w/v) COM mixture (complete supplement mixture, see below), 2 % (w/v) Bacto™ agar. Adjust pH to 7.0 with 1 N HCl or 1 N KOH (do not use NaOH) and autoclave. Pour into 100 mm diameter petri dishes, ~40 mL/dish.
4. Super KAc: 2 % (w/v) potassium acetate, 0.2 % (w/v) Bacto™ yeast extract, 0.4 % (w/v) Bacto™ peptone, 0.4 % (w/v) D-glucose, 0.087 % (w/v) COM mixture, 2 % (w/v) Bacto™ agar (BD). Adjust pH to 5.5 with 1 N HCl or 1 N KOH (do not use NaOH) and autoclave. Pour into 100 mm diameter petri dishes, ~40 mL/dish.
5. COM mixture: adenine, 0.8 g; arginine, 0.8 g; aspartic acid, 4 g; histidine, 0.8 g; leucine, 2.4 g; lysine, 1.2 g; methionine, 0.8 g; phenylalanine 2 g; threonine 8 g; tryptophan, 0.8 g; tyrosine 1.2 g; uracil, 0.8 g. Grind mixture thoroughly using a mortar and pestle or a spinning blade-type coffee grinder.
6. 1 % (w/v) potassium acetate. Sterilize by autoclaving.
7. Replica-plating block and sterile velvets.

2.2 Tetrad Dissection

1. Glusulase working solution. Glusulase® (PerkinElmer, Waltham, MA), diluted 1/10 in sterile water. Store at 4 °C.
2. Tetrad dissecting microscope. Available from Singer Instrument Co. Ltd. (www.singerinstruments.com) and from Micro Video Instruments, Inc. (www.mvi-inc.com). These are specialized instruments with a customized stage that holds a petri dish and that has click-stops in the vertical dimension to facilitate even spore placement, a long-working distance objective, and with a micromanipulator that holds a glass-fiber needle used to manipulate and dissect ascii.
3. Tetrad dissecting needles. Available from Singer Instrument Co. Ltd and from Cora Styles Lab Supplies, LLC (www.corastyles.com).
4. YEPDD agar: 2 % (w/v) Bacto™ peptone, 1 % (w/v) Bacto™ yeast extract (BD), 0.087 % (w/v) COM mixture, 2 % (w/v) Bacto™ agar (BD), 4 % D-glucose, 0.004 % adenine. Mix all ingredients except D-glucose in 0.8 × final volume, adjust pH to 5.5 with 1 N HCl, and autoclave; add D-glucose from a sterile 20 % (w/v) stock before pouring. Pour into 100 mm diameter petri dishes, ~30 mL/dish on a level surface. Make in small batches and store at room temperature (*see* **Note 2**).
5. 10 N potassium hydroxide.

2.3 Random Spores

1. 0.1 % (w/v) TWEEN® 80. Filter sterilize.
2. 5 mL round-bottom polypropylene tube (BD Falcon, 352063).
3. Sonicator, probe-type, with micro tip for use with 1 mL volumes.
4. Hemocytometer.

2.4 Sporulation in Mass Culture for DNA Analysis

1. Pre-sporulation broth (SPS): 1 % (w/v) Bacto™ peptone, 0.5 % (w/v) Bacto™ yeast extract, 1 % (w/v) potassium acetate, 0.17 % (w/v) Bacto™ yeast nitrogen base without amino acids, 1 % (w/v) ammonium sulfate, 0.5 % (w/v) potassium hydrogen pthalate. Adjust pH to 5.5 with 10 N KOH. Make fresh and autoclave 1 day before use.
2. Sporulation medium (KAc): 1 % (w/v) potassium acetate supplemented with nutrients (Table 1) according to auxotrophic requirements of the strains to be used and with 0.001 % (v/v) polypropylene glycol 2000 (Sigma-Aldrich, www.sigmaaldrich.com) as an anti-clumping agent. Warm to 30 °C before using.
3. 2.8 L triple-baffled Fernbach flask (Bellco, Vineland NJ, # 2551-02800). Warm the Fernbach flask to 30 °C before using.
4. 50 % (w/v) glycerol + 0.5 % (w/v) sodium azide.
5. Spheroplasting solution + glycerol: 1 M sorbitol; 50 mM potassium phosphate; 5 mM EDTA; 20 % (w/v) glycerol pH 7.5.

Table 1
Supplements for sporulation media

Nutrient	Stock concentration (w/v)	mL/L in KAc
Adenine	0.5 % in 0.05 M HCl	1.6
Arginine	2.0 %	0.2
Histidine	2.0 %	0.2
Isoleucine	1.0 %	0.6
Leucine	1.0 %	1.2
Lysine	1.5 %	0.4
Methionine	2.0 %	0.2
Threonine	6.0 %	1.0
Tryptophan	1.0 %	0.4
Tyrosine	0.25 %	2.0
Valine	3.0 %	1.0
Uracil	0.2 % in 1 % Na_2CO_3	1.0

2.5 Preparation of DNA from Meiotic Cells

1. Zymolyase solution: 10 mg/mL Zymolyase 100T (MP Biomedical, Solon, OH), 1 M sorbitol, 50 mM potassium phosphate, 5 mM EDTA, 5 % (w/v) D-glucose, pH 7.5. Store frozen (−20 °C) in aliquots and use once.
2. Spheroplasting solution: 1 M sorbitol, 50 mM potassium phosphate, 5 mM EDTA, 1 % 2-mercaptoethanol pH 7.5. Add 2-mercaptoethanol just before use.
3. Lysis solution: 100 mM NaCl, 50 mM Tris–HCl, 50 mM EDTA, pH 8.0.
4. 20 % (w/vol) SDS.
5. Proteinase K (Invitrogen, Carlsbad, CA) 20 mg/mL in 10 mM Tris–HCl, 20 mM calcium chloride pH 7.5, 50 % (v/v) glycerol.
6. 5 M potassium acetate.
7. TE: 10 mM Tris–HCl, 1 mM EDTA, pH 7.5.
8. DNase-free RNase.
9. 3 M sodium acetate.

2.6 Gel Electrophoresis, Blotting, and Hybridization with Radioactive Probe

1. SeaKem GTG agarose (Lonza, Basel, Switzerland).
2. TBE: 90 mM Tris base, 90 mM boric acid, 2 mM EDTA. Prepare as 10× stock.
3. 5× loading buffer: 15 % (w/v) Ficoll 400, 25 mM Tris–HCl, pH 7.4, 100 mM EDTA, 0.1 mg/mL Orange G dye (Sigma-Aldrich).

4. 0.25 N HCl
5. Alkaline transfer buffer: 1.5 M NaCl, 0.5 N NaOH.
6. Zeta-Probe blotting membrane (BioRad, Hercules, CA)
7. SSPE: 150 mM NaCl, 10 mM NaH_2PO_4, 1 mM EDTA, pH 7.4. Prepare as 20× stock.
8. Nonfat dried milk: 10 % (w/v) Carnation nonfat dried milk, 0.1 % (w/v) sodium azide. Store at 4 °C.
9. Carrier DNA: 5 mg/mL fish sperm DNA in TE.
10. Prehybridization buffer: 2× SSPE, 1 % (w/v) SDS, 0.5 % nonfat dried milk. Make fresh from concentrated stock solutions.
11. Hybridization buffer: 10 % (w/v) dextran sulfate, 2× SSPE, 1 % (w/v) SDS, 0.5 % nonfat dried milk. Dissolve dextran sulfate in water (8/10 final volume) at 65 °, then add other components from stocks. Make fresh each time, hold at 65 °C.
12. α-^{32}P dCTP, 6,000 Ci/mmol, 20 mCi/mL (Perkin Elmer NEN, Waltham, MA). Use within a few days of the assay date.
13. High Prime DNA labeling kit for use with dCTP (Roche Applied Science, www.roche-applied-science.com).
14. Sephadex G-50 MicroSpin column (GE Healthcare, www.gelifesciences.com).
15. Heat-seal bags and heat-sealer.
16. Shaking 65 °C water bath.

3 Methods

3.1 Sporulation for Genetic Analysis

3.1.1 Sporulation on Plates

1. Using a freshly grown single colony, make a ca. 2.5 cm^2 patch onto YPD agar. Incubate at 30 °C overnight (*see* **Note 3**).
2. Replica-plate patch to KAc or Super KAc agar plates, the latter if a larger number of spores are required. Incubate between room temperature and 30 °C for 2 days–1 week, until ascii are present (*see* **Note 4**).

3.1.2 Simple Sporulation in Liquid

1. Use a portion of a single colony to inoculate a 5 mL YPD culture; incubate with aeration overnight.
2. Dilute overnight culture 1/50 in 5 mL and incubate at 30 °C with aeration for 4 h. Cells should be in mid-log phase (~0.5–1×10^7 cells/mL).
3. Pellet cells by centrifugation, wash in 5 mL 1 % potassium acetate, pellet again, and resuspend in 5 mL 1 % potassium acetate.
4. Incubate with aeration, between room temperature and 30 °C, for 2 days–1 week (*see* **Note 4**).

3.2 Tetrad Dissection

Before tetrads can be dissected to separate spores, the ascus outer wall must be digested.

1. If sporulation was done on plates: scrape a small blob of spores (~2 μL) from the spore patch, using an inoculating loop or flat-end toothpick.
2. If sporulation was done in liquid: pellet spores from 50 μL of sporulation culture.
3. Resuspend/mix spores in 50 μL of glusulase working solution (*see* **Note 5**). Incubate 5–10 min at room temperature.
4. Add 0.5 mL sterile water, mix gently. Incubate at room temperature, 20–30 min.
5. Incubate at 4 °C or on ice, 1–24 h (*see* **Note 6**).
6. Using an inoculating loop with the loop bent at a 45° angle from the stem, pick up an aliquot of digested ascii, and gently spread the liquid on the loop in a single streak at the bottom or in the middle of a tetrad dissecting plate, YPD, or YPDD (*see* **Note 7** and Fig. 1a).
7. Using the needle of the dissecting microscope, pick up an ascus by gently touching the needle to the surface of the agar over or near the ascus, and then removing the needle from the surface. The needle will pick up a thin layer of water; if you are lucky (or skilled), the ascus will be trapped in it. Repeat until it is. Move the stage to the position that you wish to deposit the first spore, and briefly touch the needle to the agar surface to deposit the ascus (again, if lucky—repeat as needed).
8. Disrupt the ascus by touching it repeatedly with the needle. If you are fortunate, a single spore will break off, allowing you to move the remaining three spores to the next position, and repeating the process until all four spores are separated and place in position. Efficient ascus disruption is a skill that is best learned by experimentation and repetition. Methods include: repeatedly dropping the needle on top of the ascus; finding a "cutting edge" of the needle and using it to break the ascus apart; placing the needle directly on the ascus and gently tapping the base of the microscope so that the vibrations disrupt the ascus. Care is called for—overly vigorous manipulation or vibration of the needle will create bubbles that are difficult to distinguish from spores. If difficulty is encountered in picking up or dropping spores, the needle may need cleaning. This is done by removing it from the micromanipulator, dipping the tip in a drop of 10 N potassium hydroxide for a few minutes, then rinsing with water.
9. Incubate the dissection plate for 2–3 days at 30 °C to allow germination and colony growth. If phenotypic analysis is to be done by direct replica-plating of dissection plates, keep in mind

that colonies will expand about twofold in diameter upon replica plating, and that extended incubation can produce colonies that are difficult to replica-plate.

10. Score markers by replica-plating spore colonies to appropriate selective solid media.
11. Map distances between markers can be calculated using the following formula [6]:

$$X = \frac{6N + T}{2(P + T + N)}$$

where X = map distance (in Morgans) and P, T, and N denote frequencies of parental ditype, tetratype, and nonparental ditype tetrads, respectively. A very useful online map distance calculator, along with formulae and explanations, is available at the Stahl Lab Web site [12], and an Excel spreadsheet implementation of this Web site is available upon request from the author.

3.3 Random Spores

1. Prepare and collect spores as for tetrad analysis (Subheading 3.2, **steps 1** and **2**) and resuspend in 50 μL glusulase working solution. Zymolyase can also be used (*see* **Note 5**). Incubate overnight at 30 °C or 3–4 h at 37 °C (*see* **Note 6**).
2. Add 1.0 mL 0.1 % TWEEN® 80 and transfer to a 5 mL round-bottom polypropylene tube. Cool on ice.
3. Place tube with spores in an ice-water bath. Sonicate 3 or 4 times, 30 s sonication followed by 1 min cooling (*see* **Note 8**).
4. Examine spores in hemacytometer to determine the proportion of single spores and spore concentration. If less than 90 % of spores are present as single spores, repeat sonication.
5. Dilute spores appropriately in sterile water and plate by spreading on appropriate permissive and selective media.

3.4 Sporulation in Mass Culture for DNA Analysis

1. Four days beforehand, streak diploid strains on YPD plates and incubate at 30 °C.
2. Two days beforehand, inoculate a YPD culture with a single colony and incubate with aeration at 30 °C.
3. About 18 h before the desired time of starting sporulation, dilute the overnight culture between 1/500 and 1/2,000 into 230 mL SPS in a 2 L Erlemeyer flask (*see* **Note 9**). Incubate with vigorous shaking (~300 rpm) at 30 °C.
4. Grow cultures to an OD_{600} of 1.35–1.4 (~2×10^7 cells/mL) (*see* **Note 10**).
5. Pellet culture at 3,200 × *g*, room temperature, 3 min. Resuspend in 230 mL prewarmed KAc.
6. Take a 30 mL sample (0 h) and process for freezing (Subheading 3.4, **step 9**).

7. Pellet the remaining culture at 3,200 × *g*, room temperature, 3 min. Resuspend in 400 mL KAc and transfer to 2.8 L Fernbach flask. Incubate with vigorous aeration (350 RPM) at 30 °C.
8. At desired time points, take 30 mL samples and process for freezing.
9. To freeze samples: add 8.4 mL 50 % glycerol, 0.5 % sodium azide to kill cells and stop further recombination reactions. Pellet cells, 3,200 × *g*, 4 °C, 2 min. Thoroughly remove supernatant.
10. Resuspend pellet in 1 mL spheroplasting solution + glycerol. Transfer to a 1.5 mL microfuge tube.
11. Pellet cells briefly (30 s) in microfuge, remove supernatant thoroughly, and freeze immediately on dry ice. Store at −80 °C.

3.5 Preparation of DNA from Meiotic Cells

This method for preparing DNA will not preserve recombination intermediates, but works well for analysis of DSBs and of recombination products.

1. Thaw frozen samples on ice. Resuspend in 450 μL spheroplasting solution.
2. Add 50 μL zymolyase solution. Incubate 5–15 min at room temperature, until cells appear round in the microscope and lyse when put into a drop of 1 % SDS.
3. Spin briefly (~20 s) in microfuge. Remove supernatant completely.
4. Resuspend spheroplasts in 0.5 mL of lysis solution. Add, in the following order: 25 μL 20 % SDS; 25 μL proteinase K. Mix by gentle inversion after each addition.
5. Incubate at 65 °C for 1–2 h.
6. Add 0.2 mL 5 M potassium acetate. Mix by inverting gently and place on ice, 15 min.
7. Spin 15–30 min in microfuge at 4 °C. Gently remove supernatant, taking care to avoid carrying over precipitate (*see* **Note 11**). If necessary, spin as before to remove all remnants of the precipitate.
8. Add 0.7 mL isopropanol and mix by gentle inversion. The DNA should precipitate as a large clump (*see* **Note 12**).
9. Allow precipitate to settle, then gently decant supernatant. If pellet is loose, centrifuge 2 s in a microfuge. Dry pellet briefly by inverting tube over a paper towel, removing remaining supernatant with a sterile cotton swab, if necessary.
10. Resuspend pellet in 0.3 mL TE. Add DNase-free RNase to 0.1 μg/mL. Incubate at 30 °C for 30 min.

11. Add 1/10 volume (i.e. 30 μL) 3 M sodium acetate, mix, then 2/3 volume (i.e. 0.2 mL) isopropanol. Invert gently to mix, recover DNA precipitate and dry as in Subheading 3.5, **step 9**.
12. Resuspend in 50–100 μL TE. Yield should be ~30 μg.

3.6 Gel Electrophoresis, Blotting, and Hybridization

While many different methods for gel electrophoresis, blotting to membranes, radioactive probe preparation, and hybridization to membranes may produce good results, the following methods produce excellent resolution and signal:background ratios, allowing detection of products at levels as low as 0.1 % of total. Critical aspects of this protocol are the use of maximum specific activity radioactive label and hybridization in heat-seal bags. The latter is particularly important—hybridization in glass tubes can lead to unacceptably high background, as it is almost impossible to pour off the hybridization solution and pour in the first wash solution without drying some probe onto the membrane.

This protocol uses radioactive probe. It is assumed that the user understands and will use appropriate precautions and practices for working with radioactive material, as directions for these are beyond the scope of this chapter.

3.6.1 Gel Electrophoresis and Blotting

The following are for digests where fragments to be detected are >2 kb.

1. Gel: 0.5 % SeaKem GTG agarose in TBE, with TBE as running buffer.
2. Restriction enzyme digests use 0.5–1 μg of DNA in 20 μL final volume, using restriction digestion buffer recommended by the supplier. When digests are complete, add 1/5 volume 5× loading buffer and load onto gel.
3. Electrophoresis is for 24 h at 1.9 V/cm, with buffer recirculation (*see* **Note 13**).
4. If desired, stain DNA in gel and photograph.
5. Rinse gel with water.
6. Wash gel with >2 volumes 0.25 N HCl, with gentle agitation.
7. Rinse gel with water.
8. Transfer gel contents to Zeta-Probe membrane by capillary or vacuum transfer, using Alkaline transfer buffer.
9. Rinse membrane with 0.1× SSPE, blot and dry thoroughly at room temperature.

3.6.2 Hybridization with Radioactive Probe

1. Place membrane in heat-seal bag.
2. Boil for 2 min, enough carrier DNA stock to make a final concentration of 100 μg/mL in prehybridization buffer. Use ~0.1 mL of prehybridization buffer per cm^2 of membrane

(40 mL for a 15 × 25 cm gel). Add boiled carrier DNA to prehybridization buffer, then add to membrane in bag. Press out bubbles, seal bag, and incubate at least 4 h in 65 °C shaking water bath.

3. Prepare probe as per instructions in the High Prime kit, but either adding 50 μCi α-^{32}P dCTP to a standard reaction (using 25 ng of input DNA), or using 25 μCi α-^{32}P dCTP with a half-sized reaction (using 12.5 ng of input DNA). Separate labeled probe from unincorporated nucleotide with a MicroSpin column, as instructed by manufacturer.
4. Mix probe with enough carrier DNA stock to make a final concentration of 200 μg/mL in the final hybridization mix. Use ~0.05 mL per cm^2 of membrane (20 mL for a 15 × 25 cm gel) of hybridization buffer. Boil probe + carrier, 2 min. Mix with hybridization buffer.
5. Cut open bag, pour off prehybridization buffer, and press out remaining prehybridization buffer. Add radioactive hybridization solution, press out bubbles, seal, and incubate in 65 ° shaking water bath overnight.
6. Remove membrane from bag, and wash in 500 mL of the following solutions with gentle shaking, 30 min per wash at room temperature: (a) 2× SSPE, 0.1 % SDS; (b) 0.2× SSPE, 0.1 % SDS, preheated to 55 °; (c) 0.2× SSPE, 0.1 % SDS, preheated to 55 ° (*see* **Note 14**).
7. Remove membrane from last wash and blot with blotting paper to remove remaining liquid. If subsequent stripping and rehybridization is anticipated, wrap membrane in plastic film before it dries completely; otherwise, air dry before exposing to film or phosphorstorage screen.

4 Notes

1. Thicker plates support more robust growth and sporulation; plates used for tetrad dissection are poured thinner than normal to allow better needle clearance.
2. On occasion, YPD and YEPDD plates stored at 4 °C develop small bubbles that are similar in size and appearance to yeast spores, making tetrad dissection difficult. As a precaution, plates intended for tetrad dissection are stored at room temperature.
3. In some strain backgrounds, diploids will initiate sporulation spontaneously if stored on YPD plates for >2 days, resulting in a colony with mixed genetic constitution. For this reason, diploids used for genetic analysis should not be stored on plates, but should be isolated each time as needed, or stored frozen

at −80 °C and struck for single colonies on YPD agar 2 days before needed.

4. Strain backgrounds differ widely in terms of sporulation efficiency, sporulation time, and conditions that produce optimum sporulation. For strains of the SK1 background, high levels of sporulation are achieved on KAc plates after only 1.5 days incubation at 30 °C. Other strain backgrounds (i.e. S288c) sporulate best when incubated on plates for more extended times at room temperature. Some strain backgrounds sporulate better on plates than in liquid; others the opposite. Empirical testing of different conditions may be necessary to optimize sporulation for your particular strain background.

5. Gluculase is intestinal juice from the snail *Helix pomatia*, and contains both β-glucoronidase and β-glucoronide sulfatase. Zymolyase, a mixture of β-glucan hydrolases from *Arthrobacter luteus*, is also used to digest ascal sac walls, usually as a ~1 mg/mL solution in 1 M sorbitol. Gluculase is recommended because a working solution can be stored at 4 °C indefinitely, and because experience has found it to yield dissectable ascii over a broader range of digestion conditions. However, cell walls of stationary phase cells are digested more readily by Zymolyase, making it the enzyme of choice for preparing random spores from a culture with a high fraction of unsporulated cells.

6. Yeast strains vary widely in terms of ascus susceptibility to digestion; empirical testing of incubation times may be needed to yield ascii where enough cell wall has been removed to allow easy spore separation, but not enough so that tetrads fall apart so easily that intact tetrads cannot be picked up. Varying the time of incubation in Subheading 3.2, **steps 3** and **4** may be necessary; once digestion mixes have been moved to 4 °C, they can be stored up to 24 h. On occasion, a strain background is encountered where spores themselves are unusually sensitive to gluculase digestion. If such a strain is encountered, ascii can be digested by spreading the diluted digestion mix from Subheading 3.2, **step 4** directly on dissection plates, and storing plates overnight at 4 °C before dissecting.

7. For most yeast strains, YPD agar plates yield spore germinant colonies that can be readily genotyped by replica-plating. Strains of the SK1 background tend to form germinant colonies on YPD agar that are quite clumpy; the additional glucose in YPDD plates reduces this problem considerably. Wet plates make dissection difficult—plates should be aged at least 2 days before use. When spreading ascii, avoid excessive force or spreading back and forth, both of which will disrupt the ascii, which are now held together by surface tension and remnants of the ascal sac.

8. Power setting of the sonicator should be determined empirically. We use the greatest power that does not induce foaming; with our instrument (Misonix XL2005) this is about 40 % of maximum. Spores are quite resistant to sonication, but care should be taken to avoid heating by performing sonication on ice.
9. Appropriate dilution should be determined empirically. We usually inoculate three or four flasks with different dilutions of the same overnight culture, so that one culture will be at the correct density the following morning.
10. Different spectrophotometers have different source-cuvette-detector geometries, and with some this OD (actually, a measure of light-scattering measurement) may not reflect the same cell density. Spectrophotometers with at least 1 cm distance between the cuvette and the detector should be used, as these will give an OD_{600} of ~1.4 at the desired density. Other spectrophotometers will require empirical calibration. In our hands, SK1 cultures grown in SPS to OD_{600} of 1.4 are at the transition between exponential growth and stationary phase; the corresponding transition point will need to be empirically determined if other strains are used.
11. The precipitate formed at this stage contains insoluble denatured proteins and potassium dodecyl sulfate, and will interfere with restriction enzyme digestion if carried over.
12. Isopropanol is used, instead of ethanol, to prevent precipitation of any remaining SDS.
13. These electrophoresis conditions will place fragments of ~4.5 kb in the center of a 25 cm-long gel (Fig. 2). Greater voltage gradients will reduce resolution of larger DNA fragments.
14. Higher wash temperatures or lower SSPE concentrations, which are commonly used, bring the wash solution dangerously close to the melting temperature of the short probe segments that are produced by random priming and the radioactive decay that occurs during hybridization. This can result in unwanted loss of signal.

Acknowledgements

This work was supported by the Intramural Research Program at the Center for Cancer Research, National Cancer Institute, National Institutes of Health. I thank Hardeep Kaur for images used in the figures and Robert Shroff for comments on the manuscript.

References

1. Winzeler EA, Shoemaker DD, Astromoff A et al (1999) Functional characterization of the *S. cerevisiae* genome by gene deletion and parallel analysis. Science 285:901–906
2. Gresham D, Ruderfer DM, Pratt SC et al (2006) Genome-wide detection of polymorphisms at nucleotide resolution with a single DNA microarray. Science 311:1932–1936
3. Neiman AM (2011) Sporulation in the budding yeast *Saccharomyces cerevisiae*. Genetics 189:737–765
4. Morin A, Moores AW, Sacher M (2009) Dissection of *Saccharomyces cerevisiae* asci. J Vis Exp 27:e1146
5. Singer Instruments – Cerevisiae dissection video, singerinstruments.com/index.php?option=com_content&task=view&id=164&Itemid=975
6. Perkins DD (1949) Biochemical mutants in the smut fungus *Ustilago maydis*. Genetics 34:607–626
7. Fogel S, Mortimer R, Lusnak K et al (1979) Meiotic gene conversion: a signal of the basic recombination event in yeast. Cold Spring Harb Symp Quant Biol 43(Pt 2):1325–1341
8. Papazian HP (1952) The analysis of tetrad data. Genetics 37:175–188
9. Saccharomyces Genome Database (2013) www.yeastgenome.org
10. Borts RH, Lichten M, Hearn M et al (1984) Physical monitoring of meiotic recombination in *Saccharomyces cerevisiae*. Cold Spring Harb Symp Quant Biol 49:67–76
11. Oh SD, Jessop L, Lao JP et al (2009) Stabilization and electrophoretic analysis of meiotic recombination intermediates in *Saccharomyces cerevisiae*. Methods Mol Biol 557:209–234
12. Stahl Lab Online Tools, *molbio.uoregon.edu/~fstahl/*
13. Jessop L, Allers T, Lichten M (2005) Infrequent co-conversion of markers flanking a meiotic recombination initiation site in *Saccharomyces cerevisiae*. Genetics 169:1353–1367

Chapter 3

PCR Mutagenesis and Gap Repair in Yeast

Mark Weir and Jill B. Keeney

Abstract

Random mutagenesis of a gene and subsequent screening for phenotypic properties is a basic genetic procedure for studying gene function. When working with yeast, cloning mutated alleles into a yeast vector can be done directly due to the efficiency of homologous recombination. Here we give sample protocols for introducing random mutations to a gene using PCR and transformation of the resulting product into yeast. After screening resultant colonies for the desired phenotype, the plasmid containing the selected mutation is easily rescued to bacteria for sequencing. A reliable protocol for rescuing yeast plasmids to bacteria is given.

Key words Yeast, Mutagenesis, PCR, Homologous recombination

1 Introduction

Introduction of mutations to a specific gene or gene region and subsequent assaying of effects on gene function is a basic method for dissecting gene layout and functionality. Mutations can be designed individually (site-directed) or introduced randomly. If little is known about the functional regions of a protein, the creation of a library of random mutations to the gene and in vivo screening of the library for a specific phenotype can potentially define functional regions. A common method for introducing ran dom mutations is mutagenic PCR using a relatively low-fidelity *Taq* polymerase enzyme in an amplification reaction targeting the gene or region of interest [1, 2]. Depending on the size of the target region, and the level of mutagenesis desired, variation of PCR conditions (i.e. altering dNTP concentrations or adding $MnCl_2$) can increase the frequency of mutagenesis in the target region. The resulting PCR products are then cloned into a vector for subsequent analysis.

Due to the efficiency of homologous recombination in yeast, the cloning step can be achieved by co-transformation of a gapped plasmid with the PCR product directly into yeast, resulting in efficient

Jeffrey S. Smith and Daniel J. Burke (eds.), *Yeast Genetics: Methods and Protocols*, Methods in Molecular Biology, vol. 1205, DOI 10.1007/978-1-4939-1363-3_3, © Springer Science+Business Media New York 2014

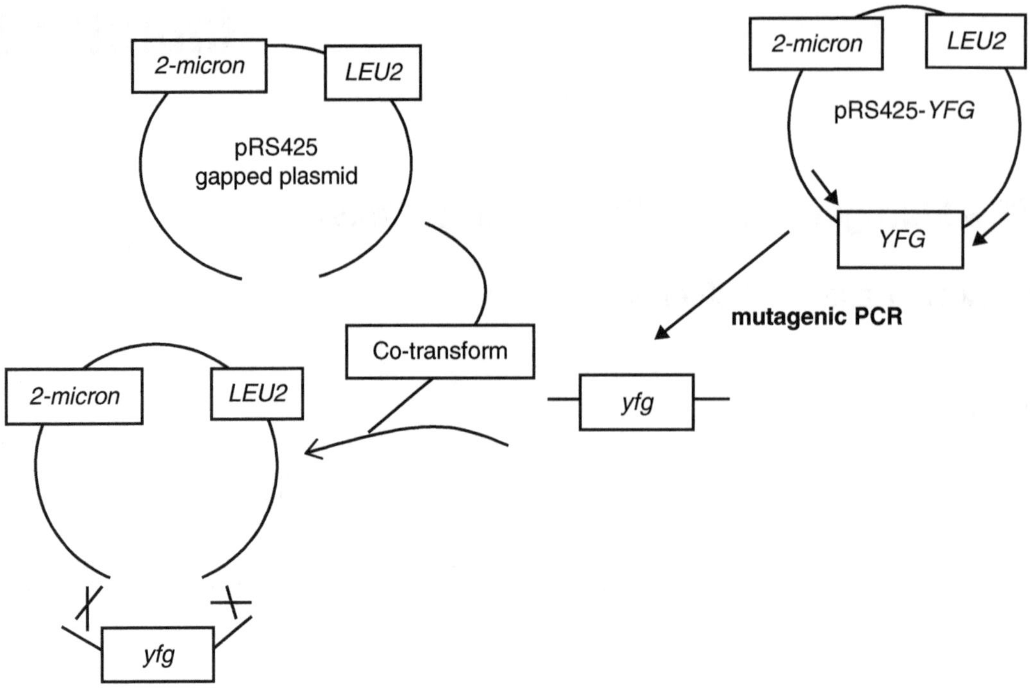

Fig. 1 Using plasmid DNA as template, a mutagenic PCR product of Your Favorite Gene (*YFG*) is produced using primers flanking the region of interest (short *black arrows*). The resulting products contain randomly introduced mutations. The transformation plasmid is gapped, leaving ends that overlap the PCR product. Upon co-transformation of the gapped plasmid and the PCR product, yeast homologous recombination mechanisms repair the gap with the PCR product, resulting in a library of yeast clones each containing a unique mutation. The resulting colonies are then directly screened for the desired phenotype. Plasmid is rescued and sequenced from selected clones

repair of the plasmid gap by the PCR product [3–5] (*see* Fig. 1). The resulting yeast clones are then screened for the desired mutant phenotype. Finally, the plasmid is rescued from clones containing the desired phenotype and sequenced to determine the genotype.

The optimum conditions for generating the desired rate of mutations are best determined empirically and numerous articles and reviews are available for guidance [6–8]. This chapter gives detailed protocols for mutagenizing a relatively small yeast gene (target region ~1,000 bp), cloning of the mutagenized target by homology-driven gap repair, and subsequent rescue of desired plasmids for sequencing.

2 Materials

1. Yeast rich medium (YPD, liquid) and selective medium (liquid and plates).
2. *Taq* polymerase.

3. *Taq* polymerase buffer (50 mM KCl, 10 mM Tris–HCl, pH 8.3, and 1.5 mM $MgCl_2$) or as supplied by manufacturer.
4. PCR primers (*see* Subheading 3.1).
5. 1 mM dNTPs, 10 mM dCTP, 10 mM dTTP.
6. 10 mM $MgCl_2$.
7. 1 M lithium acetate (LiAc).
8. 100× TE.
9. 44 % polyethylene glycol (PEG-3350) in water and filter sterilized.
10. 50 % glycerol in 0.1 M LiAc/1× TE (10 mM Tris–HCl pH 7.6, 1 mM EDTA), filter sterilized.
11. 0.1 M LiAc/1× TE (10 mM Tris–HCl pH 7.6, 1 mM EDTA).
12. PEG/LiAc solution (40 % PEG3350, 0.1 M LiAc, 1× TE (10 mM Tris–HCl pH 7.6, 1 mM EDTA)).
13. QIAprep spin column and reagents (Qiagen) or similar.

3 Methods

3.1 Primer and Plasmid Design

PCR primers are designed to flank the region of the gene to be mutated, or the entire gene, if desired. The targeted region should have two restriction sites between the primer annealing sites which do not cut elsewhere in the plasmid. These enzymes will be used for gapping the plasmid to minimize wild-type sequence available for the gap repair process. In order to eliminate wild-type sequence entirely, we use primers to the multiple cloning site of the pRS vector series [9, 10]. The template DNA for mutagenic PCR is a pRS plasmid containing the desired gene insert, while the gapped plasmid is the parent vector (no insert) cut within the cloning site. We have used this strategy successfully with a 1.5 kb insert in pRS425, using forward primer sequence 5′-GTTGTAAAACGACGGCCAGT-3′ and reverse primer sequence 5′-AAGGGAACAAAAGCTGGAGC-3′. The parent pRS425 plasmid was linearized with the restriction endonucleases *Xho*I and *Hind*III.

3.2 Mutagenic PCR and Preparation of Gapped Plasmid

1. Digest the plasmid containing the gene of interest with an enzyme that cuts at a unique site outside the target region; use this DNA as the template DNA. Since the template plasmid usually has the same selectable marker as the gapped plasmid used for co-transformation, linearizing the template plasmid reduces the background of transformations from the template DNA.
2. Set up a standard PCR reaction to confirm that primers yield a product under standard conditions. Our standard reaction contains 2–20 ng of template DNA, 0.2 μM each forward and reverse primer, 200 μM each dNTP, 50 mM KCl, 10 mM Tris–HCl, pH 8.3, and 1.5 mM $MgCl_2$ and 5 units *Taq* polymerase.

Then set up a mutagenic PCR reaction containing 2–20 ng of template DNA, 0.2 μM each forward and reverse primer, 0.2 μM dATP and dGTP, 1 mM dCTP and dTTP, 10 mM Tris–HCl, pH 8.3, and 5 mM $MgCl_2$ and 5 units *Taq* polymerase (*see* **Note 1**). PCR parameters are the same for both standard and mutagenic reactions using the primer sequences given in 3.1: preheat 95 °C 3 min, then cycle 94 °C 1 min, 46.7 °C 1 min, 72 °C 2 min for 30 cycles (*see* **Note 2**).

3. Store the reactions at 4 °C. Check for PCR product by agarose gel electrophoresis and proceed with transformation the same day, preferentially; no later than the next day (*see* **Note 3**).
4. Digest the plasmid to be gapped with restriction enzymes that cut between the primer sequences (*see* **Note 4**).

3.3 Transformation into Yeast by Lithium Acetate (LiAc) Transformation [11]

1. Inoculate 5 mL of appropriate selective medium with one colony of desired yeast strain and incubate at 30 °C in rotator 16–18 h.
2. Incubate 100 mL sterile YPD overnight at 30 °C.
3. In the morning, pour contents of inoculant into the prewarmed YPD and record an initial $O.D._{600}$ using YPD as a blank. The initial OD_{600} should be 0.1–0.2.
4. Grow to an OD_{600} of 0.7–1.0
5. Divide the sample into two (2) 50 mL conical tubes. Centrifuge tubes at 360 × *g* for 10 min in a tabletop centrifuge. Suspend pellets in 50 mL sterile water.
6. Centrifuge tubes at 360 × *g* for 10 min; suspend each pellet in sterile 25 mL LiAc/TE (0.1 M LiAc, 10 mM Tris–HCl pH 7.6, 1 mM EDTA).
7. Combine contents of both tubes and centrifuge at 360 × *g* for 10 min. Suspend in 5 mL sterile LiAc/TE.
8. Add 50 % glycerol (diluted in sterile LiAc/TE to a final concentration of 15 % glycerol and mix well).
9. Aliquot (100 μL) into microcentrifuge tubes and store at −70 °C.
10. Per 100 μL of competent yeast, add 20 μg boiled carrier DNA, 100–200 ng gapped plasmid, and 100–400 ng mutagenic PCR product (*see* **Note 5**). It is also recommended to do a reaction containing gapped plasmid only (no mutagenic PCR product) to confirm that addition of the PCR product increases transformation efficiency.
11. Incubate 5–10 min, room temperature.
12. Add 500 μL of PEG/LiAc/TE (40 % PEG3350, 0.1 M LiAc, 10 mM Tris–HCl pH 7.6, 1 mM EDTA). Mix well and incubate 30–45 min at 30 °C.
13. Heat shock 15 min at 42 °C.

14. Centrifuge briefly in a microfuge to pellet yeast. Suspend in 200 µL sterile H_2O and plate on appropriate selective medium (*see* **Note 6**).
15. Incubate plates at 30 °C 3–4 days until single colonies appear. Screen colonies directly for the desired mutant phenotype (*see* **Note 7**).

3.4 Plasmid Rescue from Yeast Using Qiagen Spin Columns (See Note 9)

(Adapted from online protocol http://labs.fhcrc.org/gottschling/Yeast%20Protocols/yplas.html.)

1. Grow 2.0 mL yeast culture in selective medium, 20–24 h, 30 °C. (Be sure that growth is thick; should be milk shake consistency, not chicken broth consistency.)
2. Centrifuge 1.0 mL of culture in screw cap microfuge tubes, with O-ring by bringing microfuge briefly to full speed. Aspirate supernatant.
3. Add remaining 1.0 mL of culture to pellet. Centrifuge as in **step 2**. Aspirate.
4. Suspend yeast in 250 µL of Qiagen buffer P1.
5. Add about 250 µL of glass beads (0.5 mm).
6. Beat for 3 min using a Disruptor Genie (Scientific Industries) or similar cell disruptor.
7. Add 250 µL of buffer P2 and gently mix well (Do not let lysis reaction continue for more than 5 min.)
8. Add 350 µL of buffer N3 and mix immediately, but gently.
9. Centrifuge for 10 min, 18,000 × *g* (full speed on most microfuges).
10. Apply supernatant to QIAprep spin column in a collection tube.
11. Centrifuge column for 45 s on full speed. Discard flow-though.
12. Wash column with 750 µL of Buffer PE and centrifuge 45 s, full speed.
13. Discard flow through and centrifuge for an additional 45 s.
14. Place column in a clean tube and add 50 µL of buffer EB.
15. Let sit for 1 min then centrifuge for 1 min, 45 s.
16. Transform 1–2 µL of DNA per aliquot of competent *E. coli*, plating on the appropriate selective medium for bacteria (*see* **Note 8**).
17. Pick bacterial colonies and proceed with bacterial plasmid DNA isolation and sequencing (*see* **Note 9**).

4 Notes

1. For a given set of mutagenic conditions and primer pair, PCR conditions may need to be optimized for temperature and $MgCl_2$ concentration to get a sufficient yield of PCR product.

We have found that a small change in $MgCl_2$ concentration (i.e. from 4 to 6 mM) impacts the yield of PCR product.

2. In a set of 136 randomly selected colonies, this mutagenesis protocol yielded a total of 215 mutations in a 600 bp region containing the reading frame of the gene of interest. The mutagenesis rate (total # mutations/plasmids sequenced was 1.58). The frequency obtained was: 43 % no mutations, 18 % single mutations, 27 % double mutations, 12 % triple mutations, and 11 % of clones with four or more mutations.
3. The product does not need to be purified if a single band is obtained in the PCR reaction.
4. We have not found it necessary to purify the digested plasmid.
5. Co-transformation should be done within a day following PCR. We have found that PCR product stored for longer periods does not yield mutant colonies, probably due to degradation of PCR product ends.
6. The number of resultant colonies will vary depending on transformation conditions and cell competency. It is recommended to plate a range of volumes to determine the best volume for optimal colony screening.
7. If no phenotypic variants are found, the plasmid can be rescued from 10 to 12 randomly selected mutants to confirm that mutants were indeed generated. If none are found, then the mutagenic PCR conditions will need to be altered to obtain a higher frequency of mutagenesis.
8. Generally, this yields 10–100 colony forming units (CFU). Using up to 20 μL will increase the yield of transformants.
9. Pick 3 and compare by RE analysis; occasionally rescued plasmid has a deletion.

References

1. Leung DW, Chen E, Goeddel DV (1989) A method for random mutagenesis of a defined DNA segment using a modified polymerase chain reaction. J Methods Cell Mol Biol 1:11–15
2. Fromant M, Blanquet S, Plateau P (1995) Direct random mutagenesis of gene-sized DNA fragments using polymerase chain reaction. Anal Biochem 224:347–353
3. Boulton SJ, Jackson SP (1996) Saccharomyces cerevisiae Ku70 potentiates illegitimate DNA double- strand break repair and serves as a barrier to error-prone DNA repair pathways. EMBO J 15:5093–5103
4. Muhlrad D, Hunter R, Parker R (1992) A rapid method for localized mutagenesis of yeast genes. Yeast 8:79–82
5. Tsukamoto Y, Kato J, Ikeda H (1997) Silencing factors participate in DNA repair and recombination in *Saccharomyces cerevisiae.* Nature 388:900–903
6. Mccullum EO, Williams BA, Zhang J et al (2010) Random mutagenesis by error-prone PCR. Methods Mol Biol 634:103–109
7. Labrou NE (2010) Random mutagenesis methods for in vitro directed enzyme evolution. Curr Protein Pept Sci 11:91–100

8. Vartanian JP, Henry M, Wain-Hobson S (1996) Hypermutagenic PCR involving all four transitions and a sizeable proportion of transversions. Nucleic Acids Res 24:2627–2631
9. Sikorski RS, Hieter P (1989) A system of shuttle vectors and yeast host strains designed for efficient manipulation of DNA in *Saccharomyces cerevisiae*. Genetics 122:19–27
10. Brachmann CB, Davies A, Cost GJ et al (1998) Designer deletion strains derived from *Saccharomyces cerevisiae* S288C: a useful set of strains and plasmids for PCR-mediated gene disruption and other applications. Yeast 14:115–132
11. Schiestl RH, Gietz RD (1989) High efficiency transformation of intact yeast cells using single stranded nucleic acids as a carrier. Curr Genet 16:339–346

Chapter 4

PCR-Mediated Epitope Tagging of Genes in Yeast

Radhika Mathur and Peter Kaiser

Abstract

Epitope tagging of genes is a powerful technique facilitating assays for gene function, determination of subcellular distribution of proteins, affinity purification, study of protein interaction with other proteins, DNA or RNA, and any other antibody-based approach in the absence of protein-specific antibodies. Here, we describe a one-step PCR-based strategy for insertion of epitope tags at the chromosomal locus. This method takes advantage of efficient homologous recombination in yeast. PCR amplified tags are directed to desired chromosomal loci with the help of primer-encoded flanking homologous sequences enabling selective epitope tagging of genes of interest.

Key words *Saccharomyces cerevisiae*, Epitope tagging, Transformation, PCR-mediated gene modification

1 Introduction

Saccharomyces cerevisiae is one of the most extensively studied eukaryotic model organisms in molecular and cellular biology. One of the characteristics that facilitate its use in rigorous genetic and biochemical analyses is highly active homologous recombination. A gene of interest can be efficiently and accurately replaced or tagged by transformation with a heterologous DNA fragment bearing as little as 40 nucleotides homology at its ends to the target sequence [1, 2].

Epitope tagging of open reading frames in yeast is a routine molecular biology procedure and is a valuable tool for studying proteins. The strategy requires a pair of primers that contain within their 5′ region, around 40-nucleotides of homology to the genomic region of interest and around 20-nucleotide homology at their 3′end to the tag cassette (Fig. 1). A large array of cassette plasmids with a wide variety of tags (Table 1) in combination with different selection markers is now readily available for use in yeast [3, 4]. These plasmids mostly comprise of the same set of linker regions flanking the tag cassette such that a small set of primers can be used

Jeffrey S. Smith and Daniel J. Burke (eds.), *Yeast Genetics: Methods and Protocols*, Methods in Molecular Biology, vol. 1205, DOI 10.1007/978-1-4939-1363-3_4, © Springer Science+Business Media New York 2014

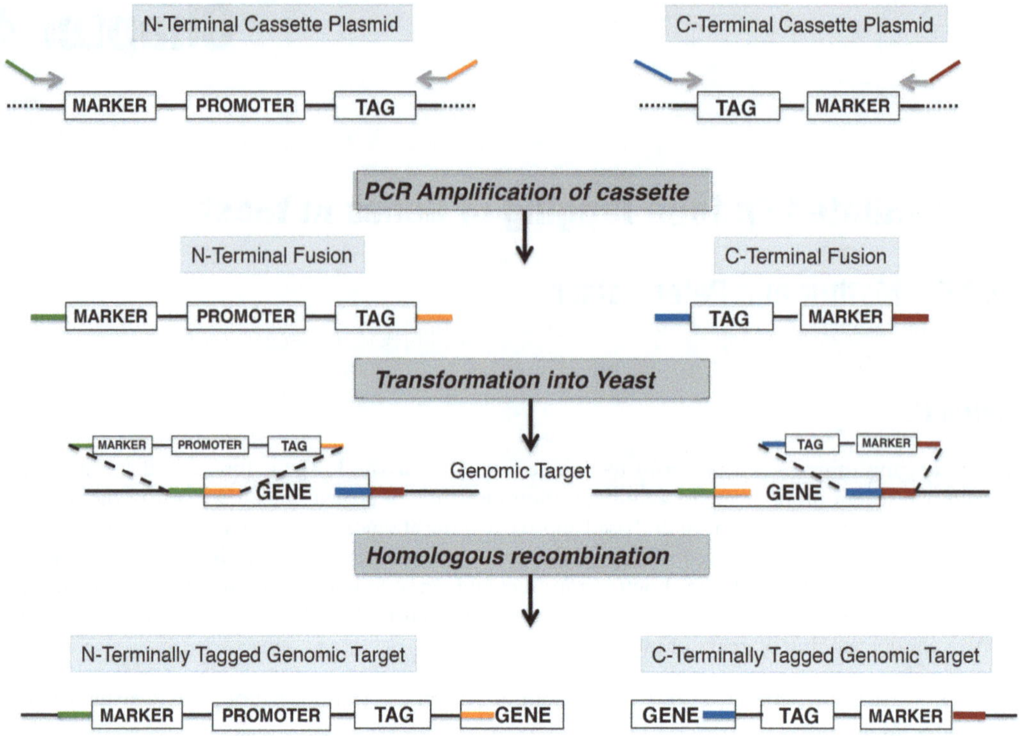

Fig. 1 The scheme of PCR-based epitope tagging

for an assortment of genome manipulations [3, 5]. Epitope tags are added either at the carboxyl (C) or amino (N) terminus of a protein. N-terminal tags typically rely on the use of heterologous promoters, though two-step strategies exist to tag genes at their N-termini under the control of their natural promoters [6].

Tag cassettes are PCR-amplified using these homology-bearing primers, and the PCR products are transformed directly into yeast. Transformants are selected based on the auxotrophic or antibiotic-resistance marker linked to the epitope tag. Successful epitope tagging is confirmed by immunoblotting to detect the protein of interest fused to the epitope tag.

2 Materials

Prepare all solutions in distilled water and store at room temperature, unless otherwise indicated.

2.1 PCR Amplification of Tag Cassette

1. TaKaRa ExTaq DNA Polymerase Kit (TaKaRa Bio Inc., Japan).
2. 3 M Sodium acetate, pH 5.2.
3. 100 % Ethanol (200 Proof).
4. 70 % Ethanol.

Table 1
List of tags available for epitope tagging in yeast

3HA (Hemagglutinin)
GFP (S65T) green fluorescent protein
Ds-Red
yEGFP
ProA
TEV-ProA
TEV-ProA-7His
TEV-GST-6His
HBH (RGS6XHis-Biotin-6XHis)
HTB (RGS6XHis-TEV-Biotin)
Biotin
RGS18XHis
6XHis
S-tag
TAP
T7
Strep-tag II
FLAG
HSV
V5
VSV-G
GST (Glutathione S-transferase)
13Myc

Most plasmids cassettes with the above tags and various selection markers are available through Addgene (www.addgene.org/) and EUROSCARF (http://web.uni-frankfurt.de/fb15/mikro/euroscarf/plasmid.html)

5. 1 % Agarose gel made in 1× TAE: 40 mM Tris acetate, 1 mM Ethylenediaminetetraacetic acid (EDTA).
6. 1× TE: 10 mM Tris–HCl, 1 mM EDTA, pH 7.5 (sterilized by autoclaving).

2.2 Yeast Transformation

1. YEP medium: 1 % (w/v) yeast extract, 2 % (w/v) tryptone, 0.006 % (w/v) adenosine and uracil (sterilized by autoclaving).
2. 1 M Lithium acetate (LiOAc), pH 7.5 (filter sterilized).
3. 10× TE: 100 mM Tris–HCl, 10 mM EDTA, pH 7.5 (sterilized by autoclaving).

4. 62.5 % Polyethylene glycol 3350 (PEG) (filter sterilized).
5. Sheared salmon sperm DNA (10 mg/mL) Stored at –20 °C.
6. Dimethyl sulphoxide, >99.9 % purity.

2.3 Western Blotting

1. YEP media: 1 % (w/v) yeast extract, 2 % (w/v) tryptone, 0.006 % (w/v) adenosine and uracil (sterilized by autoclaving).
2. Urea Buffer: 8 M Urea, 300 mM NaCl, 100 mM Tris–HCl (pH 7.5), 0.2 % SDS, 10 mM sodium pyrophosphate, 5 mM EDTA, 5 mM EGTA, 50 mM NaF, 0.1 mM orthovanadate.
3. 100 mM Phenylmethanesulfonylfluoride (PMSF) in isopropanol. Should be stored at 4 °C.
4. Protease Inhibitors: Aprotenin, Pepstatin, Leupeptin (each 1 mg/mL). Stored at 4 °C.
5. Glass beads, 0.5 mm (Biospec Products, Inc., Bartlesville, OK).
6. Antifoam A (Sigma-Aldrich).
7. 4× SDS sample buffer: 250 mM Tris–HCl pH 6.8, 8 % (w/v) SDS, 300 mM DTT, 30 % (v/v) glycerol, 0.02 % (w/v) bromophenol blue.
8. 0.45 μm nitrocellulose PVDF membrane.
9. 10× TBS-T (Tris-buffered saline with Tween): 1.37 M NaCl, 27 mM KCl, 250 mM Tris–HCl pH 7.4, 0.2 % (v/v) Tween 20.
10. Blocking buffer: 5 % (w/v) nonfat dry milk in 1× TBS-T.
11. 10× TBS: 1.37 M NaCl, 27 mM KCl, 250 mM Tris–HCl pH 7.4.
12. 2 % Sodium azide.

3 Methods

3.1 PCR Amplification of Tag Cassette

1. Perform PCR to amplify tag module using TaKaRa ExTaq DNA Polymerase Kit (*see* **Note 1**). Prepare a 300 μL reaction containing 30 μL 10× Polymerase Buffer, 2.5 mM $MgCl_2$, 1 μM each primer, 0.2 mM each dNTP, 0.1 μg plasmid DNA (containing the tag cassette) and 7.5 U of Taq DNA polymerase.
2. Divide the amplification mix from above into 3 or 6 PCR tubes with 100 μL or 50 μL each, respectively.
3. Use the following amplification conditions: Denaturation at 94 °C for 5 min, followed by 5 cycles of 94 °C for 1 min, 45 °C for 1 min, 72 °C for 2 min, and 22 cycles of 94 °C for 1 min, 60 °C for 1 min, 72 °C for 2 min and conclude with an extension step at 72 °C for 10 min (*see* **Note 2**).
4. Combine the PCR reactions into one microfuge tube (about 300 μL total volume).

5. Analyze 5 µL of PCR product by electrophoresis on a 1 % agarose gel to test efficiency of the PCR reaction.
6. Add 30 µL 3 M sodium acetate and 660 µL 100 % ethanol to the remaining PCR product and incubate at −20 °C for 20 min to precipitate DNA (*see* **Note 3**).
7. Centrifuge in a microfuge at 20,000×*g* for 10 min to pellet DNA. Carefully decant supernatant. Add 1 mL 70 % ethanol. Mix and spin again briefly. Carefully decant supernatant. Air-dry the pellet for 10 min and dissolve PCR product in 20 µL 1× TE.

3.2 Yeast Transformation

This protocol has been adapted from [7] and [8].

All steps at room temperature unless otherwise stated.

1. Grow a starter culture of yeast overnight with shaking at 30 °C in 5 mL of appropriate media (*see* **Note 4**).
2. Next day, dilute the culture to A_{600} ~ 0.3 to start 10 mL of secondary culture of yeast and grow to mid-log phase (A_{600} ~ 0.8) (*see* **Note 5**).
3. Collect cells by centrifugation at 2,000×*g* for 3 min. Gently resuspend pellet in 2 mL LiOAc/ TE (10 mM Tris–HCl, 1 mM EDTA, 100 mM LiOAc) and then spin again for 3 min. Remove the buffer and gently resuspend cells in 100 µL LiOAc/ TE and transfer to a fresh 1.5 mL centrifuge tube.
4. Add 5 µL of denatured salmon sperm DNA (*see* **Note 6**) and 5 µL of ethanol-precipitated PCR product from Subheading 3.1, **step 7** (*see* **Note 7**). Gently mix by flicking the tube, and then incubate at room temperature without shaking for 5 min.
5. Prepare 280 µL 50 % PEG in LiOAC/TE solution for each transformation aliquot. For preparing 1 mL of this solution, add 100 µL 10× TE, 100 µL 1 M LiOAc, and 800 µL 62.5 % PEG.
6. Add 50 % PEG solution to the transformation reaction, mix by inverting the tube and incubate at 30 °C for 30–60 min without shaking (*see* **Note 8**).
7. Add 43 µL DMSO and mix immediately by inverting the tube.
8. Heatshock in a water bath at 42 °C for 5 min.
9. Add 1 mL 1× TE and centrifuge at 5,000×*g* for 30 s to pellet the cells. Resuspend the cell pellet in 200 µL 1× TE and plate cells on selection plates (*see* **Note 9**).
10. Transformants usually appear after 2 days on plates incubated at 30 °C. Streak transformants onto fresh selection plates (*see* **Note 10**).

3.3 Western Blotting

To confirm successful epitope tagging 5–10 transformants should be analyzed by Immuno blotting. The rate of precise integration at the desired open reading frame varies depending on the locus, but is usually around 70 %. Any protein extraction and Western blotting procedure will work to detect successful insertion of the epitope tag. Below we describe extraction using denaturing conditions, which preserves protein integrity better than many other protocols.

1. Grow transformants overnight with shaking in 1 mL of appropriate media. Also, use the parental strain without epitope tag as negative control sample. Next day, start 5–10 mL secondary culture of yeast and grow it up to A_{600} ~ 0.8. Collect cells by centrifugation (20,000 × *g*, 1 min, room temperature), or by filtration. Quickly freeze the cell pellet and store it at −80 °C or process as below.
2. Lyse cell pellets in 200 μL of Urea Buffer. Add protease inhibitors to buffer: 1 mM PMSF and 1 μg/mL each—aprotenin, leupeptin, and pepstatin (*see* **Note 11**) with 200 μL of glass beads and 1 μL of Antifoam A. Perform lysis by rapid shaking using FastPrep FP120 (Qbiogene, Carlsbad, CA) at setting 4.5, three times for 40 s each, at 4 °C (*see* **Note 12**).
3. Separate the lysate from glass beads (*see* **Note 13**). Centrifuge the lysate at 20,000 × *g* for 10 min at room temperature to remove cell debris and transfer the clarified supernatant to a fresh microfuge tube.
4. Quantify protein concentration and dilute samples to 4 M Urea using 2× SDS sample buffer. Separate 20–40 μg of protein lysate on a standard SDS polyacrylamide gel.
5. Transfer proteins onto PVDF membrane, block the membrane at room temperature with blocking buffer for 40 min with shaking and finally incubate the blot overnight with shaking at 4 °C with appropriate antibody, diluted at manufacture recommended dilution in blocking buffer with 0.02 % sodium azide (*see* **Note 14**).
6. Wash the blot twice with 1× TBS-T (*see* **Note 15**) and then incubate with secondary antibody (diluted in blocking buffer according to manufacturer's recommendations) for 1 h at room temperature with shaking.
7. Wash the blot twice with 1× TBS-T and twice with 1× TBS and then proceed to developing the blot using a chemiluminescent substrate.

4 Notes

1. Any other Taq polymerase kit can also be used.
2. The first set of PCR cycles at 45 °C annealing temperature, ensures annealing of the short linker region of the primer to the tag cassette. The latter cycles at annealing temperature of 60 °C, allow for optimal annealing of the whole primer to the generated PCR product template.

3. Precipitation of PCR DNA is an optional step but helps to increase the efficiency of yeast transformation.
4. Generally yeast strains should be grown in YEP media containing 2 % dextrose while strains requiring selection to retain plasmids should be grown in corresponding minimal media containing 2 % dextrose. Temperature-sensitive strains should be grown at room temperature.
5. For each transformation reaction, use 10 mL of yeast culture. Make sure to reserve one reaction for a control transformation.
6. To enhance transformation efficiency, heat sheared salmon sperm DNA in 100 °C water bath for 10 min and then put on ice before use.
7. A "no PCR" control transformation should also be performed with salmon sperm DNA but no PCR product.
8. To increase transformation efficiency, cells can be incubated at room temperature over night at this stage. Temperature-sensitive strains should be incubated at room temperature for a minimum of 60 min.
9. If you select for G418, hygromycin, or zeocin resistance, incubate cells in 1 mL YEP-Dex media for at least 4 h at 30 °C or at room temperature (for Zeocin selection incubate overnight at room temperature) to allow for expression of the selectable marker. If using microfuge tubes make sure to avoid popping of the caps due to CO_2 generation by growing yeast cells. Cap protector for microfuge tubes or screw-cap tubes can be used.
10. Antibiotic-containing plates can have high background from transiently transformed cells, containing nonintegrated PCR products. Replica plate the transformants onto a new selection plate after incubating for 2 days at 30 °C. Transiently transformed cells will no longer form colonies because nonintegrated PCR fragments are unstable. In some cases an additional round of replica plating after 24 h is necessary.
11. Urea buffer can be made before hand and stored at room temperature. PMSF and protease inhibitors should be added fresh, prior to use.
12. Cells can be vortexed with glass beads five times for 1 min at 4 °C, with a 1 min break on ice between the runs. Many other methods also exist for breaking yeast cells, such as TCA extraction, or alkali lysis.
13. A convenient way to separate the lysate from glass beads is to poke a hole at the bottom of the tube with a 21G needle, insert the tube with the hole into a fresh tube and centrifuge carefully for 30 s at 1,000 ×*g*. The lysate will pass through the hole in the upper tube and collect in the lower tube, whereas the glass beads shall remain in the upper tube.

14. By adding sodium azide, one can increase the life of the diluted antibody. It can be used repeatedly and can be stored at −20 °C for a period of few months. Note that sodium azide should not be added to antibodies that are directly conjugated with HRP.
15. Each wash is performed for 10 min by shaking at room temperature.

References

1. Baudin A, Ozier-Kalogeropoulos O, Denouel A et al (1993) A simple and efficient method for direct gene deletion in *Saccharomyces cerevisiae*. Nucleic Acids Res 21:3329–3330
2. Schneider BL, Seufert W, Steiner B et al (1995) Use of polymerase chain reaction epitope tagging for protein tagging in *Saccharomyces cerevisiae*. Yeast 11:1265–1274
3. Longtine MS, McKenzie A 3rd, Demarini DJ et al (1998) Additional modules for versatile and economical PCR-based gene deletion and modification in *Saccharomyces cerevisiae*. Yeast 14:953–961
4. Janke C, Magiera MM, Rathfelder N et al (2004) A versatile toolbox for PCR-based tagging of yeast genes: new fluorescent proteins, more markers and promoter substitution cassettes. Yeast 21:947–962
5. Tagwerker C, Zhang H, Wang X et al (2006) HB tag modules for PCR-based gene tagging and tandem affinity purification in *Saccharomyces cerevisiae*. Yeast 23:623–632
6. Booher KR, Kaiser P (2008) A PCR-based strategy to generate yeast strains expressing endogenous levels of amino-terminal epitope-tagged proteins. Biotechnol J 3:524–529
7. Hill J, Donald KA, Griffiths DE (1991) DMSO-enhanced whole cell yeast transformation. Nucleic Acids Res 19:5791
8. Gietz RD, Schiestl RH (2007) High-efficiency yeast transformation using the LiAc/SS carrier DNA/PEG method. Nat Protoc 2:31–34

Chapter 5

Manipulating the Yeast Genome: Deletion, Mutation, and Tagging by PCR

Jennifer M. Gardner and Sue L. Jaspersen

Abstract

Saccharomyces cerevisiae is an ideal model eukaryotic system for the systematic analysis of gene function due to the ease and precision with which its genome can be manipulated. The ability of budding yeast to undergo efficient homologous recombination with short stretches of sequence homology has led to an explosion of PCR-based methods to delete and mutate yeast genes and to create fusions to epitope tags and fluorescent proteins. Here, we describe commonly used methods to generate gene deletions, to integrate mutated versions of a gene into the yeast genome, and to construct N- and C-terminal gene fusions. Using a high-efficiency yeast transformation protocol, DNA fragments with as little as 40 bp of homology can accurately target integration into a particular region of the yeast genome.

Key words *Saccharomyces cerevisiae*, Budding yeast, PCR, Gene deletion, Epitope tagging, GFP/YFP/CFP/BFP2/mCherry/mKate2/mRuby2, Mutation, High efficiency transformation, Whole colony PCR, Genomic DNA isolation

1 Introduction

Gene targeting by homologous recombination is one of the most powerful and important techniques available for studies in yeast. A gene at its normal chromosomal location can be removed or replaced with an allele created in vitro, such that the only genetic difference between the initial strain and the final strain is that particular allele. Therefore, phenotypes conferred by null mutations or any other types of mutations can be analyzed. Genes can also be modified so that an epitope tag (i.e., myc, HA, or FLAG) is added, or the gene can be fused to the coding sequence for fluorescent proteins, such as green fluorescent protein (GFP). Because the epitope tag or fusion is made in the genomic context, the tagged gene is subject to native regulation. The properties of a strain containing the epitope tag or fusion can be compared to an isogenic wild-type strain that lacks the tag to study gene function, localization, and regulation.

Jeffrey S. Smith and Daniel J. Burke (eds.), *Yeast Genetics: Methods and Protocols*, Methods in Molecular Biology, vol. 1205, DOI 10.1007/978-1-4939-1363-3_5, © Springer Science+Business Media New York 2014

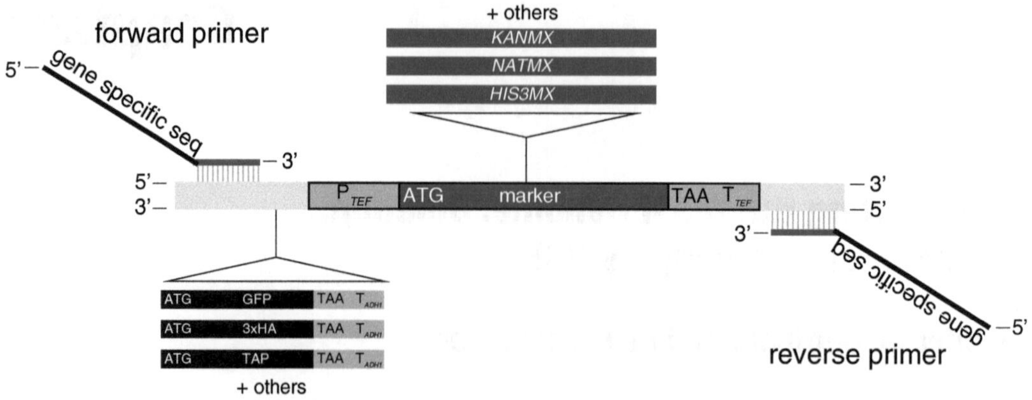

Fig. 1 Plasmids for functional analysis. Schematic showing the modular design of plasmids for functional analysis (pFA) that can be used for gene deletion and tagging. A dominant marker, such as the *S. pombe his5+* gene (*HIS3MX*), the *kan* gene from the *E. coli* transposon Tn903 (*KANMX*), or the *S. noursei nat1* gene (*NATMX*), is located between the promoter and terminator from the *A. gossypii TEF* gene. The sequence for a fusion protein (GFP), epitope tag (3xHA), or protein affinity tag (TAP) is also present in some cassettes and generally has a stop codon and a terminator from *ADH1*. Sites in the plasmid backbone of pFA allow for annealing of forward (F) and reverse (R) primers for PCR amplification of the module. In this way, a common set of primers can be used to delete a gene with different markers or create fusions to different fluorescent proteins or epitope tags. The site of integration is dependent on the gene-specific sequence in the 40–60 bp primer tails (shown in *black*) since there is no homology between the module and the yeast genome

Traditionally, gene deletions were made by one-step gene replacement using a plasmid that contains ends of the target gene where insertion is desired flanking a selectable marker [1]. Introduction of mutations generally involved generating a gene deletion followed by a "plasmid shuffle," in which a plasmid containing the mutant copy of the gene was exchanged for a plasmid containing the wild-type copy of the gene [2, 3]. Alternatively, a mutant version of the gene could be introduced into the genomic locus using a two-step gene replacement strategy involving a counterselectable marker such as *URA3* [3, 4]. Gene tagging was also done in much the same way, either through a plasmid shuffle in which a plasmid containing a tagged copy of the gene was exchanged for a plasmid containing the wild-type copy, or by two-step gene replacement [4].

Today, gene replacements and tagging are most often done by PCR, utilizing linear DNA fragments that contain ~40–60 bp of homology to the target gene and a selectable marker. A series of markers lacking homology with any region of the *S. cerevisiae* genome have been designed to facilitate rapid, efficient gene deletion or epitope tagging using PCR (Fig. 1). The key feature behind these cassettes, or modules, is that they contain a dominant drug resistance marker flanked by the *Ashbya gossypii TEF* promoter and terminator [5, 6]. This allows for expression of the marker in

S. cerevisiae and virtually eliminates the possibility for integration at random sites in the yeast genome since there are no homologous sequences. Therefore, 40–60 bp extensions that match the gene sequence of interest (often 5′ of the start and 3′ of the stop codon for deletion and the region 5′ and 3′ of the stop codon for tagging) are added to the forward and reverse PCR primers that are used to amplify these markers, and this homology is sufficient to direct the amplified cassette to the locus of interest.

Modules that contain the *kan* gene from the *Escherichia coli* transposon Tn903 (*KANMX*; confers resistance to G418), the *Streptomyces noursei nat1* gene (*NATMX*; confers resistance to nourseorthricin), the *ble* gene from the *Klebsiella pneumonia* Tn5 transposon (*BLEMX*; confers resistance to phleomycin), the *K. pneumonia hph* gene (*HYGMX*; confers resistance to hygromycin B), and the *pat* gene from *Streptomyces viridochromogenes* (*PATMX*; confers resistance to bialaphos) are available [5–7]. In addition to drug resistance markers, several modules that express a nutritional marker from a related yeast species, such as *Schizosaccharomyces pombe his5*$^{+}$ (*HIS3MX*), *Kluyveromyces lactis URA3* or *LEU2* (*KlURA3MX* or *KlLEU2*), and *Candida albicans URA3 or LYS5* (*CaURA3MX* or *CaLYS5MX*) have been designed to allow for selection using conventional medium since these complement auxotrophies present in most lab strains (*his3*, *ura3*, *leu2*, and *lys5*, respectively) [7–10]. Additional modules that contain markers not expressed using the *TEF* system have been created that allow for exchange of epitope tags, and some cassettes have been engineered to have useful features such as direct-repeat sites for marker excision [7, 11–17].

Below, we describe basic methods for gene targeting in yeast using these modules, including design and production of PCR products for the construction of deletions, the introduction of mutations or epitope tags at the genomic locus and the swapping of tags and markers. A high-efficiency transformation protocol is used for integration of PCR-derived DNA fragments with only 40 bp of homology to the target gene. Integration at the desired genomic locus can be verified by PCR using genomic DNA isolated in the rapid genomic DNA isolation protocol or by PCR of whole colonies. PCR-mediated gene disruption and tagging are essential tools for functional analysis in *S. cerevisiae*. This basic tool-kit will provide a starting point for the molecular genetic dissection of any gene.

1.1 Gene Deletions

Deletion of an entire open-reading frame (ORF) of a gene creates a null mutation, allowing for the analysis of loss-of-function phenotypes. To generate a deletion, the gene sequence from start to stop codon is removed and is generally replaced with a selectable marker. In some cases, the entire ORF cannot be removed for practical reasons such as its removal would also delete part of an overlapping gene or would remove the regulatory region of an adjacent gene. In these instances, part of the ORF is generally removed.

A basic scheme for gene deletion is depicted in Fig. 2. Primers are designed immediately upstream and downstream of gene sequence to be deleted. If the gene ORF were to be removed, the forward 5′ PCR primer (F/F1/S1) would contain 40–60 bp of DNA 5′ to the ATG, and the 3′ reverse primer (R/R1/S2) would contain 40–60 bp of DNA 3′ to the stop codon (Fig. 2a). Both primers also contain sequences to recognize the deletion module (Fig. 2b). Following PCR amplification, the 40–60 bp of gene-specific homology are sufficient to direct gene replacement at the locus of interest. Integration into the target locus is verified by PCR of genomic DNA using primers that will amplify part of the integrated marker and an adjacent area of genomic DNA, confirming that the site of integration is correct. Deletion of the ORF is also verified by PCR (Fig. 2b).

A second method for gene deletion (Fig. 3) takes advantage of the haploid yeast deletion collection containing deletions of most nonessential yeast genes. Each gene deletion is marked by a unique "barcode" and flanked by universal sequences [18]. Barcodes are short sequences of unique DNA inserted upstream and downstream of the *KANMX* module that serve as identifiers for that deletion. They can be recognized by PCR, microarray, or sequencing [19–21]. The deletion cassette, complete with barcode, can be transferred into a new strain using short, inexpensive primers using genomic DNA from a specific haploid strain from the deletion collection. Twenty-bp primers forward and reverse are designed approximately 200 bp upstream (−200F) and downstream (+200R) of the ATG and stop codon so the PCR product containing the deletion cassette has a significantly longer region of homology, which increases the efficiency of gene targeting [5, 22].

If a deletion collection exists, why would one need to remake a gene deletion? Because the deletion collection was made in a strain background (BY4743) that is not well suited for certain types of experiments (for example, most studies of meiosis use SK-1 strains and analysis of yeast lifespan in done in a version of W303 strains), deletions often need to be remade in other strain backgrounds. Furthermore, even if the BY4743 background can

Fig. 2 (continued) After transformation and selection, integration of the cassette and concurrent deletion of the chromosomal target sequence is analyzed by PCR. Using the universal R primer that recognizes P_{TEF1} (5′-GGATGTATGGGCTAAATG-3′) together with −200F and +intR, integration of the deletion module can be detected as can the presence of the wild-type gene based on the size of the PCR products. After haploids are recovered following sporulation and tetrad dissection, only a single band should be present in the new deletion strain

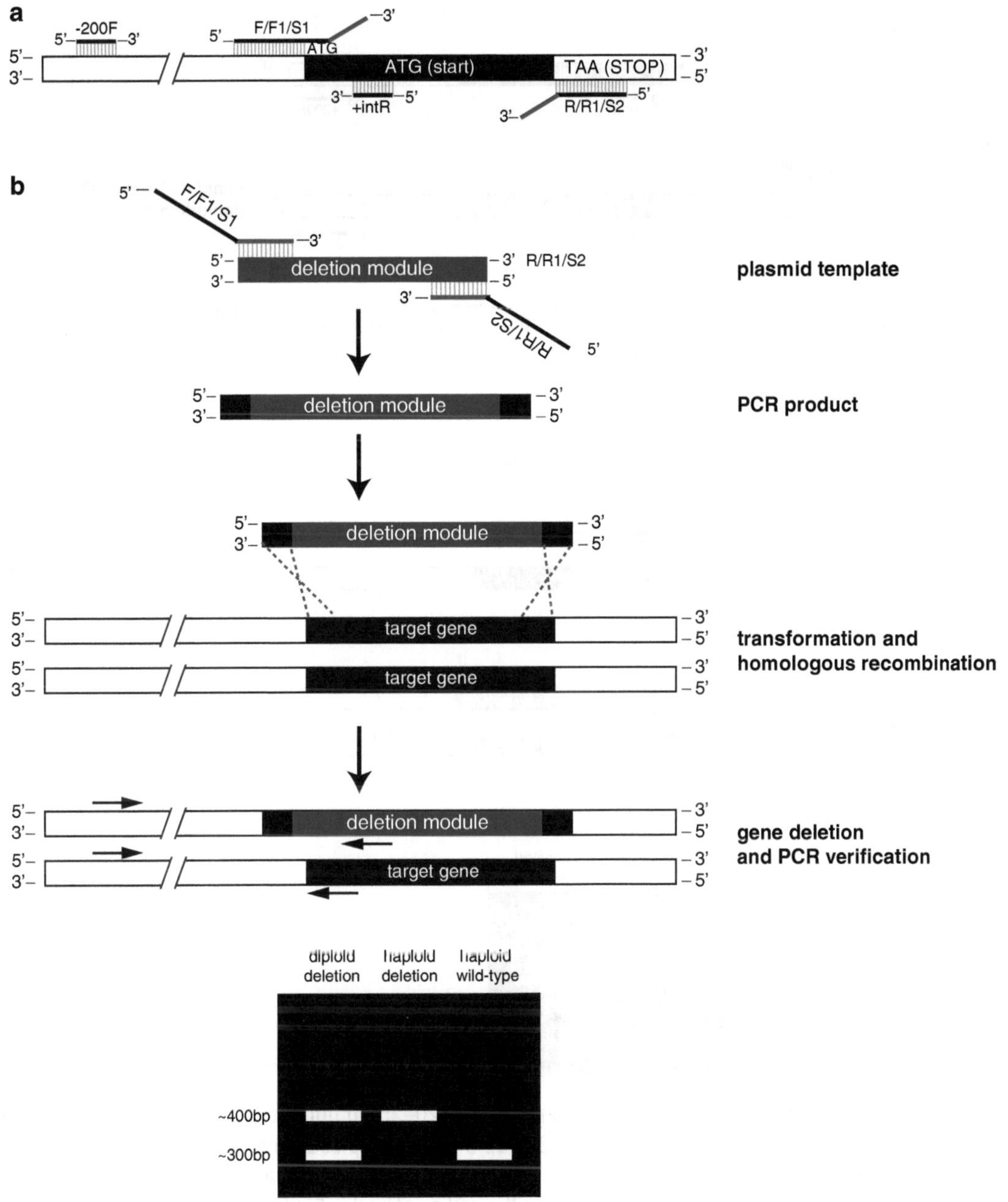

Fig. 2 De novo gene deletion. (**a**) The design of primers for gene deletion using PCR cassettes is illustrated. The forward primer (F, F1, or S1) should contain 40–60 bp of DNA 5′ to the start codon and can include the ATG, followed by a specific sequence for the deletion cassette. The reverse primer (R, R1, or S2) is the reverse complement and should contain 40–60 bp of DNA 3′ up to the stop codon and can include the stop codon, followed by a specific sequence for the deletion cassette. The sites of two additional 20–25 bp primers used for verification of the deletion are also depicted. −200F is located ~200 bp 5′ to the start codon on the coding strand and +intR is the reverse complement, ~100 bp downstream of the start codon. (**b**) PCR of the deletion template with the forward and reverse primer results in a PCR product with short regions of homology to the target gene (*black*), sufficient to direct homologous recombination and gene replacement.

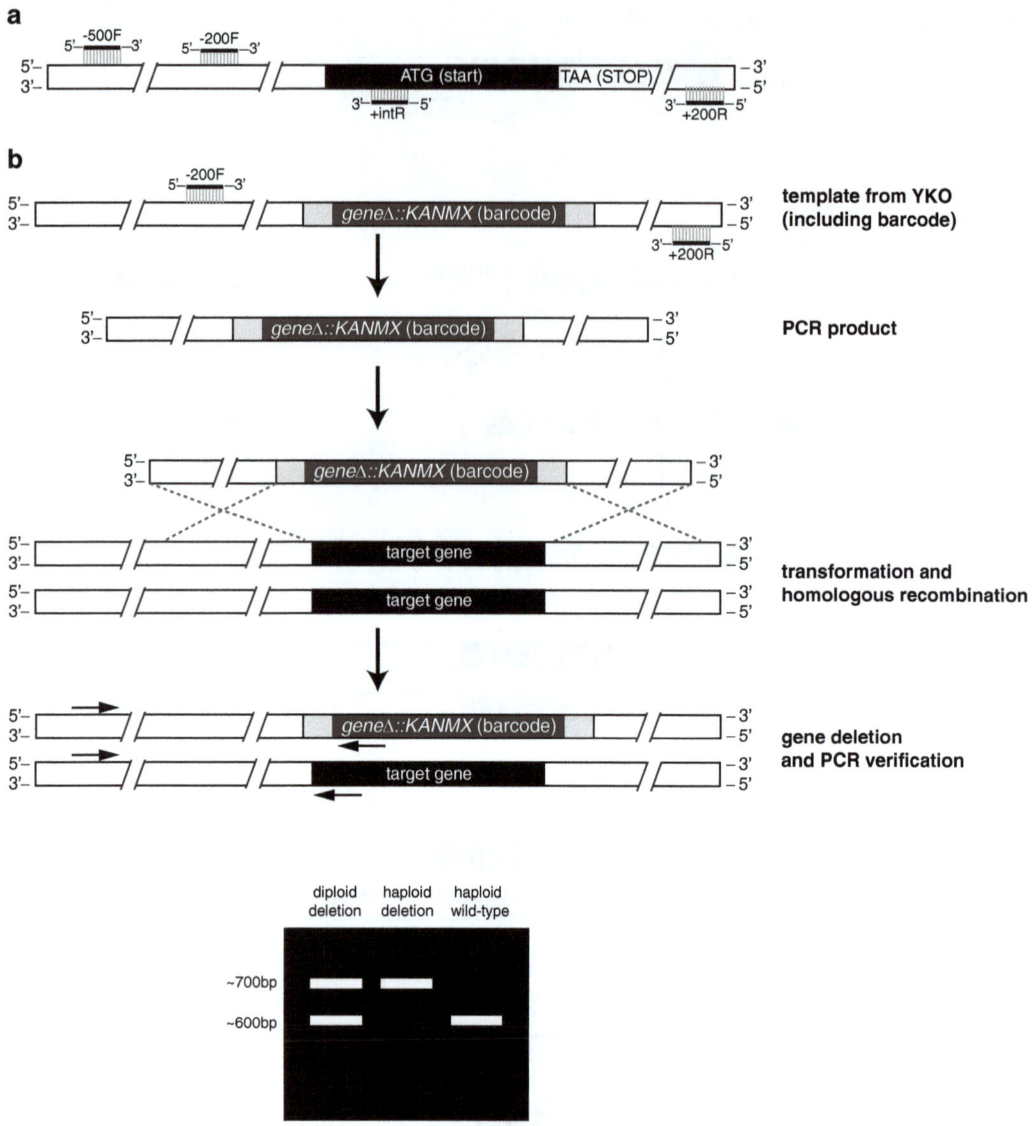

Fig. 3 Gene deletion using the deletion collection. (**a**) Short 20–25 bp primers designed ~200 bp 5′ to the start codon on the coding strand (–200F) and ~200 bp 3′ to the stop codon on the reverse strain (+200R) can be used to amplify a deletion and its associated barcode from the yeast deletion collection (YKO). Additional primers ~500 bp 5′ to the start codon on the coding strand (–500F) and ~100 bp downstream of the start codon on the reverse complement (+intR) are used for PCR verification of the deletion. (**b**) PCR using -200F and +200R results in a PCR product with ~200 bp regions of homology on both ends, which increases the frequency of accurate homologous recombination and gene replacement. Integration of the cassette and concurrent deletion of the chromosomal target sequence is analyzed by PCR using the –500F, +intR, and the universal R primer. If the deletion was done in a diploid, two products will be present: a ~600 bp band representing the intact target gene and a ~700 bp band representing the deletion. After haploids are recovered following sporulation and tetrad dissection, only the single ~700 bp band should be present in the new deletion strain

be used for experiments, it is good practice when studying a particular gene to remake the deletion in a new strain. This is because many of the strains present in the yeast deletion collection are aneuploid or contain mutations in genes other than the deletion (*see* for example, [23–25]). Making your own deletion will ensure that the deletion strain does not contain extragenic mutations and is isogenic with others used in your experiments.

1.2 Integrating Mutations

Deleting entire ORF creates a null mutant. However, in many cases it is often advantageous to study hypomorphic alleles that give only partial activity. This is particularly true for essential genes—genes that when deleted result in inviable cells. The most common type of hypomorphic allele used in yeast is the temperature-sensitive mutant, which lives at the permissive temperature of 23 °C and dies at the restrictive or non-permissive temperature of 36–37 °C [26]. Temperature-sensitive mutants are often generated during a phenotype-based screen. However, site-directed mutants created following proteomic, comparative genomic, or sequence/motif analysis can be tested for phenotypes such as temperature sensitivity, cold sensitivity, or sensitivity to various chemicals [3].

Plasmids containing new alleles of a gene are often introduced using the plasmid shuffle [2, 3]. However, because centromeric plasmids used in this method are present in ~1–4 copies per cell, there is often considerable cell-to-cell heterogeneity in phenotypes. In addition, the phenotype of a mutant on a plasmid is often not identical to the phenotype when the mutation is integrated in single copy into the genomic locus. Although the plasmid shuffle method is still advantageous for screening a large number of new mutants due to its relative ease, functional studies and genetic manipulations are best done on strains in which the mutant allele is introduced into the genomic locus, replacing the wild-type copy of the gene. This can be done using PCR-based methods.

To replace a wild-type gene with a mutant allele, the mutant gene is amplified by PCR from a DNA template and is co-transformed with a marker that contains overlapping homology and homology with the target locus (Fig. 4). In some cases the two fragments are fused in a second round of PCR to increase the frequency of positive transformants. Integration into the genomic locus occurs by one-step gene replacement and should be verified by phenotype and by PCR and DNA sequence analysis. The marker can be left in the strain so the mutation can be easily tracked; however, a variation of this method involves excision of the marker so that the only change in the strain is the mutation [7, 13, 27]. Other methods that couple the creation of mutant alleles to allele integration into the chromosome have also been described [28, 29].

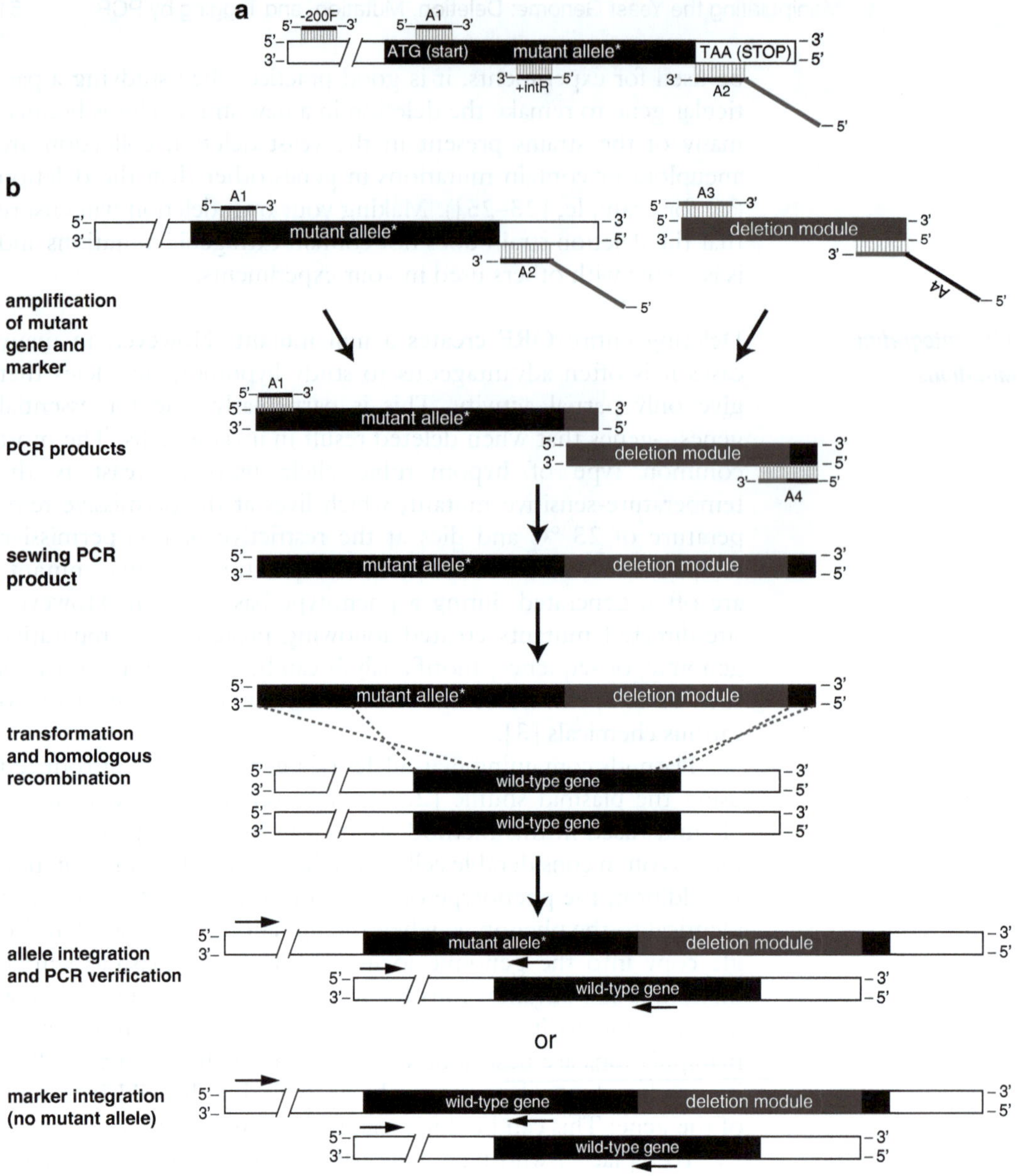

Fig. 4 Introduction of mutations by PCR. (**a**) The design of primers for introduction of mutations into the genomic locus using PCR. A 20 bp gene-specific forward primer that is at least 200 bp 5′ of the mutation (A1) and a reverse primer (A2) that contains the sequence, 5′-AGTAGCTGATTAACTCTATGATTTAAAGGGCAGTATAGCGACCAGC ATTCAC-3′ followed by 20 bp in gene specific sequence from the noncoding strand immediately after the stop codon, are used to amplify the mutant gene from a DNA template such as a plasmid or genomic DNA. (**b**) A universal forward primer (5′-ACATGGAGGCCCAGAATACCCTCCTT-3′) (A3) and a reverse primer (A4) that contains sequence from the noncoding strand downstream from the A2 primer followed by the indicated sequence from the marker (5′-CAAGGAGGGTATTCTGGGCCTCCATGT-3′) are used to amplify a module from the pFA plasmids. The resulting PCR products can be mixed and fused together in a second round of PCR using A1 and A4 primers or directly transformed into yeast. Because homologous recombination can occur along the length of the gene, some transformants will integrate the marker but not the mutation (*bottom*) into the target locus while others will integrate the allele and the marker (*top*). The −200F and +intR or other primers designed to amplify the region of the gene containing the mutation can allow for detection of the mutation by sequence analysis of the PCR product or by restriction digest if a silent restriction site is engineered into the mutant

1.3 Epitope Tagging and Construction of Fusion Proteins

The ability to epitope tag or create fusion proteins with the endogenous wild-type or mutant copy of yeast genes makes yeast one of the most powerful systems for the study of multiple biological processes. Tagging with small peptide sequences such as HA, MYC, and FLAG are useful for immunochemistry; fusions with glutathione-S-transferase (GST), protein A (proA), or a tandem affinity tag (TAP) are useful for protein purification, and fusions to GFP or red fluorescent protein (RFP) or one of their variants (i.e., BFP, CFP, YFP, mCherry, dsRed, RedStar, Venus, mKate2) are helpful for cell biological studies. Because the tag is engineered into the genomic locus, the tagged version of the gene is expressed and regulated like the wild-type gene. Furthermore, the parental strain serves as an important isogenic negative control. A number of different methods can be used to engineer protein fusions depending on the location of the tag [8, 11, 12, 16, 27, 30, 31].

A plethora of C-terminal tagging modules have been developed for virtually every epitope and fluorescent protein variant, including yeast codon optimized versions of most fluorescent proteins [11, 12, 32–39]. Using ~40–60 bp sequences immediately upstream and downstream of the stop codon as targeting sequences on the forward and reverse PCR primers (F2/S3/F5 and R1/S2/F3, respectively), the resulting PCR fragments from these modules are integrated into the genome by one-step gene replacement (Fig. 5a, b). Correct integration can be confirmed by PCR and/or by checking for expression of the tagged protein in the microscope or using western blotting.

N-terminal tagging is considerably more complex because insertion of a marker in the 5′ region of a gene will generally disrupt its expression. Therefore, N-terminal tagging of essential genes must be done in diploids since gene function is disrupted during the tagging procedure. Furthermore, the marker must be excised in order to examine the expression of native versions of both essential and nonessential genes (Fig. 5c). Specially designed tagging modules that contain direct repeat sequences on either side of the marker, such as loxP sites, have been created to allow for marker removal. In the case of loxP, excision can be stimulated by transient expression of the *cre* recombinase [30, 31]. Other systems use a *URA3* marker that can be counterselected by growing cells on 5-fluoro-orotic acid (5-FOA) [27, 40, 41].

1.4 Marker and Epitope Exchange

The fact that most PCR deletion and tagging modules share a common design is a mixed blessing. It is convenient that the same set of primers can be used with many modules. Generating new epitopes or fluorescent proteins can be done using the same general strategy, and because many permutations of markers and tags are available, strains for multiple functional studies can be created using a relatively small number of reagents. However, the homology

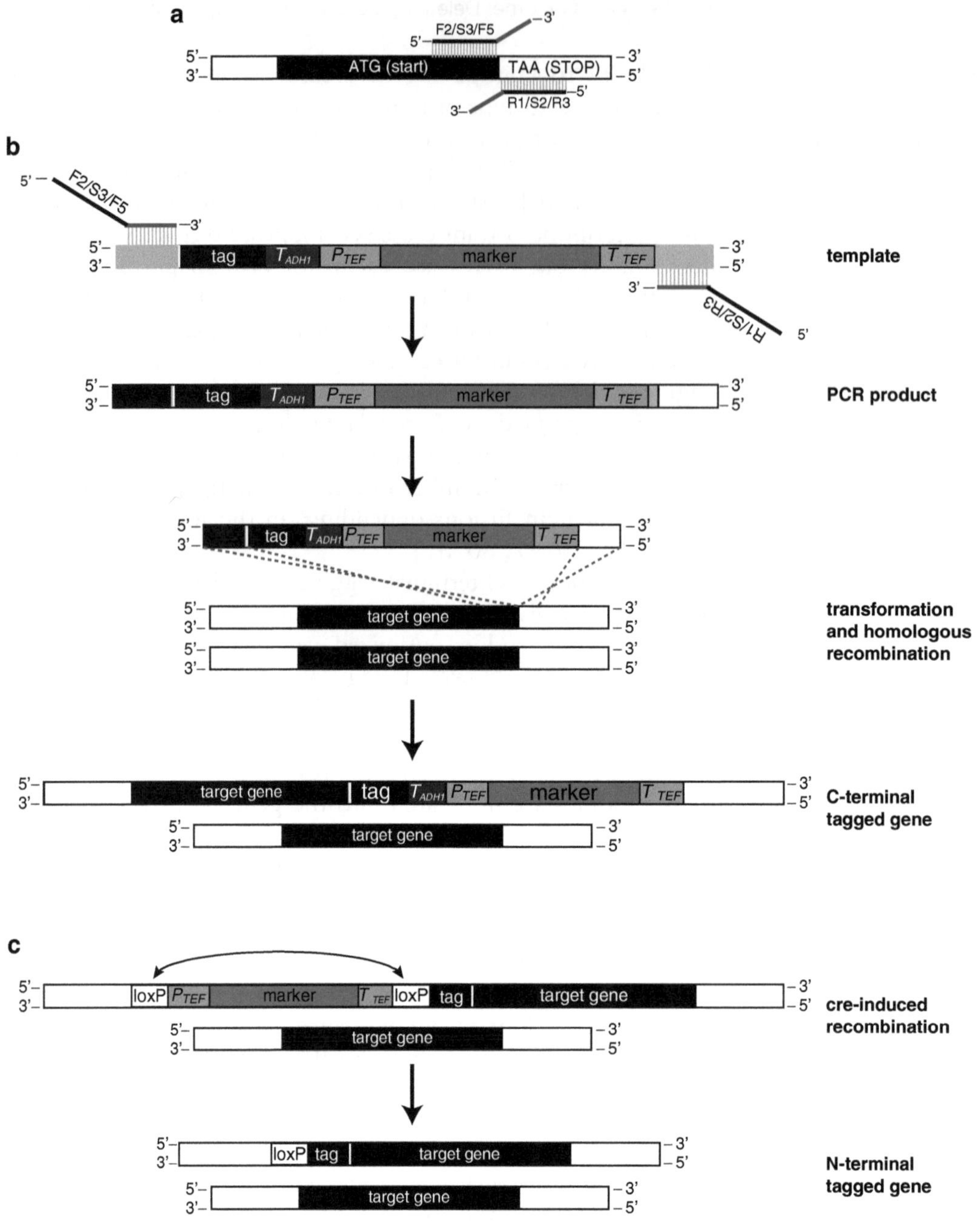

Fig. 5 Gene tagging by PCR. (**a**) The design of primers for C-terminal gene tagging using PCR cassettes is illustrated. The forward primer (F2/S3/F5) should contain 40–60 bp of DNA 5′ up to, but not including, the stop codon. The reverse primer (R1/S2/R3) should contain 40–60 bp of DNA 3′ to the stop codon followed by a specific sequence for the cassette. The same primer used for gene deletion can also be used for C-terminal tagging since the same site of integration can be used. (**b**) PCR of the tagging template with forward and reverse primers results in a PCR product with short regions of homology to the target gene, sufficient to direct homologous recombination at the C-terminus of the gene. After transformation, selection, and integration of the cassette, positive colonies are identified by western blotting, PCR and fluorescence microscopy. (**c**) Homologous recombination of the tagging cassette at the N-terminus of a gene can also be done using primers that direct integration to the 5′ end of a gene. However, because the module contains a marker gene for selection in addition to the tag, expression of the target gene is affected. Direct repeats, including loxP sites, can be used for excision of the marker, resulting in an N-terminally tagged version of a gene. This same strategy can also be used to tag internal regions

between the cassettes is problematic in generating strains with multiple deletions or tags because the modules are able to recombine with each other [12]. The simplest way to generate strains containing multiple deletions or tags is to simply select for all markers when transfroming a PCR product into a strain that already contains a deletion or tag made by PCR. There does not appear to be cross-resistance between the drug resistance markers so they can easily be used in combination [6]. In addition, most drugs are compatible with minimal medium required for nutritional selection, although some, such as G418 and bialaphos, work in modified versions of minimal medium in which ammonium acetate has been replaced with mono-sodium glutamate and proline, respectively [6, 7, 42].

Longer regions of homology such as those generated by amplifcation of genomic DNA from a deletion mutant will increase the liklihood of targeting a genomic locus rather than a previously integrated cassette [5, 22]. In addition, certain PCR modules lack the *TEF* sequences so these can be helpful to create strains that contain multiple deletions or tags (e.g., *pFA6-TRP1*, *pFA6-KlURA3*, *pFA6-KlLEU2*, *pFA6-natNT2*, *pFA6-hphNT1;* markers in these modules are expressed using their native promoter/terminator or the *TEF* promoter and the *CYC1 (NT1) or ADH1 (NT2)* terminators) [7, 11, 12]. It is important to note that in some strains such as W303, *TRP1* amplified from these modules will most often integrate into the genomic locus since this strain background contains a point mutation at *trp1* [1]. In addition, the BY4743 strain background used for the yeast deletion collection is prototrophic for *TRP1* so it cannot be used [43]. Several strategies to "recycle" markers have been developed; a gene can be deleted then the marker can be excised using loxP or by mitotic recombination of other direct-repeat sequences, allowing that marker to be used again in the same strain [7, 9, 10, 13, 14, 27]. The importance of marker recycling is illustrated in a study of the highly duplicated hexose transporter genes; deletion of 20 genes was required to inihibit growth on hexoses [44].

The homology between cassettes can be useful in building strains since markers, and even epitope tags, are easily interchanged by exploiting the homology built into the module design. A simple PCR of one marker that contains the *TEF* promoter and terminator followed by transformation and selection can convert all of the markers in the yeast that are flanked by *TEF* to the new marker. Sung and colleagues switched the *ADH1* terminator and removed the *TEF* promoter and marker gene, replacing it with that of *KlURA3* driven by its own promoter [16]. Using these epitope switching modules, a previously integrated tag can be switched to another tag and/or another marker (Fig. 6). In principle, this type of module could be used to convert an existing library such as the GFP or TAP library to an epitope of choice.

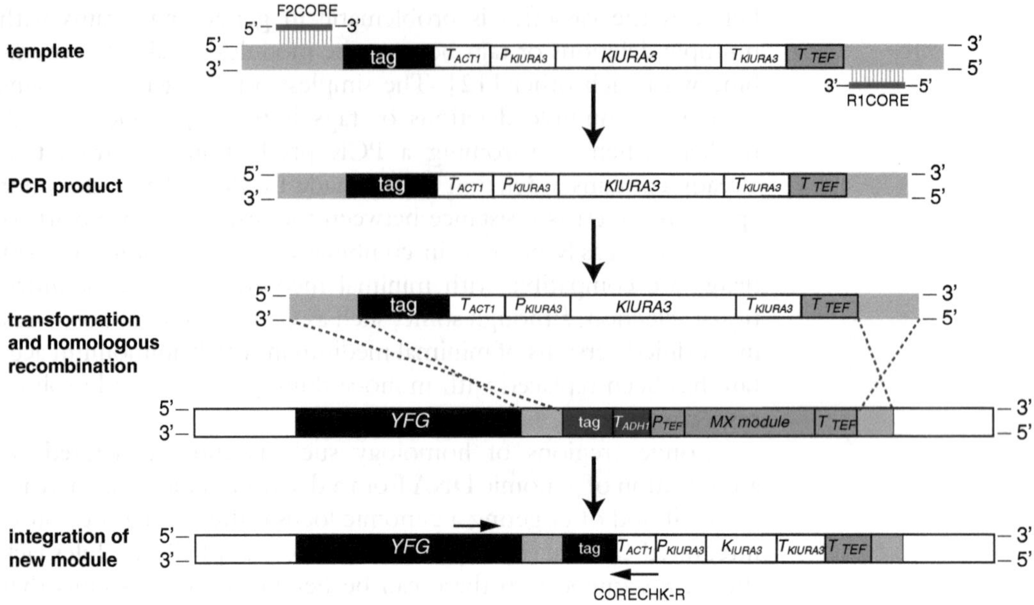

Fig. 6 Epitope exchange. The epitope switching module is similar in design to pFA plasmids, only the terminator for *ACT1* replaces that of *ADH1*. In addition, the *K. lactis URA3* gene has been inserted in place of the *TEF* promoter and *S. pombe his5+* gene. Amplification of this module using the F2CORE (5′-GGTCGACGGATCC CCGGGTT-3′) and R1CORE (5′-TCGATGAATTCGAGCTCGTT-3′) primers allows for one-step switching of any MX module in the genome. A primer located in the *ACT1* terminator (COREChk: 5′-ATACGCGCACAAAAGCAGAG-3′) allows for integration of the new module to be verified by PCR. In the case of tagged proteins, this can also be done by western blotting or microscopy

1.5 High Efficiency Yeast Transformation

Success of PCR-mediated gene disruption or fusion ultimately involves efficient uptake of DNA and recombination into the target locus. It is imperative to use a high efficiency tranformation protocol [5, 45]. Transformations using many PCR products, especially drug resistance modules, cannot be directly plated to selection medium but rather need a period of outgrowth in which the drug resistance gene or nutritional marker can be expressed [5, 6, 8, 42].

1.6 Verifying Integrants

Transformants need to be verified to ensure that the module has integrated into the intended target locus. PCR of genomic DNA or whole colony PCR analysis can be used to determine the site of integration. A universal reverse primer (5′-GGATGTATGGGC TAAATG-3′) that can be used with virtually all tagging modules, together with a gene specific reverse primer (+intR) and a gene specific forward primer (−200 F) are used to screen for integration of the module into the target locus and disruption of the target gene (in the case of gene deletion). The most common phenomenon observed when deleting a gene by PCR or conventional methods is gene duplication, which occurs in about 8 % of all transformants. Some of these events are due to duplication of the entire chromosome, while others involve duplication of a chromosome arm or smaller segment [46]. Therefore, it is vital to demonstrate both the absence of the ORF and the presence of the deletion module when generating any new deletion strain.

2 Materials

2.1 Gene Deletion De Novo by One-Step Gene Replacement

1. 10 ng/μl deletion cassette template (Table 1).
2. 25 μM F/F1/S1 and R/R1/S2 PCR primers made up in TE (10 mM Tris–HCl, pH 8.0, 1 mM EDTA). *See* Fig. 2a and Table 1 legend for primer design (*see* **Notes 1** and **2**).
3. Thermophilic DNA Polymerase (Taq) and 10× buffer supplied by manufacturer (*see* **Note 3**).
4. 25 mM dNTP mix: 25 mM dATP, 25 mM dTTP, 25 mM dCTP, 25 mM dGTP.

2.2 Gene Deletion Using the Yeast Deletion Collection

1. 10 ng/μl genomic DNA template (*see* **Notes 9** and **10** and *Rapid Isolation of Genomic DNA*).
2. 25 μM -200 F and +200 F primers made up in TE (10 mM Tris–HCl, pH 8.0, 1 mM EDTA). *See* Fig. 3a and Table 1 legend for primer design (*see* **Note 11**).
3. Taq and 10× Polymerase buffer (*see* **Note 3**).
4. 25 mM dNTP mix.

2.3 Integrating Mutations

1. 10 ng/μl DNA template (i.e., a plasmid containing a mutant version of gene or genomic DNA from yeast containing the mutation of interest; *see* **Notes 12** and **13**).
2. 10 ng/μl deletion cassette template (Table 1).
3. 25 μM A1, A2, A3 and A4 primers made up in TE. *See* Fig. 4a and **Note 14** for primer design (also *see* **Note 2**).
4. Taq and 10× Polymerase buffer (*see* **Note 3**).
5. 25 mM dNTP mix.

2.4 C-terminal Tagging

1. 10 ng/μl tagging cassette template (Table 1).
2. 25 μM F2/S3/F5 and R2/S2/R3 primers made up in TE. *See* Fig. 5a and **Notes 19–21** for primer design (also *see* **Note 2**).
3. Taq and 10× Polymerase buffer (*see* **Note 3**).
4. 25 mM dNTP mix.

Table 1
Plasmids for functional analysis

Function	Plasmid name	Marker	Approximate size PCR product (bp)	Primers and notes	Reference
Deletion modules					
Deletion	pFA6-KANMX6	*kan*	1,559	F1/R1	[12]
Deletion	pFA6-TRP1	*S.cer TRP1*	1,036	F1/R1	[12]
Deletion	pFA6-HIS3MX6	*S. pombe his5+*	1,403	F1/R1	[12]
Deletion	pFA6-NATMX4	*nat*	1,300	F/R	[6]
Deletion	pFA6-PATMX4	*pat*	1,300	F/R	[6]
Deletion	pFA6-HYGMX4	*hph*	1,600	F/R	[6]
Deletion	pFA6-HIS3X6	*S. pombe his5+*	1,400	F/R	[8]
Deletion	pFA6-KANMX4	*kan*	1,300	F/R	[5]
Deletion	pFA6-CaURA3MX4 (pAG60)	*C. albicans URA3*	1,500	F/R	[9]
Deletion	pFA6-LYS5MX4	*S. cer LYS5*	1,000	F/R	[10]
Deletion	pFA6-CaLYS5MX4	*C. albicans LYS5*	1,100	F/R	[10]
Deletion	pYM-HYGNT1	*hph*	1,600	S1/S2	[11]
Deletion	pYM-NATNT2	*nat*	1,300	S1/S2	[11]
Recyclable deletion modules					
Deletion/excision	pFA6-KANMX3	*kan*	1,300	F/R; direct repeats	[5]
Deletion/excision	pFA6-CaURA3MX3 (pAG61)	*C. albicans URA3*	1,500	F/R; direct repeats	[9]
Deletion/excision	pFA6-NATMX3	*nat*	1,300	F/R; direct repeats	[6]
Deletion/excision	pFA6-PATMX3	*pat*	1,300	F/R; direct repeats	[6]
Deletion/excision	pFA6-HYGMX3	*hph*	1,600	F/R; direct repeats	[6]
Deletion/excision	pFA6-LYS5MX3	*S. cer LYS5*	1,000	F/R; direct repeats	[10]
Deletion/excision	pFA6-CaLYS5MX3	*C. albicans LYS5*	1,200	F/R; direct repeats	[10]
Deletion/excision	pFA6-loxP-KANMX6-loxP (pUG6)	*kan*	1,700	F/R; loxP sites	[7]
Deletion/excision	pFA6-loxP-HIS3MX6-loxP (pUG27)	*S. pombe his5+*	1,550	F/R; loxP sites	[7]
Deletion/excision	pFA6-loxP-BLEMX6-loxP (pUG66)	*ble*	1,300	F/R; loxP sites	[7]
Deletion/excision	pFA6-loxP-KlURA3-loxP (pUG72)	*K. lactis URA3*	1,700	F/R; loxP sites	[7]
Deletion/excision	pFA6-loxP-KlLEU2-loxP (pUG73)	*K. lactis LEU2*	2,500	F/R; loxP sites	[7]
Deletion/excision	pFA6-loxP-LYS2-loxP (pUG-LYS2)	*S. cer LYS2*	4,500	F/R; loxP sites	[14]

Deletion/excision	pFA6-loxP-HIS3MX6-loxP (pUG-SpHIS5)	*S. pombe his5+*	1,300	F/R; loxP sites	[14]
Deletion/excision	pFA6-loxP-LYS5MX4-loxP	*S. cer LYS5*	1,100	F/R; loxP sites	[10]
Deletion/excision	pFA6-loxP-CaLYS5MX4-loxP	*C. albicans LYS5*	1,300	F/R; loxP sites	[10]
Deletion/excision	pFA6-loxP-NATMX6-loxP (pUG74)	*nat*	1,500	F/R; loxP sites	[17]
Deletion/excision	pFA6-loxP-HYGMX-loxP (pUG75)	*hph*	1,900	F/R; loxP sites	[17]
C-terminal tagging					
C-GFP	pFA6-GFP(S65T)-KANMX6	*kan*	2,504	F2/R1	[12]
C-GFP	pFA6-GFP(S65T)-TRP1	*S.cer TRP1*	1,981	F2/R1	[12]
C-GFP	pFA6-GFP(S65T)-HIS3MX6	*S. pombe his5+*	2,348	F2/R1	[12]
C-GFP	pFA6-GFP(S65T)-NATMX6	*nat*	2,200	F2/R1	[34]
C-3xHA	PFA6-3HA-KANMX6	*kan*	1,898	F2/R1	[12]
C-3xHA	pFA6-3HA-TRP1	*S.cer TRP1*	1,375	F2/R1	[12]
C-3xHA	pFA6-3HA-HIS3MX6	*S. pombe his5+*	1,742	F2/R1	[12]
C-3xHA	pFA6-3HA-NATMX6	*nat*	1,600	F2/R2	[34]
C-13xMYC	PFA6-13MYC-KANMX6	*kan*	2,325	F2/R1	[12]
C-13xMYC	pFA6-13MYC-TRP1	*S.cer TRP1*	1,802	F2/R1	[12]
C-13xMYC	pFA6-13MYC-HIS3MX6	*S. pombe his5+*	2,169	F2/R1	[12]
C-13xMYC	pFA6-13MYC-NATMX6	*nat*	2,000	F2/R1	[34]
C-GST	pFA6-GST-KANMX6	*kan*	2,465	F2/R1	[12]
C-GST	pFA6-GST-TRP1	*S.cer TRP1*	1,942	F2/R1	[12]
C-GST	pFA6-GST-HIS3MX6	*S. pombe his5+*	2,309	F2/R1	[12]
C-GST	pFA6-GST-NATMX6	*nat*	2,200	F2/R1	[34]
C-3xHA	pFA6-3HA-KANMX6 (pYM1)	*kan*	1,830	S3/S2	[32]
C-3xHA	pFA6-3HA-HIS3MX6 (pYM2)	*S. pombe his5+*	1,674	S3/S2	[32]
C-3xHA	pYM-3HA-HYGNT1 (pYM24)	*hph*	1,950	S3/S2	[11]
C-6xHA	pFA6-6HA-KANMX4 (pYM14)	*kan*	1,900	S3/S2	[11]
C-6xHA	pYM-6HA-HYGNT1 (pYM3)	*hph*	2,000	S3/S2	[11]
C-6xHA	pYM-6HA-NATNT1 (pYM17)	*nat*	1,312	S3/S2	[11]
C-6xHA	pFA6-6HA-HIS3MX6 (pYM22)	*S. pombe his5+*	1,900	S3/S2	[11]
C-6xHA	pFA6-6HA-KlTRP1 (pYM3)	*K. lactis TRP1*	1,312	S3/S2	[32]
C-3xMYC	pFA6-3MYC-KANMX6 (pYM4)	*kan*	1,851	S3/S2	[32]
C-3xMYC	pFA6-3MYC-HIS3MX6 (pYM5)	*S. pombe his5+*	1,695	S3/S2	[32]
C-3xMYC	pFA6-3MYC-KlTRP1 (pYM23)	*K. lactis TRP1*	1,400	S3/S2	[11]

(continued)

Table 1
(continued)

Function	Plasmid name	Marker	Approximate size PCR product (bp)	Primers and notes	Reference
C-9xMYC	pFA6-9MYC-KANMX4 (pYM18)	*kan*	1,900	S3/S2	[11]
C-9xMYC	pYM-9MYC-HYGNT1 (pYM20)	*hph*	2,000	S3/S2	[11]
C-9xMYC	pYM-9MYC-NATNT1 (pYM21)	*nat*	1,300	S3/S2	[11]
C-9xMYC	pFA6-9MYC-HIS3MX6 (pYM19)	*S. pombe his5+*	1,900	S3/S2	[11]
C-9xMYC	pFA6-9MYC-KlTRP1 (pYM6)	*K. lactis TRP1*	1,480	S3/S2	[32]
C-ProA	pFA6-ProA-KANMX6 (pYM7)	*kan*	2,109	S3/S2	[32]
C-TEV-ProA	pFA6-TEV-ProA-KANMX6 (pYM8)	*kan*	2,133	S3/S2	[32]
C-TEV-ProA-7HIS	pFA6-TEV-ProA-7HIS-KANMX6 (pYM9)	*kan*	2,145	S3/S2	[32]
C-TEV-ProA-7HIS	pFA6-TEV-ProA-7HIS-HIS3X6 (pYM10)	*S. pombe his5+*	1,989	S3/S2	[32]
C-TEV-GST-6HIS	pFA6-TEV-GST-6HIS-KANMX4 (pYM11)	*kan*	2,251	S3/S2	[32]
C-EGFP	pFA6-EGFP-KANMX4 (pYM12)	*kan*	2,469	S3/S2	[32]
C-HA	pFA6-1HA-KANMX4 (pYM45)	*kan*	1,800	S3/S2	[11]
C-TAP	pFA6-TAP-KANMX4 (pYM13)	*kan*	2,000	S3/S2	[11]
C-yeGFP	pFA6-yeGFP-KANMX4 (pYM12)	*kan*	2,469	S3/S2	[11]
C-yeGFP	pYM-yeGFP-HYGNT1 (pYM25)	*hph*	2,600	S3/S2	[11]
C-yeGFP	pFA6-yeGFP-HIS3MX6 (pYM44)	*S. pombe his5+*	2,300	S3/S2	[11]
C-yeGFP	pFA6-yeGFP-KlTRP1 (pYM26)	*K. lactis TRP1*	2,000	S3/S2	[11]
C-ECFP	pFA6-ECFP-KANMX4 (pYM30)	*kan*	2,469	S3/S2	[11]
C-ECFP	pFA6-ECFP-HIS3MX6 (pYM31)	*S. pombe his5+*	2,300	S3/S2	[11]
C-ECFP	pFA6-ECFP-KlTRP1 (pYM32)	*K. lactis TRP1*	2,000	S3/S2	[11]
C-EYFP	pFA6-EYFP-KANMX4 (pYM39)	*kan*	2,469	S3/S2	[11]
C-EYFP	pYM-EYFP-HYGNT1 (pYM40)	*hph*	2,600	S3/S2	[11]
C-EYFP	pFA6-EYFP-HIS3MX6 (pYM44)	*S. pombe his5+*	2,300	S3/S2	[11]
C-DsRed (yRFP)	pFA6-DsRed-KANMX4 (pYM37)	*kan*	2,500	S3/S2	[11]
C-RedStar*	pFA6-RedStar*-KANMX4 (pYM42)	*kan*	2,500	S3/S2	[11]
C-RedStar2	pFA6-RedStar2-KANMX4 (pYM43)	*kan*	2,500	S3/S2	[11]
C-PA-GFP	pYM-PAGFP-HYGNT1 (pYM48)	*hph*	2,600	S3/S2	[11]
C-ECFP	pFA6-ECFP-HIS3MX6	*S. pombe his5+*	2,348	F2/R1	[38]
C-ECFP	pFA6-ECFP-TRP	*TRP1*	1,981	F2/R1	[38]

C-ECFP	pFA6-ECFP-NATMX6	*nat*	2,200	F2/R1	[34]
C-EYFP	pFA6-EYFP-HIS3MX6	*S. pombe his5+*	2,348	F2/R1	[38]
C-EYFP	pFA6-EYFP-TRP	*TRP1*	1,981	F2/R1	[38]
C-EYFP	pFA6-EYFP-NATMX6	*nat*	2,200	F2/R1	[34]
C-TAP	pFA6-CTAP4-NATMX6	*nat*	2,300	F2/R1	[34]
C-2xProA-TEV	pFA6-2xProA-TEV-TRP1	*TRP*	2,600	F2/R1	[37]
C-FLAG	pFA6-6xGLY-FLAG-HIS3MX	*S. pombe his5+*	1,700	FG/R1	[33]
C-FLAG	pFA6-6xGLY-FLAG-KANMX6	*kan*	1,800	FG/R1	[33]
C-FLAG	pFA6-6xGLY-FLAG-HYGMX4	*hph*	1,900	FG/RG	[33]
C-3xFLAG	pFA6-6xGLY-3xFLAG-HIS3MX	*S. pombe his5+*	1,700	FG/R1	[33]
C-3xFLAG	pFA6-6xGLY-3xFLAG-KANMX6	*kan*	1,800	FG/R1	[33]
C-3xFLAG	pFA6-6xGLY-3xFLAG-HYGMX4	*hph*	1,900	FG/RG	[33]
C-Strep tag II	pFA6-6xGLY-Strep-tagII-HIS3MX	*S. pombe his5+*	1,700	FG/R1	[33]
C-Strep tag II	pFA6-6xGLY-Strep-tagII-KANMX6	*kan*	1,800	FG/R1	[33]
C-Strep tag II	pFA6-6xGLY-Strep-tagII-HYGMX4	*hph*	1,900	FG/RG	[33]
C-T7	pFA6-6xGLY-T7-HIS3MX	*S. pombe his5+*	1,700	FG/R1	[33]
C-T7	pFA6-6xGLY-T7-KANMX6	*kan*	1,800	FG/R1	[33]
C-T7	pFA6-6xGLY-T7I-HYGMX4	*hph*	1,900	FG/RG	[33]
C-His tag	pFA6-6xGLY-His-tag-HIS3MX	*S. pombe his5+*	1,700	FG/R1	[33]
C-His tag	pFA6-6xGLY-His-tag-KANMX6	*kan*	1,800	FG/R1	[33]
C-His tag	pFA6-6xGLY-His-tag-HYGMX4	*hph*	1,900	FG/RG	[33]
C-S tag	pFA6-6xGLY-S-tag-HIS3MX	*S. pombe his5+*	1,700	FG/R1	[33]
C-S tag	pFA6-6xGLY-S-tag-KANMX6	*kan*	1,800	FG/R1	[33]
C-S tag	pFA6-6xGLY-S-tag-HYGMX4	*hph*	1,900	FG/RG	[33]
C-Myc	pFA6-6xGLY-MYC-HIS3MX	*S. pombe his5+*	1,700	FG/R1	[33]
C-Myc	pFA6-6xGLY-MYC-KANMX6	*kan*	1,800	FG/R1	[33]
C-Myc	pFA6-6xGLY-MYC-HYGMX4	*hph*	1,900	FG/RG	[33]
C-VSV-G	pFA6-6xGLY-VSV-G-HIS3MX	*S. pombe his5+*	1,700	FG/R1	[33]
C-VSV-G	pFA6-6xGLY-VSV-G-KANMX6	*kan*	1,800	FG/R1	[33]
C-VSV-G	pFA6-6xGLY-VSV-G-HYGMX4	*hph*	1,900	FG/RG	[33]
C-HSV	pFA6-6xGLY-HSV-HIS3MX	*S. pombe his5+*	1,700	FG/R1	[33]
C-HSV	pFA6-6xGLY-HSV-KANMX6	*kan*	1,800	FG/R1	[33]
C-HSV	pFA6-6xGLY-HSV-HYGMX4	*hph*	1,900	FG/RG	[33]
C-V5	pFA6-6xGLY-V5-HIS3MX	*S. pombe his5+*	1,700	FG/R1	[33]

(continued)

Table 1
(continued)

Function	Plasmid name	Marker	Approximate size PCR product (bp)	Primers and notes	Reference
C-V5	pFA6-6xGLY-V5-KANMX6	*kan*	1,800	FG/R1	[33]
C-V5	pFA6-6xGLY-V5-HYGMX4	*hph*	1,900	FG/RG	[33]
C-3xE2a	pFA6-3xE2a-KANMX6	*kan*	1,600	S3/S2	[36]
C-3xE2a	pFA6-3xE2a-HYGMX6	*hph*	1,900	S3/S2	[36]
C-3xE2a	pFA6-3xE2a-NATMX6	*nat*	1,400	S3/S2	[36]
C-3xE2b	pFA6-3xE2b-KANMX6	*kan*	1,600	S3/S2	[36]
C-3xE2b	pFA6-3xE2b-HYGMX6	*hph*	1,900	S3/S2	[36]
C-3xE2b	pFA6-3xE2b-NATMX6	*nat*	1,400	S3/S2	[36]
C-yEGFP	pFA6-link-yEGFP-HIS3MX6 (pKT128)	*kan*	2,300	F5/R3	[35]
C-yEGFP	pFA6-link-yEGFP-KANMX6 (pKT127)	*S. pombe his5+*	2,500	F5/R3	[35]
C-yEGFP	pFA6-link-yEGFP-CaURA3MX (pKT209)	*C. albicans URA3*	2,500	F5/R3	[35]
C-yEVenus	pFA6-link-yEVenus-HIS3MX6 (pKT90)	*kan*	2,300	F5/R3	[35]
C-yEVenus	pFA6-link-yEVenus-KANMX6 (pKT103)	*S. pombe his5+*	2,500	F5/R3	[35]
C-yECFP	pFA6-link-yECFP-HIS3MX6 (pKT101)	*kan*	2,300	F5/R3	[35]
C-yECFP	pFA6-link-yECFP-KANMX6 (pKT102)	*S. pombe his5+*	2,500	F5/R3	[35]
C-yECFP	pFA6-link-yECFP-CaURA3MX (pKT174)	*C. albicans URA3*	2,500	F5/R3	[35]
C-ymCherry	pFA6-link-ymCherry-HIS3MX6	*kan*	2,300	F5/R3	[49]
C-ymCherry	pFA6-link-ymCherry-KANMX6	*S. pombe his5+*	2,500	F5/R3	[49]
C-ymCherry	pFA6-link-ymCherry-CaURA3MX	*C. albicans URA3*	2,500	F5/R3	[49]
C-2xDendra	pFA6-2xDendra-HIS3MX6 (pKW2207)	*S. pombe his5+*	3,000	F2/R1	[50]
C-yBFP2	pFA6-link-yBFP2-HIS3MX6	*kan*	2,300	F5/R3	[39]
C-yBFP2	pFA6-link-yBFP2-KANMX6	*S. pombe his5+*	2,500	F5/R3	[39]
C-yBFP2	pFA6-link-yBFP2-CaURA3MX	*C. albicans URA3*	2,500	F5/R3	[39]
C-yGFPγ	pFA6-link-yGFPγ-HIS3MX6	*kan*	2,300	F5/R3	[39]
C-yGFPγ	pFA6-link-yGFPγ-KANMX6	*S. pombe his5+*	2,500	F5/R3	[39]
C-yGFPγ	pFA6-link-yGFPγ-CaURA3MX	*C. albicans URA3*	2,500	F5/R3	[39]
C-mRuby2	pFA6-link-ymRuby2-HIS3MX6	*kan*	2,300	F5/R3	[39]
C-mRuby2	pFA6-link-ymRuby2-KANMX6	*S. pombe his5+*	2,500	F5/R3	[39]
C-mRuby2	pFA6-link-ymRuby2-CaURA3MX	*C. albicans URA3*	2,500	F5/R3	[39]

C-mKate2	pFA6-link-ymKate2-HIS3MX6	*kan*	2,300	F5/R3	[39]
C-mKate2	pFA6-link-ymKate2-KANMX6	*S. pombe his5+*	2,500	F5/R3	[39]
C-mKate2	pFA6-link-ymKate2-CaURA3MX	*C. albicans URA3*	2,500	F5/R3	[39]
C-mTurquoise2	pFA6-link-ymTurquoise2-CaURA3MX	*C. albicans URA3*	2,500	F5/R3	SLJ, unpublished
C-link-HIS-TEV-ProA	pFA6-link-HIS7-TEV-ProA-HIS3MX6 (pKW2194)	*S. pombe his5+*	3,000	F2/R1	[50]
N-terminal tagging					
N-6HA	pFA6-loxP-KANMX6-loxP-6HA (pOM10)	*kan*	1,912	Nt-F/Nt-R	[31]
N-6HA	pFA6-loxP-KlURA3-loxP-6HA (pOM12)	*K. lactis URA3*	1,891	Nt-F/Nt-R	[31]
N-6HA	pFA6-loxP-KlLEU2-loxP-6HA (pOM13)	*K. lactis LEU2*	2,727	Nt-F/Nt-R	[31]
N-9MYC	pFA6-loxP-KANMX6-loxP-9MYC (pOM20)	*kan*	2,080	Nt-F/Nt-R	[31]
N-9MYC	pFA6-loxP-KlURA3-loxP-9MYC (pOM22)	*K. lactis URA3*	2,059	Nt-F/Nt-R	[31]
N-9MYC	pFA6-loxP-KlLEU2-loxP-9MYC (pOM23)	*K. lactis LEU2*	2,895	Nt-F/Nt-R	[31]
N-yEGFP	pFA6-loxP-KANMX6-loxP-yEGFP (pOM40)	*kan*	2,458	Nt-F/Nt-R	[31]
N-yEGFP	pFA6-loxP-KlURA3-loxP-yEGFP (pOM42)	*K. lactis URA3*	2,437	Nt-F/Nt-R	[31]
N-yEGFP	pFA6-loxP-KlLEU2-loxP-yEGFP (pOM43)	*K. lactis LEU2*	3,273	Nt-F/Nt-R	[31]
N-TEV-ProA	pFA6-loxP-KANMX6-loxP-TEV-ProA (pOM70)	*kan*	2,149	Nt-F/Nt-R	[31]
N-ProA	pFA6-loxP-KANMX6-loxP-ProA (pOM60)	*kan*	2,125	Nt-F/Nt-R	[31]
N-TEV-GST-6HIS	pFA6-loxP-KANMX6-loxP-TEV-GST-ProA (pOM50)	*kan*	2,471	Nt-F/Nt-R	[31]
N-TEV-ProA-7HIS	pFA6-loxP-KANMX6-loxP-TEV-ProA-7HIS (pOM80)	*kan*	2,161	Nt-F/Nt-R	[31]
N-3xE2a	ploxP-KANMX6-loxP-E2a	*kan*	1,700	N-Fe/N-Re	[36]
N-3xE2a	ploxP-HYGMX6-loxP-E2a	*hph*	1,900	N-Fe/N-Re	[36]
N-3xE2a	ploxP-NATMX6-loxP-E2a	*nat*	1,500	N-Fe/N-Re	[36]
N-3xE2b	ploxP-KANMX6-loxP-E2b	*kan*	1,700	N-Fe/N-Re	[36]
N-3xE2b	ploxP-HYGMX6-loxP-E2b	*hph*	1,900	N-Fe/N-Re	[36]
N-3xE2b	ploxP-NATMX6-loxP-E2b	*nat*	1,500	N-Fe/N-Re	[36]
N-EGFP	ploxP-KANMX-loxP-EGFP	*kan*	2,399	N-F/N-R	[30]
N-yEGFP	ploxP-KANMX-loxP-yEGFP	*kan*	2,399	N-F/N-R	[51]
N-yEYFP	ploxP-KANMX-loxP-yEYFP	*kan*	2,399	N-F/N-R	[51]
N-yECFP	ploxP-KANMX-loxP-yECFP	*kan*	2,399	N-F/N-R	[51]

(continued)

Table 1
(continued)

Function	Plasmid name	Marker	Approximate size PCR product (bp)	Primers and notes	Reference
Switching plasmids					
S-GFP	pFA6-GFP(S65T)-ACT1T-KlURA3	*K. lactis URA3*	2,400	F2CORE/R1CORE	[16]
S-TAP	pFA6-TAP-ACT1T-KlURA3	*K. lactis URA3*	2,400	F2CORE/R1CORE	[16]
S-GST	pFA6-GST-ACT1T-KlURA3	*K. lactis URA3*	2,400	F2CORE/R1CORE	[16]
S-13MYC	pFA6-13MYC-ACT1T-KlURA3	*K. lactis URA3*	2,300	F2CORE/R1CORE	[16]
S-3HA	pFA6-3HA-ACT1T-KlURA3	*K. lactis URA3*	1,800	F2CORE/R1CORE	[16]
S-FLAG	pFA6-FLAG-ACT1T-KlURA3	*K. lactis URA3*	1,800	F2CORE/R1CORE	[16]

Primers suggested for use with each plasmid are indicated. Since primers are designed to recognize the multiple cloning site of the pFA plasmid, they can often, but not always, be used interchangeably (e.g., the S1 and S2 primer can be used in place of F/F1 and R/R1 for deletion). Primer sequences: F1, 5′-(40 bp up to ATG)-CGGATCCCCGGTTAATTAA-3′; R1, 5′-(40 bp after stop reverse complement)-GAATTCGAGCTCGTTTAAAC-3′; F2, 5′-(40 bp before stop)-CGGATCCCCGGGTTAATTAA-3′; S1, 5′-(40 bp up to ATG)-CGTACGCTGCAGGTCGAC-3′; S3, 5′-(40 bp before stop)-CGTACGCTGCAGGTCGAC-3′; S2, 5′-(40 bp after stop reverse complement)-ATCGATGAATTCGAGCTCG-3′; F, 5′-(40 bp up to ATG)-CAGCTGAAGCTTCGTACGC-3′; R, 5′-(40 bp after stop reverse complement)-GCATAGGCCACTAGTGGATCTG-3′; N-F, 5′-(40 bp up to ATG)-CGGCCGCCAGGG-3′;N-R, 5′-(40 bp after ATG reverse complement)-TTTGTACAATTCATCCATACCATG-3′; FG, 5′-(40 bp before stop)-GGGGGAGGCGGGGGTGGA-3′; RG, 5′-(40 bp after stop reverse complement)-GAATTCGAGCTCGTTTCGA-3′; N-Fe, 5′-(40 bp up to ATG)-ATGTGCAGGTCGACAACCCTTAAT-3′; N-Re, 5′-(40 bp after ATG reverse complement)-GCGGCCGCATAGGCCACT-3′; F5, 5′-(40 bp before stop)-GGTGACGGTGCTGGTTTA-3′; R3, 5′-(40 bp after stop reverse complement)-TCGATGAATTC-GAGCTCG-3′; Nt-F, 5′-(40 bp up to ATG)-ATGTGCAGGTCGACAACCCTTAAT-3′; Nt-R, 5′-(40 bp after ATG reverse complement)-GCGGCCGCATAGGCCACT-3′. The following plasmids for *cre*-induced recombination have been described: p*URA3-GAL1-cre* (pSH47), p*HIS3-GAL1-cre* (pSH62), p*TRP1-GAL1-cre* (pSH63), p*ble-GAL1-cre* (pSH65) [13], YEp351-*cre-cyh* [14], pNAT-GAL1-cre (pSH66), p*KAN-GAL1-cre* (pSH67), p*HYG-GAL1-cre* (pSH69), p*LEU2-GAL1-cre* (pSH68) [17]. Most plasmids are available for a nominal fee from AddGene or Euroscarf. Others can be obtained by contacting the referenced authors

2.5 Marker and Epitope Exchange

1. 10 ng/μl deletion cassette template (Table 1).
2. 25 μM PR78 (5′-CCTTGACAGTCTTGACGTGC-3′) and PR79 (5′-CGCACTTAACTTCGCATCTG-3′) primers made up in TE for marker exchange [6].
3. 10 ng/μl switching plasmid template (Table 1).
4. 25 μM F2CORE (5′-GGTCGACGGATCCCCGGGTT-3′) and F1CORE (5′-TCGATGAATTCGAGCTCGTT-3′) primers made up in TE for epitope exchange [16].
5. Taq and 10× Polymerase buffer (*see* **Note 3**).
6. 25 mM dNTP mix.

2.6 High Efficiency Yeast Transformation

1. 1 M lithium acetate (LiOAc): dissolve 51 g LiOAc in 450 ml ddH_2O. Adjust the volume to 500 ml. Filter-sterilize or autoclave.
2. 10× TE: Mix 10 ml 0.5 M EDTA, 50 ml 1 M Tris–HCl pH 7.6, and 440 ml ddH_2O. Filter-sterilize or autoclave.
3. LiOAc Mix: Mix 50 ml 1 M LiOAc, 50 ml 10× TE, and 400 ml ddH_2O. Filter-sterilize or autoclave.
4. PEG Mix: Heat 50 ml ddH_2O to a near boil. To this add, 40 g PEG 3350, 10 ml 10× TE, and 10 ml 1 M LiOAc. Stir and heat until PEG dissolves. Adjust the volume to 100 ml and filter-sterilize with a 0.45 μm filter unit (*see* **Note 27**).
5. YPD plates: 1 % yeast extract, 2 % peptone, 2 % dextrose, 2 % Bacto-Agar (YPD).
6. Selection media: YPD supplemented with 200 mg/L G418 disulfate salt (Life Technologies) for selection of *KANMX* transformants, with 100 mg/L nourseothricin (Werner Bioagents) for selection of *NATMX* transformants, with 300 mg/L hygromycin B (Sigma or Roche) for selection of *HYGMX* transformants, with 7.5 mg/L phleomycin (Sigma or InvivoGen) for *BLEMX* transformants; 0.7 % yeast nitrogen base without amino acids with ammonium sulfate, 2 % dextrose, 0.2 % amino acid drop-out mix (Sunrise or Bio101), 2 % Bacto-Agar for selection of transformants complementing an auxotrophy; 0.17 % yeast nitrogen base without amino acids and without ammonium sulfate, 0.1 % proline, 2 % dextrose, 2 % Bacto-Agar supplemented with 200 mg/L bialaphos (Toku-E) for selection of *PATMX* transformants (glufosinate can also be used at 600–800 mg/L). Antibiotics should be filter-sterilized and added to autoclaved medium that has cooled to ~60–65 °C. Pour plates when medium is ~55 °C. Let plates dry for at least one day prior to use.

2.7 Rapid Isolation of Genomic DNA

1. TSENT: 2 % Triton X-100, 1 % SDS, 1 mM EDTA, 100 mM NaCl, 10 mM Tris–HCl pH 8.0.
2. 3 M sodium acetate, pH 5.2 (NaOAc).

3. 100 % ethanol (room temperature).
4. 70 % ethanol (ice cold).

2.8 Whole Colony PCR Verification of Integrants

1. 0.02 N NaOH.
2. 25 μM universal reverse (5′-GGATGTATGGGCTAAATG-3′), +intR, and −200 F primers made up in TE. *See* Figure legends for design of +intR and −200 F primers.
3. Taq and 10× Taq buffer (*see* **Note 39**).
4. 25 mM dNTP mix.

3 Methods

3.1 Gene Deletion De Novo by One-Step Gene Replacement

1. Set up the following PCR reaction in a 0.2 ml PCR tube: (*see* **Note 4**)

   1 μl deletion cassette template (approximately 10 ng)

   4 μl 25 μM F/F1/S1 primer

   4 μl 25 μM R/R1/S2 primer

   10 μl 10× Polymerase Buffer

   0.5 μl 25 mM dNTPs

   0.5 μl Polymerase

   80 μl H_2O
2. Vortex to mix then centrifuge briefly.
3. PCR program in the thermocycler: (*see* **Notes 5** and **6**).

   1 cycle: 94 °C, 5 min

   10 cycles: 94 °C, 30 s; 52 °C, 1 min; 68 °C, 2 min

   20 cycles: 94 °C, 30 s; 52 °C, 1 min; 68 °C, 2 min +5 s per cycle

   1 cycle: 72 °C, 10 min

   Pause: 4 °C
4. Check for PCR product by agarose gel electrophoresis. Generally, a single band is visible using 5 μl of the PCR reaction. The size of the band will depend on the cassette used for deletion (Table 1).
5. Transform 35–50 μl of the PCR reaction into yeast using the *High Efficiency Transformation Protocol* (*see* **Notes 7** and **8**).
6. Plate the entire transformation onto YPD and incubate overnight at 30 °C (*see* **Note 8**).
7. Replica plate to selective medium and incubate 2–3 days at 30 °C.
8. Verify integration using PCR.

3.2 Gene Deletion Using the Yeast Deletion Collection

1. Prepare genomic DNA from the desired deletion mutant from the haploid yeast deletion collection using the *Rapid Isolation of Genomic DNA* Protocol.
2. Set up the following PCR reaction in a 0.2 ml PCR tube: (*see* **Note 4**)

   1 µl genomic DNA template (approximately 10 ng)

   4 µl 25 µM −200 F primer

   4 µl 25 µM +200 F primer

   10 µl 10× Polymerase Buffer

   0.5 µl 25 mM dNTPs

   0.5 µl Polymerase

   80 µl H_2O
3. Continue starting at **Step 2** of Subheading 3.1.

3.3 Integrating Mutations

1. Prepare genomic DNA from the desired mutant using the *Rapid Isolation of Genomic DNA* Protocol to generate a DNA template, or prepare plasmid DNA containing mutation of interest (*see* **Notes 12** and **13**).
2. Set up the following PCR reaction in a 0.2 ml PCR tube: (*see* **Note 15**)

   1 µl mutant DNA template (approximately 10 ng)

   4 µl 25 µM A1 primer

   4 µl 25 µM A2 primer

   10 µl 10× Polymerase Buffer

   0.5 µl 25 mM dNTPs

   0.5 µl Polymerase

   80 µl H_2O
3. Set up the following PCR reaction in a separate 0.2 ml PCR tube: (*see* **Note 4**)

   1 µl deletion cassette template (approximately 10 ng)

   4 µl 25 µM A3 primer

   4 µl 25 µM A4 primer

   10 µl 10× Polymerase Buffer

   0.5 µl 25 mM dNTPs

   0.5 µl Polymerase

   80 µl H_2O
4. Vortex both reactions to mix then centrifuge briefly

5. PCR program in the thermocycler: (*see* **Note 16**)

   1 cycle: 94 °C, 5 min

   10 cycles: 94 °C, 30 s; 52 °C, 1 min; 68 °C, 2 min

   20 cycles: 94 °C, 30 s; 52 °C, 1 min; 68 °C, 2 min + 5 s per cycle

   1 cycle: 72 °C, 10 min

   pause: 4 °C
6. Check for PCR products by agarose gel electrophoresis. Generally, a single band is visible using 5 μl of the PCR reaction. The size of the band will depend on the cassette used and the size of the mutant gene.
7. Transform 35 μl of BOTH PCR reactions into yeast using the *High Efficiency Transformation* Protocol (*see* **Notes 17** and **18**).
8. Plate the entire transformation onto YPD and incubate overnight at 23 °C or 30 °C (*see* Subheading 4).
9. The next day, replica plate to selective medium and incubate 2–3 days at 23 °C or 30 °C.
10. Verify integration using PCR.

3.4 C-terminal Tagging

1. Set up the following PCR reaction in a 0.2 ml PCR tube: (*see* **Note 4**)

   1 μl tagging cassette template (approximately 10 ng)

   4 μl 25 μM F2/S3/F5 primer

   4 μl 25 μM F1/S2/F3 primer

   10 μl 10× Polymerase Buffer

   0.5 μl 25 mM dNTPs

   0.5 μl Polymerase

   80 μl H_2O
2. Continue starting at **Step 2** of Subheading 3.1 (*see* **Notes 22–25**).

3.5 Marker Exchange

1. Set up the following PCR reaction in a 0.2 ml PCR tube: (*see* **Note 4**)

   1 μl deletion DNA template (approximately 10 ng)

   4 μl 25 μM PR78 primer

   4 μl 25 μM PR79 primer

   10 μl 10× Polymerase Buffer

   0.5 μl 25 mM dNTPs

   0.5 μl Polymerase

   80 μl H_2O
2. Continue starting at **Step 2** of Subheading 3.1 (*see* **Note 26**).

3.6 Epitope Exchange

1. Set up the following PCR reaction in a 0.2 ml PCR tube:

   1 μl switching DNA template (approximately 10 ng)

   4 μl 25 μM F2CORE primer

   4 μl 25 μM R1CORE primer

   10 μl 10× Polymerase Buffer

   0.5 μl 25 mM dNTPs

   0.5 μl Polymerase

   80 μl H_2O

2. Continue starting at **Step 2** of Subheading 3.1 (*see* **Note 26**).

3.7 High Efficiency Yeast Transformation Modified from [47] for PCR Products)

1. Grow up 50 ml of cells overnight in YPD or SC at the appropriate temperature (*see* **Notes 8, 18, 22** and **28**).
2. Measure OD_{600} in the morning.
   - If OD_{600} is >1.0, dilute cells back to 0.1 and grow 4–6 h
   - If OD_{600} is 0.2–1.0, cells can be used immediately or diluted for use later in the day
   - If OD_{600} is less than 0.2, continue growing cells
3. Centrifuge down cells in 50 ml conical for 3–5 min at 3,000 × *g*.
4. Meanwhile, remove salmon sperm DNA (10 mg/ml; Applied Biosystems) from −20 °C and boil for 5 min. Immediately place on ice (*see* **Note 29**).
5. Pour off medium and resuspend cell pellet in 5 ml sterile 1× TE by vortexing.
6. Centrifuge down cells 3 min at 3,000 × *g*.
7. Pour off 1× TE and resuspend pellet in 5 ml LiOAc mix by vortexing.
8. Centrifuge down cells 3 min at 3,000 × *g*.
9. Pour off LiOAc mix and resuspend pellet in 0.2–1 ml LiOAc mix (*see* **Note 30**).
10. In a 1.5 ml eppendorf tube mix:

    1–5 μg DNA in 10–50 μl (*see* **Notes 7, 17** and **31**)

    10 μl 10 mg/ml salmon sperm DNA, freshly boiled

    100 μl cells in LiOAc mix

11. Add 700 μl PEG mix to each.
12. Vortex briefly to resuspend cells.
13. Incubate for 30 min at room temperature.
14. Add 48 μl DMSO to each.
15. Vortex briefly to mix.
16. Incubate for 15 min at 42 °C.

17. Centrifuge down for 1 min at 5,000 × *g* in microfuge.
18. Aspirate off liquid.
19. Add 200 μl YPD (*see* **Note 32**).
20. Spread onto a YPD plate and incubate overnight at 23 °C or 30 °C (*see* **Note 33**).
21. The next day, replica plate to selective medium and incubate at appropriate temperature until colonies appear, generally 2–4 days (*see* **Notes 34** and **35**).

3.8 Rapid Isolation of Genomic DNA (Modified from [48])

1. Grow strain to saturation in medium of choice.
2. Harvest 1.5 ml of overnight culture in screw-cap microfuge tube (*see* **Note 36**).
3. Resuspend in 200 μl TSENT by vortexing.
4. Add 0.3 g (about 150 μl worth) of acid washed glass beads (Biospec) to the resuspended cells.
5. Add 200 μl phenol–chloroform–IAA (25:24:1).
6. Screw on lid and vigorously vortex 5 min.
7. Centrifuge for 5 min at full speed in microfuge.
8. Transfer upper aqueous phase to a new eppendorf tube.
9. Ethanol precipitate one time:
   (a) Add 10 μl 3 M NaOAc pH 5.2 and 500 μl 100 % room temperature ethanol
   (b) Vortex to mix
   (c) Centrifuge 5 min at full speed in microfuge
   (d) Remove supernatant, being careful not to disturb the pellet
   (e) Add 500 μl cold 70 % EtOH to pellet and briefly vortex
   (f) Centrifuge 5 min at full speed in microfuge
   (g) Remove supernatant, being careful not to disturb the pellet
   (h) Centrifuge again briefly and remove residual liquid with a pipet
   (i) Let air dry 5–15 min
10. Resuspend in 50 μl TE (*see* **Notes 37** and **38**).

3.9 Whole Colony PCR Verification of Integrants

1. Using a P200 pipet tip, scrape approximately 10 % of a single colony (about the size of a match tip, around 5×10^5 cells) from a fresh plate and resuspend in 25 μl freshly made 0.02 N NaOH in an Eppendorf tube by twirling (*see* **Notes 40** and **41**).
2. Vortex briefly and boil for 5 min.
3. Immediately place tube on ice.

4. Vortex each sample vigorously for 10 s.
5. Set up the PCR reactions as follows:

   17 μl ddH_2O

   2.5 μl 10× Taq Buffer

   0.5 μl 10 mM dNTPs

   1 μl 25 μM F primer

   1 μl 25 μM R primer

   1 μl Taq
8. Add mix to PCR tubes containing 2 μl of DNA from your yeast.
9. Vortex briefly to mix.
10. Run the following PCR program in the thermocycler:

    1 cycle: 94 °C, 5 min

    35 cycles: 94 °C, 1 min; 50 °C, 1 min; 72 °C, 2 min

    1 cycle: 72 °C, 10 min

    Pause: 4 °C
11. Load the entire reaction onto agarose gel and visualize band after electrophoresis. The size of the product will depend on the positions of the PCR primers.

4 Notes

1. Typically the primers are designed to replace the open reading frame of a yeast gene with a selectable marker. However, in some cases only a portion of the gene can be deleted because removal of the entire ORF also removes regulatory regions or coding sequence of additional genes. Each primer ends with a universal sequence that is designed to amplify various selectable markers from plasmid templates (*see* Table 1 and Fig. 2a).
2. Oligonucleotides for PCR-mediated gene disruption are typically 60–80 bp in length. Many companies recommend HPLC or PAGE purification to increase the yield and purity of full-length primer.
3. Both Taq and high fidelity proofreading polymerases are suitable for this application. Premixed PCR master mix kits have also been successfully used.
4. Due to a high GC content in *NATMX* and *HYGMX* templates, the reaction must be supplemented with DMSO to 5 % when these templates are used [6].
5. The 68 °C extension temperature helps to preserve the life of the polymerase and results in greater full-length product yield.

6. A short annealing time helps reduce nonspecific background often observed when amplifying cassettes with direct repeats [9].
7. It is not necessary to clean up the PCR reaction unless multiple PCR products are observed. Precipitating the DNA prior to transformation is also not necessary.
8. The region of homology is short and so the efficiency of gene replacement is low, and therefore one must use a high efficiency transformation protocol. However, since the dominant drug resistance markers do not have sequence homology with the yeast genome, integration will usually occur at the target locus [5, 6, 8]. Note if one tries to use this procedure with a selectable marker derived from the *S. cerevisiae* genome, the PCR product will frequently integrate into the marker locus [8, 22, 45]. This is common when using the *TRP1* marker in W303, for example. *S. pombe his5*$^+$ present in *HIS3MX* is only 59 % identical to *HIS3* [8], and *K. lactis URA3* and *LEU2* are only 73 % and 77 % homologous to *S. cerevisiae URA3* and *LEU2* [7]. Because there are no long stretches of sequence homology, these cassettes can be used for deletion in most strain backgrounds and will not undergo recombination with the *S. cerevisiae* gene.
9. Generally, 1 μl of a 1:50 dilution of genomic DNA from the *Rapid Isolation of Genomic DNA* Protocol is used.
10. To use a marker other than *KANMX*, switch the marker in the yeast deletion strain using the *Marker Exchange* Protocol. Make and amplify genomic DNA from the new strain.
11. The standard oligonucleotides used for amplification from genomic DNA are ~20 bp long and do not need to be purified. The ~200 bp of homology generated using this method significantly enhances the targeting efficiency [5, 22]; longer regions of homology can be generated by changing the positions for the forward and reverse primers.
12. Mutants can be created using site directed mutagenesis of a plasmid, using kits such as the QuikChange Mutagenesis kit (Agilent) or the GeneArt Site-Directed Mutagenesis System (Life Technologies). It is helpful to not only introduce the desired mutation, but also to either destroy or create a restriction enzyme site so that the mutant allele can be easily distinguished from the wild-type gene without the need for sequencing. Many DNA analysis programs include a feature that allows silent restriction enzyme sites to be identified and engineered along with the mutation into PCR primers used to create the new allele.
13. Although it is possible to amplify a mutant gene from genomic DNA, it is sometimes useful to transfer the mutant gene onto a plasmid. This can be done using gap repair [3].

14. The position of the A1 primer does not need to be at the start codon, as depicted in Fig. 4. Rather, it needs to be approximately 200 bp 5′ to the mutation. Increasing the region of homology on the 5′ side of the mutation will increase the number of transformants that contain the mutant allele since there is a greater chance for recombination in this region. For large genes, this may not be feasible due to the size of the PCR product.
15. A proof-reading polymerase should be used for amplification of mutant allele in order to allow for error-free amplification of the mutant gene.
16. The extension time may have to be adjusted if long mutant alleles are amplified.
17. Co-transformation of PCR products will usually result in transformants, since yeast can successfully recombine both products. The PCR products can be stitched together in a PCR reaction using A1 and A4 primers.
18. Because it is likely that the mutant allele will result in a fitness disadvantage, there will be a selective pressure for recombination events that result in the wild-type allele if haploids are used. Therefore, construction of mutants should be done in diploids, so that a wild-type copy of the gene is present.
19. The forward primer used for C-terminal tagging must be in-frame with the fusion protein or the epitope tag.
20. The stop codon does not need to be included in the reverse primer. Each tagging module contains a stop codon following the epitope or fusion protein, as well as a transcription terminator (Fig. 1).
21. The same reverse primer R1/S2/R3 can be used for both C-terminal tagging and the construction of de novo gene deletions.
22. It is good practice to transform PCR products for C-terminal tagging into diploid yeast strains in the event that the resulting gene fusion is nonfunctional or only partially functional. Upon sporulation, haploid strains in both mating types can be recovered, which is useful for subsequent strain constructions. In addition, any random mutations that may be introduced during the transformation can be out-crossed [46].
23. Correct integration of PCR fragments used for tagging or gene fusion can be accomplished by direct visualization under the microscope for fluorescent fusion proteins or by western blotting and protein mobility on SDS-PAGE using commercially available antibodies to detect the tagged proteins. The parental untagged strain is a negative control.

24. The most rigorous way to ensure that a PCR fragment has integrated at the desired region is using PCR. A primer that recognizes the cassette and gene specific primers will give a PCR fragment only if the tagging module has integrated into the gene of interest.
25. It is important to show, if at all possible, that the protein fusion to your favorite gene functions normally. This is best accomplished by complementing all mutant phenotypes associated with deletions of your favorite gene. Growth on rich medium at 30 °C does not necessarily mean the fusion protein is fully functional.
26. Select for the new marker following transformation. Verify that the markers have switched by scoring for sensitivity to the original marker. Integration of the new module can also be verified by PCR using reverse primers that specifically recognize each module. KANMX-R, 5′-GGCCGGGTGACCCGGCGG GG-3′; HIS3MX-R, 5′-GGAGTCAATAATTTCATCGCTG CC-3′;NATMX-R,5′-GCCTCCATGTCGCTGGCCGGG-3′; HYG-MX-R, 5′-CATGCCCCTGAGCTGCGCACG-3′; CaU RA3MX-R,5′-CCTCGACATCATCTGCCC-3′;KlURA3-R,5′-CAGACCGATCTTCTACCC- 3′; KlLEU2-R, 5′-AGTTATC CTTGGATTTGG- 3′; CoreCHK-R, 5′-ATACGCGCACAA AAGCAGAG-3′.
27. PEG Mix can be autoclaved, but care must be taken to ensure the PEG is at the proper concentration. In addition, it is important to store the PEG Mix in a tightly capped container to prevent evaporation of water, which will increase the PEG concentration. Small variations in PEG concentration will reduce transformation efficiency. It is for this reason that large batches of PEG mix are not made and stored.
28. Since nonhomologous end-joining is believed to be repressed in diploids, proper targeting to the homologous chromosomal locus is higher in diploids. Gene deletions and tagging should always be done in diploids followed by sporulation and tetrad dissection to ensure that there is only one insertion of the marker and that additional collateral mutations have not been induced at other loci. Johnston et al. estimate that 5–10 % of all haploid transformants contain a mutation that results in an observable growth defect not related to the targeted DNA [46].
29. It is not necessary or desirable to boil the carrier DNA every time. Keep small aliquots at −20 °C and boil after three or four thaws. Keep on ice when out of the freezer.
30. The volume in which to resuspend cells is dependent on cell number. Because cells grown in SD or SC medium are at a lower density than cells grown in YPD or other rich media, smaller volumes are used. If more cells are needed for multiple

transformations, the protocol is easily scaled up by simply growing a larger culture and then dividing that culture into multiple 50 ml conical tubes. Alternatively, larger cultures can be harvested using 250 ml or 500 ml centrifuge bottles; in this case, wash volumes are increased accordingly.

31. It is always helpful to include a no DNA control when performing transformations.
32. Some strains are particularly sensitive to any remaining PEG mix so it is useful to wash the cells once in sterile water or medium prior to plating. Although sterile water can also be used for cell resuspension, transformants will appear more quickly if YPD is used; this is particularly important when working with non-wild-type strains. The residual amino acids in YPD do not affect selection on drop-out plates.
33. Allow the transformed cells to recover prior to plating to selective medium containing drugs. The recovery time allows for expression of the drug resistance gene. To ensure that transformants are clonal, it is best to plate the transformation to YPD overnight (recovery), then replica plate the lawn of cells to selective medium the next day. The entire transformation can be plated on a single plate when using an integrating plasmid or a PCR product. Transformants will generally be visible in 3–4 days at 30 °C. Approximately 5–50 transformants are expected.
34. Background is often high with a PCR-based transformation. These are thought to be abortive transformants that have not integrated the marker. Replica plating to a second selection plate can reduce the background.
35. Transformants should be colony purified by streaking for single colonies on a selective plate.
36. After removing medium, the cell pellet can be flash frozen in liquid N_2 and stored at −80 °C indefinitely.
37. This DNA is suitable for analysis and amplification by PCR. Due to the presence of RNA and other contaminants, additional cleanup steps are required for restriction digests, Southern and Northern blot analysis and quantitative real-time PCR.
38. For verifying integration events, use the following PCR reaction conditions:

    39.25 μl ddH_2O

    5 μl 10× Polymerase Buffer

    0.25 μl 25 mM dNTPs

    2 μl 25 μM F primer

    2 μl 25 μM R primer

    0.5 μl Taq

1 μl genomic DNA (approximately 10 ng)

And run the following PCR program:

1 cycle: 94 °C, 5 min

35 cycles: 94 °C, 1 min; 50 °C, 1 min; 72 °C, 2 min

1 cycle: 72 °C, 10 min

pause: 4 °C

Note that extension time and annealing temperature may need to be adjusted for primer pair.

39. This method works best if Taq polymerase is used.
40. Following transformation, potential positives are streaked for single colonies on selective media, then patches are made from single colonies on YPD for analysis. Whole colony PCR works best using freshly patched cells that have not been refrigerated.
41. Be careful not to transfer any agar since this will inhibit the PCR reaction.

Acknowledgements

We are indebted to participants of the Cold Spring Harbor Yeast Genetics and Genomics Course for their insights into yeast cell manipulation. We thank members of the Jaspersen lab for comments on the manuscript. S.L.J. is supported by the Stowers Institute for Medical Research and the American Cancer Society (RSG-11-030-01-CSM).

References

1. Rothstein RJ (1983) One-step gene disruption in yeast. Methods Enzymol 101:202–211
2. Elledge SJ, Davis RW (1988) A family of versatile centromeric vectors designed for use in the sectoring-shuffle mutagenesis assay in *Saccharomyces cerevisiae*. Gene 70:303–312
3. Cormack B, Castano I (2002) Introduction of point mutations into cloned genes. Methods Enzymol 350:199–218
4. Rothstein R (1991) Targeting, disruption, replacement, and allele rescue: integrative DNA transformation in yeast. Methods Enzymol 194:281–301
5. Wach A, Brachat A, Pohlmann R et al (1994) New heterologous modules for classical or PCR-based gene disruptions in *Saccharomyces cerevisiae*. Yeast 10:1793–1808
6. Goldstein AL, McCusker JH (1999) Three new dominant drug resistance cassettes for gene disruption in *Saccharomyces cerevisiae*. Yeast 15:1541–1553
7. Gueldener U, Heinisch J, Koehler GJ et al (2002) A second set of *loxP* marker cassettes for Cre-mediated multiple gene knockouts in budding yeast. Nucleic Acids Res 30:e23
8. Wach A, Brachat A, Alberti-Segui C et al (1997) Heterologous *HIS3* marker and GFP reporter modules for PCR-targeting in *Saccharomyces cerevisiae*. Yeast 13:1065–1075
9. Goldstein AL, Pan X, McCusker JH (1999) Heterologous *URA3MX* cassettes for gene replacement in *Saccharomyces cerevisiae*. Yeast 15:507–511
10. Ito-Harashima S, McCusker JH (2004) Positive and negative selection *LYS5MX* gene replacement cassettes for use in *Saccharomyces cerevisiae*. Yeast 21:53–61

11. Janke C, Magiera MM, Rathfelder N et al (2004) A versatile toolbox for PCR-based tagging of yeast genes: new fluorescent proteins, more markers and promoter substitution cassettes. Yeast 21:947–962
12. Longtine MS, McKenzie A 3rd, Demarini DJ et al (1998) Additional modules for versatile and economical PCR-based gene deletion and modification in *Saccharomyces cerevisiae*. Yeast 14:953–961
13. Gueldener U, Heck S, Fielder T et al (1996) A new efficient gene disruption cassette for repeated use in budding yeast. Nucleic Acids Res 24:2519–2524
14. Delneri D, Tomlin GC, Wixon JL et al (2000) Exploring redundancy in the yeast genome: an improved strategy for use of the cre-loxP system. Gene 252:127–135
15. Chee MK, Haase SB (2012) New and Redesigned pRS Plasmid Shuttle Vectors for Genetic Manipulation of *Saccharomyces cerevisiae*. G3 (Bethesda) 2:515–526
16. Sung MK, Ha CW, Huh WK (2008) A vector system for efficient and economical switching of C-terminal epitope tags in *Saccharomyces cerevisiae*. Yeast 25:301–311
17. Hegemann JH, Heick SB (2011) Delete and repeat: a comprehensive toolkit for sequential gene knockout in the budding yeast *Saccharomyces cerevisiae*. Methods Mol Biol 765:189–206
18. Shoemaker DD, Lashkari DA, Morris D et al (1996) Quantitative phenotypic analysis of yeast deletion mutants using a highly parallel molecular bar-coding strategy. Nat Genet 14:450–456
19. Winzeler EA, Shoemaker DD, Astromoff A et al (1999) Functional characterization of the *S. cerevisiae* genome by gene deletion and parallel analysis. Science 285:901–906
20. Smith AM, Heisler LE, Mellor J et al (2009) Quantitative phenotyping via deep barcode sequencing. Genome Res 19:1836–1842
21. Smith AM, Durbic T, Kittanakom S et al (2012) Barcode sequencing for understanding drug-gene interactions. Methods Mol Biol 910:55–69
22. Manivasakam P, Weber SC, McElver J et al (1995) Micro-homology mediated PCR targeting in *Saccharomyces cerevisiae*. Nucleic Acids Res 23:2799–2800
23. Hughes TR, Roberts CJ, Dai H et al (2000) Widespread aneuploidy revealed by DNA microarray expression profiling. Nat Genet 25:333–337
24. Lehner KR, Stone MM, Farber RA et al (2007) Ninety-six haploid yeast strains with individual disruptions of open reading frames between *YOR097C* and *YOR192C*, constructed for the *Saccharomyces* genome deletion project, have an additional mutation in the mismatch repair gene *MSH3*. Genetics 177:1951–1953
25. Copic A, Latham CF, Horlbeck MA et al (2012) ER cargo properties specify a requirement for COPII coat rigidity mediated by Sec13p. Science 335:1359–1362
26. Hartwell LH, Culotti J, Reid B (1970) Genetic control of the cell-division cycle in yeast. I. Detection of mutants. Proc Natl Acad Sci U S A 66:352–359
27. Reid RJ, Lisby M, Rothstein R (2002) Cloning-free genome alterations in *Saccharomyces cerevisiae* using adaptamer-mediated PCR. Methods Enzymol 350:258–277
28. Langle-Rouault F, Jacobs E (1995) A method for performing precise alterations in the yeast genome using a recycable selectable marker. Nucleic Acids Res 23:3079–3081
29. Maeder CI, Maier P, Knop M (2007) A guided tour to PCR-based genomic manipulations of *S. cerevisiae* (PCR-targeting). Methods Microbiol 36:55–78
30. Prein B, Natter K, Kohlwein SD (2000) A novel strategy for constructing N-terminal chromosomal fusions to green fluorescent protein in the yeast *Saccharomyces cerevisiae*. FEBS Lett 485:29–34
31. Gauss R, Trautwein M, Sommer T et al (2005) New modules for the repeated internal and N-terminal epitope tagging of genes in *Saccharomyces cerevisiae*. Yeast 22:1–12
32. Knop M, Siegers K, Pereira G et al (1999) Epitope tagging of yeast genes using a PCR-based strategy: more tags and improved practical routines. Yeast 15:963–972
33. Funakoshi M, Hochstrasser M (2009) Small epitope-linker modules for PCR-based C-terminal tagging in *Saccharomyces cerevisiae*. Yeast 26:185–192
34. Van Driessche B, Tafforeau L, Hentges P et al (2005) Additional vectors for PCR-based gene tagging in *Saccharomyces cerevisiae* and *Schizosaccharomyces pombe* using nourseothricin resistance. Yeast 22:1061–1068
35. Sheff MA, Thorn KS (2004) Optimized cassettes for fluorescent protein tagging in *Saccharomyces cerevisiae*. Yeast 21:661–670
36. Tamm T (2009) Plasmids with E2 epitope tags: tagging modules for N- and C-terminal PCR-based gene targeting in both budding and fission yeast, and inducible expression vectors for fission yeast. Yeast 26:55–66
37. Gadal O, Strauss D, Braspenning J et al (2001) A nuclear AAA-type ATPase (Rix7p) is required

for biogenesis and nuclear export of 60S ribosomal subunits. EMBO J 20:3695–3704

38. Gadal O, Strauss D, Petfalski E et al (2002) Rlp7p is associated with 60S preribosomes, restricted to the granular component of the nucleolus, and required for pre-rRNA processing. J Cell Biol 157:941–951
39. Lee S, Lim WA, Thorn KA (2013) Improved blue, green, and red fluorescent protein tagging vectors for *S. cerevisiae*. PLoS One 8: e67902
40. Schneider BL, Seufert W, Steiner B et al (1995) Use of polymerase chain reaction epitope tagging for protein tagging in *Saccharomyces cerevisiae*. Yeast 11:1265–1274
41. Moqtaderi Z, Struhl K (2008) Expanding the repertoire of plasmids for PCR-mediated epitope tagging in yeast. Yeast 25:287–292
42. Webster TD, Dickson RC (1983) Direct selection of *Saccharomyces cerevisiae* resistant to the antibiotic G418 following transformation with a DNA vector carrying the kanamycin-resistance gene of Tn903. Gene 26:243–252
43. Brachmann CB, Davies A, Cost GJ et al (1998) Designer deletion strains derived from *Saccharomyces cerevisiae* S288C: a useful set of strains and plasmids for PCR-mediated gene disruption and other applications. Yeast 14: 115–132
44. Wieczorke R, Krampe S, Weierstall T et al (1999) Concurrent knock-out of at least 20 transporter genes is required to block uptake of hexoses in *Saccharomyces cerevisiae*. FEBS Lett 464:123–128
45. Baudin A, Ozier-Kalogeropoulos O, Denouel A et al (1993) A simple and efficient method for direct gene deletion in *Saccharomyces cerevisiae*. Nucleic Acids Res 21:3329–3330
46. Johnston M, Riles L, Hegemann JH (2002) Gene disruption. Methods Enzymol 350: 290–315
47. Gietz RD, Woods RA (2002) Transformation of yeast by lithium acetate/single-stranded carrier DNA/polyethylene glycol method. Methods Enzymol 350:87–96
48. Hoffman CS, Winston F (1987) A ten-minute DNA preparation from yeast efficiently releases autonomous plasmids for transformation of *Escherichia coli*. Gene 57:267–272
49. Slaughter BD, Schwartz JW, Li R (2007) Mapping dynamic protein interactions in MAP kinase signaling using live-cell fluorescence fluctuation spectroscopy and imaging. Proc Natl Acad Sci U S A 104:20320–20325
50. Onischenko E, Stanton LH, Madrid AS et al (2009) Role of the Ndc1 interaction network in yeast nuclear pore complex assembly and maintenance. J Cell Biol 185:475–491
51. Hailey DW, Davis TN, Muller EG (2002) Fluorescence resonance energy transfer using color variants of green fluorescent protein. Methods Enzymol 351:34–49

Chapter 6

Preparation of Yeast Cells for Live-Cell Imaging and Indirect Immunofluorescence

Lucy F. Pemberton

Abstract

In spite of their small size, the cellular morphology, structure, and protein localization of yeast cells can be successfully imaged. A detailed protocol for preparing yeast cells for live-cell imaging is described, including techniques to immobilize yeast for time-lapse microscopy. Protocols for indirect immunofluorescence are outlined, including strategies for fixation, cell wall digestion, and the use of primary and secondary antibodies conjugated to fluorescent moieties. Alternative approaches to these techniques are discussed, highlighting the advantages and disadvantages where possible. Using these protocols, investigation of yeast cell structure and protein localization will continue to yield important insights into yeast cell biology and regulation.

Key words *Saccharomyces cerevisiae*, Fluorescence microscopy, Spheroplast, Live-cell imaging, Concanavalin A, Agar pad

1 Introduction

Fluorescence microscopy is a well-established technique allowing the localization of different structures and individual macromolecules within the cell. In the yeast *S. cerevisiae* both indirect immunofluorescence and live-cell fluorescent imaging are readily accomplished. The relative ease with which yeast proteins can be tagged within the genome, or deletion mutants constructed and even purchased, makes yeast an attractive organism to study different dynamic processes such as cell division. In addition, the availability of different fluorescent moieties, the development of sensitive cameras, and advances in electronic light and confocal laser scanning microscopy have helped overcome challenges inherent to yeast microscopy, which are a consequence of its relatively small size. Protocols will be described for the preparation of slides for live-cell imaging of cellular proteins with fluorescent tags. We will also describe protocols for indirect immunofluorescence of fixed cells using antibodies to native proteins or expressed tags. Both techniques

Jeffrey S. Smith and Daniel J. Burke (eds.), *Yeast Genetics: Methods and Protocols*, Methods in Molecular Biology, vol. 1205, DOI 10.1007/978-1-4939-1363-3_6, © Springer Science+Business Media New York 2014

are useful; live-cell imaging allows dynamic studies, in addition to avoiding cell fixation and permeabilization, thus helping to preserve cellular morphology. However, setting up time-lapse microscopy requires expensive equipment, and gathering and analyzing this type of data is time consuming. Immunofluorescence provides a snapshot and allows the use of specific antibodies, often bypassing the need for genetic manipulation or plasmid expression. However, the preparation of slides takes much longer, and can be prone to problems with fixation and permeabilization. After the preparation of slides outlined here, cells can be viewed with a standard upright fluorescence microscope. The expression of fluorescent tags in living cells also lends itself to more complicated microscopy using confocal scanning technologies, such as spinning disk microscopy, FRAP, and FLIP, in addition to high-throughput approaches. The pros and cons of these techniques are discussed at length in earlier protocols [1–4]. We do not discuss the choice of microscope or data analysis for these methods as they are dependent on the specific questions being asked and are addressed more thoroughly elsewhere [1–3, 5].

2 Materials

2.1 Live-Cell Imaging

2.1.1 Simple Method for Imaging Live Cells

1. Cover slips, 22 × 22 mm (Fisher).
2. Microscope slides: For live-cell imaging plain frosted slides (Fisher).

2.1.2 Preparation of Con A-Coated Slides

1. Concanavalin A (Sigma).
2. Microscope slides: For live-cell imaging plain frosted slides (Fisher).
3. Cover slips, 22 × 22 mm (Fisher).
4. Nail polish: We prefer colored polish because it is easier to see where it has been applied.

2.1.3 Preparation of Agarose Pads for Imaging

1. Seakem Gold agarose (Fisher).
2. Difco Yeast nitrogen base (YNB) without amino acids (Fisher).
3. 10× complete supplement mixture (CSM) in 20 % glucose: Take 0.8 g of CSM powder (Sunrise Science Products) and 20 g of dextrose (Fisher) and dissolve in 100 mL of dH_2O. Filter sterilize the solution and store in foil at 4 °C. Be careful as this medium becomes easily contaminated.
4. VALAP: A mixture of equal amounts of Vaseline, lanoline, and paraffin wax flakes (Fisher). Mix together equal amounts (by weight) in a glass beaker and heat on a hot plate set to low, mixing occasionally. When the liquid is homogeneous, aliquot

into small glass bottles that can be easily reheated. Allow to cool, cap, and store at room temperature. VALAP needs to be re-melted on the hot plate before use; never leave unattended and do not allow it to overheat.

2.2 Indirect Immunofluorescence in Yeast

1. Formaldehyde: 37 % solution. Formaldehyde eventually breaks down to formic acid, so small amounts should be purchased and not be kept for extended periods (Fisher).
2. Phosphate buffer: 13.6 KH_2PO_4 (anhydrous), 2.1 g KOH is dissolved in 1 L of H_2O, the pH should be 6.5.
3. SPC buffer: 218.6 g sorbitol (1.2 M), 17.4 g K_2HPO_4 (anhydrous), and 7 g citric acid (monohydrate) are dissolved in 1 L of H_2O. This solution is filter sterilized and kept at 4 °C as it easily becomes contaminated.
4. Several different combinations of enzymes for digestion of the cell wall are described. We routinely use 10 % glusulase (Perkin Elmer) and 0.1 mg/mL zymolyase 20T (US Biologicals) in SPC buffer. Other protocols include 0.4 mg/mL zymolyase 100T (US Biologicals) in SPC buffer.
5. Microscope slides: For indirect immunofluorescence, we use teflon-coated slides. The 15-well multitest slides use less antibody per well (MP).
6. Poly-L-lysine (molecular weight 400,000, Sigma): Add 50 mg to 10 mL of water, and dissolve to make a 0.5 % stock. Store the stock in aliquots at −20 °C. The thawed aliquots may be re-frozen. We also use poly-L-lysine that is purchased as a 0.1 % w/v liquid stock (Sigma).
7. Two Coplin jars: Tall enough to hold microscope slides (Fisher).
8. PBS: Phosphate-buffered saline (PBS) 10×: dissolve 80 g NaCl, 2 g KCl, 14.4 g $Na_2HPO_4{\cdot}H_2O$, and 2.4 g KH_2PO_4 in 1 L of deionized H_2O, pH to 7.4 with HCl. PBS is sterilized by autoclaving and stored at RT. PBS is dissolved to 1× before use. The indirect immunofluorescence protocols described here use 1× PBS, 2 % milk, and 0.1 % Tween-20. For 100 mL, take 10 mL of 10× PBS, 90 mL H_2O, 100 μL Tween-20, and 2 g of Carnation Nonfat Dried Milk. Mix together until milk has dissolved. To make PBS, 2 % BSA, and 0.1 % Tween-20 replace the milk with 2 g BSA (BSA Fraction V, Fisher).
9. Antibodies: For secondary antibodies we routinely use Rabbit anti-Mouse FITC FC region and Goat anti-Rabbit Cy3 (Jackson Immunoresearch). Primary antibodies can be either monoclonal or polyclonal. Detect TAP or PrA tags with Rabbit affinity purified antibody to Mouse IgG (MP Biologicals). Antibodies need to be titrated when first used to determine the optimum concentration.

10. Mounting medium: *p*-Phenylenediamine (Sigma) functions as an anti-bleaching agent. Take care since this chemical is carcinogenic. Dissolve 10 mg of *p*-phenylenediamine in 1 mL of phosphate-buffered saline (PBS) by vortexing. Add 9 mL of glycerol and mix well. Store the mounting medium in foil at −20 °C, or −80 °C for long-term storage. The medium should be discarded when it turns brown. DAPI can also be added to the medium at 0.1 μg/mL concentration if required. Some researchers prefer to use commercially prepared mounting medium (Molecular Probes).
11. DAPI: Dissolve sufficient 4′,6-diamidino-2-phenylindole dihydrochloride (DAPI; Molecular Probes) in water to prepare a 1 mg/mL (1,000×) solution. Store in the dark for several years at −20 °C. DAPI can also be added at 100 ng/mL to the mounting medium.
12. Cover slips (Corning, No. 1 weight, 24×60 mm).

3 Methods

General. All protocols require first growing yeast in the appropriate liquid medium. Where possible it is advisable to grow yeast for immunofluorescence in rich medium such as YPD, as this results in more efficient cell wall digestion. For indirect immunofluorescence, cells should not be overgrown prior to fixation with an OD_{600} of <1.0. If cells are saturated after overnight culture they should be diluted back to at least ~0.4 OD_{600} and allowed to double again. If possible avoid "pink" yeast strains that are *ade1* or *ade2* mutants, as they accumulate a fluorescent intermediate in adenine biosynthesis in the vacuole that can interfere with imaging. If *ade1* or *ade2* strains are used, the production of the intermediate can be eliminated by growing cells in medium containing adenine (0.1 mg/mL) for 18–24 h.

3.1 Live-Cell Imaging

Three methods for preparing logarithmically growing cells for live-cell imaging are described. For pelleting steps requiring centrifugation, the RCF should be kept to approx 500×*g* as faster speeds can lead to the deformation of yeast cells and internal organelles. A number of fluorescent moieties can be used in yeast cells and the relative advantages and disadvantages are discussed elsewhere [1]. The choice of fluorescent moiety is up to the individual but in experiments using one fluorophore, the enhanced S65T mutant GFP (EGFP) usually gives the most intense and stable signal and is therefore often the best choice [6, 7]. If dual labeling is necessary, effective combinations are CFP and YFP, or GFP and mCherry. In our lab we routinely use two tandem copies of EGFP fused to the protein C terminus, leading to a stronger GFP signal

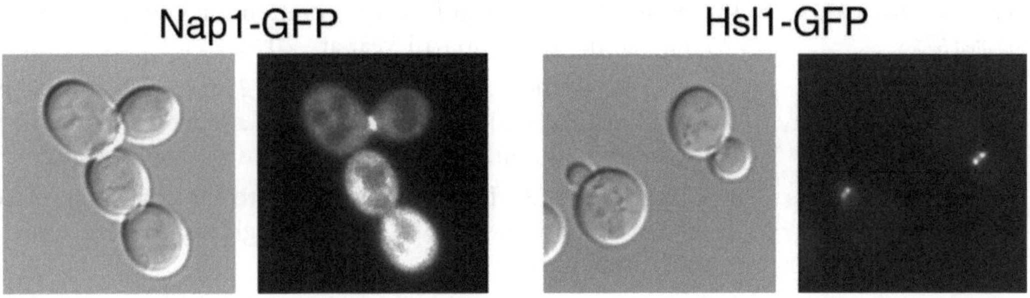

Fig. 1 Wild-type yeast cells expressing plasmids encoding either the histone chaperone Nap1 or the kinase Hsl1 fused to two tandem copies of EGFP. The corresponding DIC image is also shown. Nap1 is localized to the cytoplasm and bud neck in cells at G2/M whereas Hsl1 is restricted to the bud neck

and also increasing the molecular weight of small proteins above the diffusion limit of the nuclear pore complex, which is important for nuclear transport studies [8] (Fig. 1). It has been noted that EGFP is brighter when cells are cultured in neutral or slightly basic medium (*see* **Note 1**, ref. 9).

3.1.1 Simplest Method

Cells growing in media can be pipetted directly onto a slide with no preparation. In addition, individual colonies on plates can be similarly viewed to determine whether they are expressing GFP.

1. Take 4 μL of cells in growth media and pipette into the center of a plain glass slide. Alternatively, add a colony of yeast cells to 4 μL of H_2O onto a slide and use a pipette tip to break up the colony and mix into the liquid.
2. Place a 22 × 22 mm cover slip on top of the cells. Before use, wipe cover slips with a Kimwipe to remove small specks of dust, which can prevent the cover slip from lying flat on the droplet.
3. Apply a little pressure to the cover slip; too much pressure will lead to the cells bursting, and too little results in the cells floating around. The cover slip remains in place by surface tension and these slides remain useable for about 20 min, after which they dry out.

Immobilized Cells. It is more common to immobilize yeast cells for live-cell imaging experiments, using either concanavalin A (con A)-treated slides, or on agar pads. These substrates prevent the cells from moving during the experiment. Con A is a lectin that binds to carbohydrate on the yeast cell surface. Agar pads are composed of synthetic medium solidified with agar. The con A technique is faster and easier, and avoids the background fluorescence sometimes detected with agar pads. The agar pads are more effective for longer time-lapse experiments as they provide nutrients to the living cells. The production of con A-coated cover slips and agar pads is described below.

3.1.2 Use of Con A-Coated Slides for Imaging

1. Prepare a 0.5 mg/mL stock solution of con A in water (this can be stored up to 1 year at −20 °C).
2. Add 5 μL of the con A solution onto a 22 × 22 mm cover slip, spread out the solution with the side of a pipet tip, and air-dry for around 20 min, in a box to prevent dust from sticking.
3. Once dry add 3 μL of cell suspension to the cover slip, place the cover slip upside down on a clean glass slide, and apply a small amount of pressure.
4. Seal the cover slip to the slide using nail polish.

3.1.3 Use of Agarose Pads for Imaging

The protocol for the production of agarose pads is based on that described by the Yeast Resource Center, University of Washington, and described in their youtube video, http://www.youtube.com/watch?v=ZrZVbFg9NE8. Pads can also be prepared from gelatin (*see* **Note 2**).

1. To prepare agarose, add 6.7 g YNB without amino acids to 900 mL of H_2O and autoclave. Microwave 0.5 g of agarose into 45 mL of the YNB solution, aliquot into Eppendorf tubes (0.45 mL/tube), and store in the fridge.
2. Make a separate stock of 10× CSM-complete mix (to provide amino acids) in 20 % glucose, filter sterilize the solution through a 0.22 μm filter to remove particles, and sterilize.
3. Melt the 0.45 mL aliquot of agarose in a boiling heat block, transfer to a 65 °C block, add 50 μL of the amino acid/glucose mixture to the agarose, vortex the solution, and keep at 65 °C.
4. Tape around the ends of a glass microscope slide with Scotch tape to act as spacers. Add a 30 μL drop of agar between the spacers and spread with a tip to make a round drop. Place a second slide over the first by touching the slides together at one end first, and slowly lowering the second slide on top of the first. Flatten the drop between the slides; after slight pressure between the two slides the pad of agar will become oval shaped.
5. Place the slide on a Kimwipe on top of a frozen ice block for about 60 s to solidify. Do not let the agarose freeze. Slide the two slides apart and the pad should remain in the center of one slide if this process is successful. Allow the pad to dry for about 3 min before use.
6. 3 μL of cell suspension in the YNB media is added to the pad and a cover slip. The cover slip can be sealed to the slide by first applying VALAP around the edges of the pad. First melt the VALAP on a heat block and using a cotton swab, apply a thin line around the edge of the pad. Then place the cover slip on top before the VALAP solidifies.

3.2 Indirect Immunofluorescence in Yeast

These protocols are based on protocols from Kilmartin and Adams [10], and Rout and Kilmartin [11] and Silver, P.A. [12].

1. Grow yeast to an OD_{600} less than or equal to 1 as indicated above.
2. Add 1/10th volume of 37 % w/v formaldehyde directly to the culture (0.5 mL to a 5 mL culture). Mix the yeast by gentle swirling. Yeast should be fixed at room temperature for the appropriate time for the particular antigen. Thirty minutes is suitable for many antigens although others such as HA are formaldehyde sensitive and should be fixed for 10 min. If necessary carry out a time course to determine the optimal fixation time; cells should be gently agitated every 10 min to prevent them from settling out while fixing. Formaldehyde is purchased as a solution or prepared freshly (*see* **Note 3**).
3. After fixation pellet the yeast by centrifugation for 5 min at 500 × *g* in a clinical centrifuge at room temperature.
4. Resuspend the pellet in 1.0 mL of 100 mM potassium phosphate buffer, pH 6.5 and transfer to a microfuge tube.
5. Wash the cells three times in 1.0 mL of 100 mM potassium phosphate buffer, pH 6.5 by pelleting for 20 s at 500 × *g* in a microfuge and aspirating the supernatant. The cell pellet should be resuspended gently with a pipette. If cells have been growing in selective media, or are overgrown or are slow-growing mutants, the cell walls can be harder to digest. Cells can be pretreated with DTT to increase the efficiency of digestion. This step is not necessary for log-phase cells grown in YPD. Resuspend in 1 mL 100 mM Tris–HCl, pH 9.0, and 100 mM DTT and incubate at 30 °C for 10 min, and then pellet cells as before. It is not necessary to wash again prior to adding the enzymes.
6. Yeast cells have cell walls, and for antibodies to gain access to the interior, the walls must first be digested with enzymes. Resuspend the cell pellet in 1 mL of 1.2 M Sorbitol phosphate-citrate buffer (SPC) and pellet as before. The sorbitol provides osmotic stability to cells, which will be important when the cell wall is digested. To digest cell walls, resuspend cells in a small volume (100–200 μL depending on the pellet size) of SPC containing 10 % glusulase and 0.01 % zymolyase 20T for 90–120 min at 30 °C with occasional agitation. Do not vortex cells during or after digestion as they are fragile. There are several alternative enzyme cocktails that can be used to efficiently digest the cell wall. These include adding 20 μL of freshly made 10 mg/mL zymolyase 100T to yeast in 0.5 mL of SPC buffer, or adding 20 μL of 50 mg/mL freshly made zymolyase 20T to yeast in 0.5 mL SPC buffer. These cocktails can be incubated for 30–60 min at 30 °C. The precise cocktail is

usually determined by what enzymes are available in the lab. The more highly purified zymolyase 100T is more active and can generally be used for shorter periods but this benefit is offset by the fact that it is more expensive. In all cases cells should be viewed by phase microscopy to determine whether efficient digestion has taken place (*see* ref. 13 for examples of spheroplasted cells and *see* **Note 4**).

7. Wash digested cells as before three times in 1 mL cold SPC, resuspend the pellet in approximately two pellet volumes of SPC, and leave on ice. These cells can remain on ice for several hours and can also be frozen at –80 °C and thawed with minimal loss of integrity.
8. Poly-L-lysine-coated slides are prepared, as yeast cells do not adhere well to glass. Prior to using, the slides are wiped free of dust with a tissue. We use 15-well slides that have a teflon coating to separate wells. Into each well pipette a drop of poly-L-lysine. This can be prepared from powder as a 0.1 % stock in water, or purchased as a liquid stock. PEI can also be used in place of poly-L-lysine (*see* **Note 5**). Leave the slides for 10 min at room temperature ensuring that the poly-L-lysine does not dry out. Gently pass the slides two times through a stream of running distilled water and allow to air-dry (excess drops around the edges can be carefully aspirated away).
9. After the 15-well slides have been coated with poly-L-lysine, add 3–5 μL of cells in SPC buffer per well and let attach at room temperature for 5 min. Label slides with a pencil in the white area, as other markers may be washed off in the methanol/acetone.
10. Remove excess liquid by carefully aspirating at the side of the well. Place two slides back to back and plunge into a Coplin jar of –20 °C MeOH (chilled in a freezer for 2–3 h). Move the slide up and down 2–3 times to ensure that it is coated and leave for 5 min.
11. Remove slides and plunge into a Coplin jar of acetone at room temperature for 30 s. Again move slide up and down in Coplin jar a few times. Remove slides and shake or flick hard in air for 30 s to completely dry.
12. Using a black marker pen draw a ring around each well; this helps keep the drops of antibody confined, allows the use of less antibody, and prevents mixing of different antibodies. The dry slides are now ready for antibody incubation. Dry slides are stable at –20 °C for 1–4 weeks.
13. Prepare a humidity chamber by assembling a plastic box with a lid with two damp paper towels laid flat in the bottom. Once the slides are rehydrated they should never be allowed to dry out prior to mounting. Block the cells by adding a drop of

PBS-2 % milk-0.1 % Tween-20 to each well. The drop should be added, then aspirated, and then immediately reapplied for efficient blocking. To prevent the wells from drying, set up a vacuum aspirator with a gel loading tip on the end; the wells are aspirated with one hand and the media immediately reapplied by pipette using the other hand (this technique is shown in 14). Leave cells to block for 5 min at RT in the box.

14. Have the antibodies diluted to the correct concentration in PBS-2 % milk-0.1 % Tween-20 prior to removing blocking solution. Aspirate wells and immediately reapply 3–5 μL of antibody at the appropriate dilution. For most antibodies incubation is for 1–2 h at room temperature, or overnight at 4 °C in the humidity box is sufficient. *See* **Notes 6** and **7** for appropriate controls to use and **Note 8** for the use of fluorescently labeled chemicals to label ells.
15. Wash the slide four times at room temperature with PBS-2 % milk-0.1 % Tween-20 by applying buffer and aspirating at the edge of the well.
16. Incubate with the 2 °C antibody for 1 h at room temperature. Dilute the fluorescent antibody in PBS-2 % milk-0.1 % Tween-20. Place the humidity box in the dark for this incubation; a drawer usually suffices or wrap in aluminum foil.
17. Wash the cells as before with of PBS-2 % milk-0.1 % Tween-20. Perform three additional washes with 2 % BSA in PBS 0.1 % Tween-20. It is important to remove the milk as it appears as a hazy film under the microscope. After washing the slide is ready to mount.
18. Mount in 1 mg/mL *p*-phenylene diamine and 0.1 μg/mL DAPI/glycerol mounting medium by removing all the liquid from each well and then applying two drops of medium to the center of the slide (*see* **Note 9**). Apply a large cover slip and push down so that the medium spreads evenly over each well. Aspirate any drops that appear around the cover slip. We use a flat piece of plastic originally intended to open microfuge tubes to push out any air bubbles; a wooden applicator stick will also work. After removing excess mounting medium, the slide is sealed by painting around the edge of the cover slip with nail polish. Once the nail polish is dry the slides are ready to view; the DAPI may take 5 min to effectively stain the DNA.

Microscopy: Slides prepared above are ready for viewing. Due to their size yeasts are typically viewed using a 100× oil objective with a 1.3 or 1.4 numerical aperture. We use a Nikon 800 microscope with Intensilight, fitted with a 100× NA 1.4 PlanApo objective, and a Hamamatsu Orca R2 CCD camera controlled by Openlab. Image processing is performed using Openlab and Adobe Photoshop. There are other comparable brands of microscope and camera and Metamorph software is commonly used for image capture and manipulation. For the corresponding analysis of overall cell structure by transmitted light microscopy, differential interference contrast optics are most commonly used in yeast. Most microscopes come with a package of filters that are specific to GFP/FITC (Ex 425-475, DM 480, BA 485-535) and Cyc3/Rhodamine and Texas Red (Ex 528-553, DM 563, BA 600-660), as well as DAPI and Hoechst (Ex 330-380, DM 400, BA 435-485). These work well for basic double labeling with GFP and Cy3, but it may be necessary to invest in additional filter sets if other fluorescent moieties or double-labeling combinations are used. Confocal microscopy requires additional specialized equipment described elsewhere but the preparation of slides is the same. Kohlwein provides an excellent description of microscope setup [1].

4 Notes

1. It has been reported that EGFP fluorescence is increased in a neutral or slightly basic pH medium [9]. Therefore when using medium that is slightly acidic, 1 M Tris–HCl pH 7.5 can be added to cultures to a final concentration of 10 mM (1:100) and yeasts are incubated in this media with shaking for about 5 min. This media is replaced with PBS pH 7.4 and prepared for live-cell imaging as described above [9].
2. We have described live-cell microscopy using agarose pads, but it is also possible to make the pads using 50 μL of 25 % gelatin dissolved in synthetic media as above [15]. After application of cells and cover slip these slides are sealed with VALAP.
3. It is also possible to prepare fresh paraformaldehyde (30 %): 2.0 g paraformaldehyde solid is added to 5 mL of dH_2O in a Pyrex flask. The flask is heated on a Bunsen burner to boiling with intermittent mixing. After the addition of 0.1 mL 1 M NaOH and mixing the solution is chilled on ice and filtered with a 4.5 μm syringe filter. The solution should be used the same day.
4. Efficient removal of the cell wall is essential for the successful immunofluorescence labeling as it allows antibody accessibility, and increases the efficiency by which cells stick to the slide. As described above, yeasts that are not growing under optimal conditions are harder to convert to spheroplasts. There are several different combinations of enzymes that can be used, each requiring different time periods for optimal digestion. Therefore, to ensure efficient digestion cells should be observed by phase-contrast microscopy. Cells that have been digested efficiently change appearance and go from round, bright, and

reflective looking cells to appearing darker, grayer, and flatter. Very pale translucent appearing cells are over-digested. Images of under- or over-digested and correctly digested cells are shown in [13].

5. Cells can also be attached to slides using polyethyleneimine (PEI) instead of poly-L-lysine. A 0.5 % PEI stock is prepared in water and is stable in the dark for several weeks. The PEI is spread over the slide evenly using a pipette tip. The slide is then immediately washed by passing several times through a stream of distilled water and allowed to air-dry.

6. For indirect immunofluorescence controls should include omission of the primary antibody, which is replaced by PBS-Tween-20-1 % milk during the incubation period. This controls for cross-reactivity with the second antibody. A specificity control for the primary antibody may include using a yeast strain in parallel where the gene encoding the antigen or protein of interest has been deleted. If epitope-tagged proteins are used such as HA or Myc, it is advisable to use a yeast strain not expressing the relevant epitope tag as a control. The use of monoclonal or purified antibodies will result in lower background. When using polyclonal antibodies or rabbit immunoglobulin for the purpose of detecting TAP and protein A tags, pre-absorption of nonspecific antibodies from polyclonal antibody serum can lower the background. This is accomplished by preparing yeast as for immunofluorescence following **steps 1–7**. Use the relevant deletion strain for polyclonal antibodies, or wild-type yeast where whole rabbit IgG is used for TAP tag detection. Add the cells to an aliquot of antibody that requires pre-absorption and leave at 4 °C, occasionally mixing to resuspend the cells. The cells are then pelleted and the antibody removed; the antibody is now ready for use.

7. Double labeling for indirect immunofluorescence: We usually sequentially add the first primary antibody followed by its specific second antibody, followed by the second primary and then second secondary following the washing protocol above. The secondary antibodies can be added together as long as they do not cross-react. It is important to also label cells with only one fluorescent moiety at a time as a control to determine that the signal is not visible using the filters for the second fluorescence moiety. For example check the GFP signal with the filter set for Cy3 and vice versa to ensure that there is no signal bleed through.

8. It is also possible to label yeast with fluorescently labeled chemicals such as phalloidin to localize actin and calcofluor white to localize bud scars. Protocols for these procedures are described elsewhere [16].

9. DAPI staining is used to mark the nucleus or observe DNA content. Cells that have been fixed and permeabilized with methanol readily take up DAPI. Cells can be incubated with 1 μg/mL DAPI in H_2O, or can be mounted in medium that contains DAPI as described above. For live-cell imaging DAPI or Hoechst is not so readily taken up into cells and is eventually toxic preventing imaging for long time periods. Cells that appear very bright after DAPI staining are often dead, and therefore it is advisable to also observe the cells by bright-field microscopy. Note that cells will also incorporate these chemicals into mitochondria, leading to a spotted appearance around the cell periphery. Where it is necessary to colocalize the nucleus or nuclear periphery, some investigators co-express a histone-GFP or a nucleoporin such as Nup49 fused to an appropriate fluorophore [17, 18].

Acknowledgement

L.F.P. was funded by NIH grant GM 65385-09.

References

1. Kohlwein SD (2000) The beauty of the yeast: live cell microscopy at the limits of optical resolution. Microsc Res Tech 51:511–529
2. Thorn K (2010) Spinning-disk confocal microscopy of yeast. Methods Enzymol 470:581–602
3. Wolinski H, Natter K, Kohlwein SD (2009) The fidgety yeast: focus on high-resolution live yeast cell microscopy. Methods Mol Biol 548:75–99
4. van Drogen F, Peter M (2004) Revealing protein dynamics by photobleaching techniques. Methods Mol Biol 284:287–306
5. Rines DR, Thomann D, Dorn JF et al (2011) Live cell imaging of yeast. Cold Spring Harb Protoc; 2011; doi:10.1101/pdb.top065482
6. Reichel C, Mathur J, Eckes P et al (1996) Enhanced green fluorescence by the expression of an *Aequorea victoria* green fluorescent protein mutant in mono- and dicotyledonous plant cells. Proc Natl Acad Sci U S A 93: 5888–5893
7. Kahana JA, Silver PA (2001) Use of the *A. victoria* green fluorescent protein to study protein dynamics in vivo. Curr Protoc Mol Biol Chapter 9:Unit 9.7C
8. Mosammaparast N, Jackson KR, Guo Y et al (2001) Nuclear import of histone H2A and H2B is mediated by a network of karyopherins. J Cell Biol 153:251–262
9. Baggett JJ, Shaw JD, Sciambi CJ et al (2003) Fluorescent labeling of yeast. Curr Protoc Cell Biol Chapter 4:Unit 4.13
10. Kilmartin JV, Adams AE (1984) Structural rearrangements of tubulin and actin during the cell cycle of the yeast *Saccharomyces*. J Cell Biol 98:922–933
11. Rout MP, Kilmartin JV (1990) Components of the yeast spindle and spindle pole body. J Cell Biol 111:1913–1927
12. Silver P (2009) Indirect immunofluorescence labeling in the yeast *Saccharomyces cerevisiae*. Cold Spring Harb Protoc pdb prot5317
13. Niu W, Hart GT, Marcotte EM (2011) High-throughput immunofluorescence microscopy using yeast spheroplast cell-based microarrays. Methods Mol Biol 706:83–95
14. Keeling JW, Miller RK (2011) Indirect immunofluorescence for monitoring spindle assembly and disassembly in yeast. Methods Mol Biol 782:231–244
15. Yeh E, Skibbens RV, Cheng JW et al (1995) Spindle dynamics and cell cycle regulation of dynein in the budding yeast, *Saccharomyces cerevisiae*. J Cell Biol 130:687–700
16. Hasek J (2006) Yeast fluorescence microscopy. Methods Mol Biol 313:85–96
17. Salmon ED, Inoue T, Desai A et al (1994) High resolution multimode digital imaging system for mitosis studies in vivo and in vitro. Biol Bull 187:231–232
18. Bucci M, Wente SR (1997) In vivo dynamics of nuclear pore complexes in yeast. J Cell Biol 136:1185–1199

Chapter 7

Single Yeast Cell Imaging

Heimo Wolinski and Sepp D. Kohlwein

Abstract

Microscopic imaging techniques play a pivotal role in the life sciences. Here we describe labeling and imaging methods for live yeast cell imaging. Yeast is an excellent reference organism for biomedical research to investigate fundamental cellular processes, and has gained great popularity also for large-scale imaging-based screens. Methods are described to label live yeast cells with organelle-specific fluorescent dyes or GFP-tagged proteins, and how cells are maintained viable over extended periods of time during microscopy. We point out common pitfalls and potential microscopy artifacts arising from inhomogeneous labeling and depending on cellular physiology. Application and limitation of bleaching techniques to address dynamic processes in the yeast cell are described.

Key words *Saccharomyces cerevisiae*, Confocal microscopy, Fluorescence, Photobleaching, FRAP, FLIP

1 Introduction

Imaging techniques play a pivotal role in biomedical research, at the subcellular, cellular, and organismal levels. A wide range of specific labeling techniques that are available for living cells and widespread application of the green fluorescent protein (GFP) allow to visualize complex biological processes in three dimensions and over time, thus giving unprecedented insights into dynamic cellular processes and physiology. Advances in technologies, including microscope setups, optical devices and lenses, laser light sources, and more sensitive detection systems allow highly sophisticated imaging techniques and reaching optical resolution way beyond the diffraction limit [1–4]. Mass storage capabilities and ever-increasing computer power and speed are the basis for quantitative image processing and large-scale imaging-based screens of (yeast) mutant collections or drugs that impact cellular morphology [5, 6]. Current technology allows unprecedented imaging quality in terms of acquisition time, towards high-resolution quantitative data. However, because of the great impact of imaging techniques ('an image is worth a 1,000 words') it is crucial to understand the

Jeffrey S. Smith and Daniel J. Burke (eds.), *Yeast Genetics: Methods and Protocols*, Methods in Molecular Biology, vol. 1205, DOI 10.1007/978-1-4939-1363-3_7, © Springer Science+Business Media New York 2014

imaging process and to get a clear picture of potential pitfalls and limitations [6, 7]. Current technologies also allow imaging-based screens of cells as small as yeast [6, 8–13], which is an organism of great experimental potential for biomedical research. Numerous human disease genes have orthologues in yeast [14], and most of the metabolic and central cellular processes are highly conserved between yeast and men, making yeast with its vast repertoire of experimental techniques—including microscopy—an ideal reference organism [15].

This chapter describes basic staining and imaging protocols for yeast cells, using confocal imaging. Techniques and protocols described below were established on Leica TCS4d, SP2 and SP5 upright and inverted confocal microscopes, but are applicable on any fluorescence microscope with appropriate excitation and emission filter settings, and photo bleaching capabilities. Due to the small size of yeast cells in the range of 5–8 μm—only one order of magnitude above the wavelength of visible light—an optimized microscope setup is essential, including the choice of lenses with highest possible numerical aperture. Live yeast cell imaging requires careful assessment of the optimum growth and immobilization conditions during (long-term) observation to avoid stress-induced artifacts, which may result in misinterpretation of imaging data. It needs to be considered that labeling with organelle-specific fluorescent dyes is critically dependent on the growth phase during which the examination is performed: fixed cells are typically easier to label homogeneously since the transport barrier of the plasma membrane that controls the efficient uptake/internalization of the dye is inactivated—at the cost of nonviable cells. In contrast, uptake of fluorescent dyes in living cells may depend on the activity of pleiotropic drug resistance pumps, making it necessary to optimize the labeling protocols according to the growth phase of the cells. Here we describe protocols for multilabeling of cells using green-fluorescent protein (GFP) fusion proteins and reference dyes such as MitoTracker™ Red as well as rather new dyes, such as LD540 to label lipid droplets [16] and Draq5™, which is an efficient red-emitting fluorophore for labeling DNA, in yeast cells.

As with all fluorescence imaging techniques, bleaching can be a nuisance but potentially avoided/reduced by the appropriate sample preparation protocol; it may, however, also be used advantageously to assess the dynamics of cellular processes and subcellular interactions in the course of (qualitative) FRAP (fluorescence recovery after photo bleaching) and FLIP (fluorescence loss in photo bleaching) experiments that will be discussed. These more sophisticated types of experiments clearly require additional attention to obtain reliable results. Especially the careful selection of the bleaching area and optimization of the region of interest (ROI) need to be considered to avoid potential artifacts.

This chapter describes cell preparation and labeling techniques that are well established in our laboratory. Potential pitfalls are pointed out and practical solutions to these problems discussed in Subheading 4. The rather simple experimental setups and protocols should facilitate their application in basic and advanced (yeast) imaging courses.

2 Materials

2.1 Preparation of Growth Media

1. Rich medium (YPD): Bacto™ yeast extract, d(+)-glucose monohydrate, Bacto™ peptone, Bacto™ agar, sterile glass flasks or glass tubes, distilled water, autoclave, sterile filter (0.2 μm).
2. SC medium: ammonium sulfate, yeast nitrogen base w/o amino acids, amino acid and purine and pyrimidine concentrates.

2.2 Cultivation of Yeast Cells

1. Rotary shaker.
2. 6- and 12 Well microtiter plates.
3. Thermomixer (Eppendorf, Inc.).

2.3 Immobilization of Yeast Cells Using Agar Sheets

1. Bacto™ agar.
2. 15 mL plastic tubes.
3. Microscope glass slides (25 × 75 mm).
4. Autoclave.

2.4 Immobilization of Yeast Cells Using Fluorescently Labeled Agar Sheets

1. Bacto™ agar.
2. 15 mL plastic tubes.
3. Microscope glass slides (25 × 75 mm).
4. Autoclave.
5. Water-bath with temperature control.
6. Stock solutions of fluorescence dyes.

2.5 Labeling Yeast Organelles for Simultaneous Detection of GFP Fusion Proteins

2.5.1 Labeling of Nuclei

1. Draq5™ 5 mM stock solution (Biostatus, Ltd.).
2. 37 % formaldehyde (Sigma, Inc.).
3. 1 M sorbitol in Aqua Dest Laboratory Water (Servoprax).
4. 50 mM Tris/HCl pH 7.4.

2.5.2 Labeling of Mitochondria

1. MitoTracker Red CM-H2XRos (Life Sciences, Inc.) stock solution: 1 mg/mL in DMSO.

2.5.3 Labeling of Lipid Droplets Using LD540

1. LD540 [16] stock solution: 1 mg/mL in DMSO.

2.5.4 Labeling of Lipid Droplets Using Nile Red

1. Nile Red [17], (Sigma, Inc.) stock solution: 1 mg/mL in DMSO, 37 % formaldehyde (Sigma, Inc.), 1 M sorbitol in aqua dest., 50 mM Tris/HCl pH 7.4.

2.5.5 Labeling of Vacuoles

1. FM4-64 (Life Sciences, Inc.) stock solution: 1 mg/mL in DMSO.

2.6 Application of Photobleaching Techniques

Confocal microscope with spectral detection and acousto optical tunable filter (AOTF) for region of interest scanning.

2.6.1 Optimization of the Imaging Setup

1. Cells expressing GFP fusion proteins.
2. High numerical aperture water (NA 1.2), Glycerol (NA 1.3) or oil (NA 1.4) immersion objectives

2.6.2 Evaluation of the Lateral and Axial Bleach Patterns

1. Cells expressing GFP fusion proteins.
2. 1 M sorbitol in aqua dest.
3. 50 mM Tris/HCl pH 7.4.
4. 37 % formaldehyde (Sigma, Inc.).

2.6.3 Setup for Fluorescence Recovery After Photobleaching (FRAP)

1. Cells expressing GFP fusion proteins.

2.6.4 Setup for Fluorescence Loss in Photobleaching (FLIP)

1. Cells expressing GFP fusion proteins.

2.7 Four-Dimensional Live Cell Imaging of Yeast Cells

1. Cells expressing GFP fusion proteins or fluorescently labeled cells.
2. Programmable microscope stage.
3. Objective heater system (Bioptechs Inc.).
4. Solid agar media slides.
   (a) 63× or 100× NA 1.4 oil immersion objective.

2.8 Imaging of Low-Abundance Fusion Proteins

1. High numerical aperture water or glycerol immersion objectives with correction collar.

3 Methods

3.1 Preparation of Growth Media

1. Weigh Bacto™ yeast extract (1 % w/v) and Bacto™ peptone (2 % w/v) in an appropriate flask and fill up with distilled water to 9/10 of the final volume. Alternatively, for SC media weigh in ammonium sulfate (0.5 % w/v), Yeast Nitrogen Base w/o

ammonium sulfate and amino acids (0.17 % w/v) in an appropriate flask and fill up with distilled water to 9/10 of the final volume.

2. Autoclave at 121 °C for 20 min.
3. Add 1/10 volume of 10× glucose solution (2 % w/v final concentration) by sterile filtration to the media, mix well and dispense into sterile flasks or glass tubes. For SC media, add concentrates of amino acids and bases, as required to satisfy the auxotrophic requirements of the strains.
4. For solid media plates add 2 % Bacto™ agar to the solution prior to autoclaving (*see* **Note 1**).

3.2 Cultivation of Yeast Cells

1. Prepare a pre-culture by inoculating cells from storage agar plates into culture flasks containing 5 mL YPD for 12 h at 30 °C on a rotary shaker (180 rpm).
2. To obtain yeast cells in stationary growth phase, inoculate 100 mL fresh YPD in a 500 mL flask with 50 μL of the preculture and cultivate cells for 72 h on a rotary shaker (180 rpm). To obtain logarithmically growing yeast cells inoculate fresh YPD medium with 1/100 volume of a stationary-phase culture and cultivate for 4–8 h at 30 °C on a rotary shaker. For small-scale cultivation, grow cells in 6- or 12-well microtiter plates in 5–2 mL of medium under shaking and temperature control (30 °C) on a thermomixer (*see* **Note 2**).

3.3 Immobilization of Yeast Cells Using Agar Sheets

1. Weigh in Bacto™ agar in an appropriate flask and fill up with distilled water to the final volume to obtain a 2 % (w/v) solution.
2. Autoclave at 121 °C for 20 min.
3. Pour 5 mL of the liquid agar into a 15 mL plastic tube. Pour the agar on a microscope slide. Take care that the edges of the slide are fully covered with agar.
4. Allow the agar to solidify for 20 min at RT.
5. Concentrate a fluorescently labeled cell suspension (*see* Subheading 3.5) by centrifugation at 2,000 ×*g* for 3 min using a tabletop centrifuge.
6. Apply 1 μL of the cell pellet to the middle of the agar sheet. Alternatively, for multiple sample imaging, cut with a razor blade vertical and horizontal slits ~5 mm apart, into the agar. Apply cell suspensions to each resulting field (*see* **Note 3**) and mount the cells with a large coverslip (50 × 24 mm) suitable for confocal imaging (0.17 mm thickness).
7. Remove the agar on the slide left and right from the coverslip and mount the slide on the microscope stage.

3.4 Immobilization of Yeast Cells Using Fluorescently Labeled Agar Sheets

1. Prepare (medium) agar (*see* Subheading 3.1).
2. Transfer 5 mL of melted (medium) agar into a 15 mL plastic tube and cool to approx. 60 °C in a water bath (the agar should remain liquid).
3. Add the corresponding fluorescence dye (*see* **Note 4**) in the same concentration as used for liquid media to the warm (medium) agar. Vortex thoroughly but gently to avoid air bubbles.
4. Pour the fluorescently labeled agar on a microscope slide. Take care that the edges of the slide are fully covered with agar.
5. Allow polymerization of the agar for 20 min at RT.
6. Follow **steps 5–7** of Subheading 3.3.

3.5 Labeling of Yeast Organelles for Simultaneous Detection of GFP Fusion Proteins

3.5.1 Labeling of Nuclei

1. Cultivate yeast cells to the desired growth stage (*see* Subheading 3.2).
2. Concentrate cells by centrifugation at 2,000×g for 3 min using a tabletop centrifuge; remove surplus liquid.
3. Wash cells 2× with 1 mL of 1 M sorbitol (*see* **Note 5**).
4. Centrifuge cells; remove surplus liquid.
5. Add 1 mL of 3.7 % formaldehyde in 1 M sorbitol (stationary phase cells). Alternatively, add 1 mL of 1 % formaldehyde in 0.5–1 M sorbitol (logarithmic phase cells). Vortex. Fix cells by continuous rotation using a tube rotator for 10–15 min (stationary phase cells) or for 3–5 min (logarithmic cells) at RT (*see* **Notes 6** and **7**). Centrifuge cells; remove surplus liquid.
6. Wash cells 3× with 1 mL of 1× PBS pH 7.4.
7. Prepare Draq5™ working solution by adding 0.2 μL of Draq5™ stock solution (final concentration 1 μM) to 1 mL of 1× PBS pH 7.4 in a 1.5 mL tube (*see* **Note 8**).
8. Add 1 mL of the staining solution to the pellet; vortex.
9. Stain for 1–3 min (*see* **Note 9**).
10. Centrifuge cells; remove surplus liquid.
11. Mount 1 μL of the cell pellet on an agar sheet (*see* Subheading 3.3) or standard microscope slide.
12. Immediately image cells with filters for GFP ($\lambda_{exc}/\lambda_{em}$ 488/500–535 nm) and Draq5™ ($\lambda_{exc}/\lambda_{em}$ 633/640–700 nm); Fig. 1a (*see* **Note 30**).

3.5.2 Labeling of Mitochondria

1. Cultivate yeast cells to the desired growth stage (*see* Subheading 3.2).
2. Add 1 μL of MitoTracker Red CM-H2XRos stock solution (final concentration 1 μg/mL) to 1 mL of the cell suspension (*see* **Note 10**).

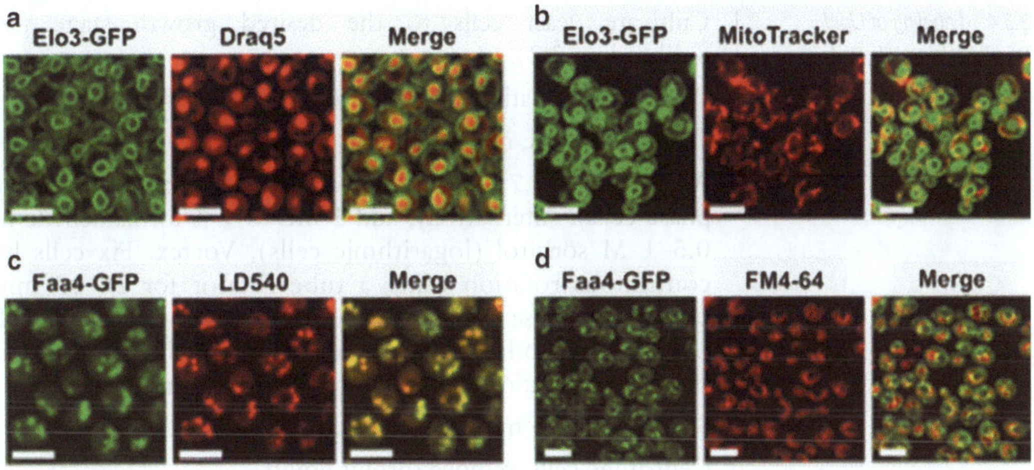

Fig. 1 Simultaneous imaging of GFP fusion proteins and red-emitting reference dyes in yeast cells. (**a**) Endoplasmic reticulum (Elo3-GFP, *green*) and nuclei (Draq5™, *red*); (**b**) endoplasmic reticulum (Elo3-GFP, *green*) and mitochondria (MitoTracker Red CM-H2XRos, *red*); (**c**), surface (phospholipid monolayer) of lipid droplets (Faa4-GFP, *green*) and vacuolar membranes (FM4-64, *red*); (**d**), surface (phospholipid monolayer) of lipid droplets (Faa4-GFP, *green*) and LD neutral lipid core (LD540, *red*). Elo3 is a fatty acid elongase in the endoplasmic reticulum, and Faa4 is an acyl-CoA synthetase localizing predominantly to the surface of lipid droplets, under the experimental conditions. Bar = 10 μm

3. Stain for 15–30 min by continuous rotation using a tube rotator for 15 min at RT.
4. Centrifuge cells; remove surplus liquid.
5. Mount 1 μL of the cell pellet on an agar sheet (*see* Subheading 3.3) or standard microscope slide (*see* **Note 11**).
6. Image cells with filters for GFP ($\lambda_{exc}/\lambda_{em}$ 488/500–535 nm) and MitoTracker ($\lambda_{exc}/\lambda_{em}$ 543/550–650 nm or 561/570–650 nm); Fig. 1b (*see* **Note 30**).

3.5.3 Laboling of Lipid Droplets Using LD540

1. Cultivate yeast cells to the desired growth stage (*see* Subheading 3.2).
2. Transfer 1 mL of the cell suspension to a 1.5 mL tube.
3. Add 1 μL of LD540 stock solution (final concentration 1 μg/mL) to the cell suspension (*see* **Note 12**).
4. Stain for 10–15 min.
5. Centrifuge cells; remove surplus liquid.
6. Mount 1 μL of the cell pellet on an agar sheet (*see* Subheadings 3.3 and 3.4) or standard microscope slide (*see* **Note 11**).
7. Image cells with filters for imaging of GFP ($\lambda_{exc}/\lambda_{em}$ 488/500–535 nm) and LD540 ($\lambda_{exc}/\lambda_{em}$ 543/550–650 nm or 561/565–650 nm); Fig. 1c (*see* **Note 30**).

3.5.4 Labeling of Lipid Droplets Using Nile Red

1. Cultivate yeast cells to the desired growth stage (*see* Subheading 3.2).
2. Wash cells 2× with 1 mL 1 M sorbitol (*see* **Note 5**).
3. Centrifuge cells; remove surplus liquid.
4. Add 1 mL of 3.7 % formaldehyde in 1 M sorbitol (stationary phase cells). Alternatively, add 1 mL of 1 % formaldehyde in 0.5–1 M sorbitol (logarithmic cells). Vortex. Fix cells by continuous rotation using a tube rotator for 10–15 min (stationary phase cells) or for 5 min (logarithmic cells) at RT (*see* **Notes 7** and **13**). Add 1 μL of Nile Red stock solution (final concentration 1 μg/mL) to the cell culture.
5. Stain for 10–15 min.
6. Centrifuge cells; remove surplus liquid.
7. Mount 1 μL of the cell pellet on an agar sheet (*see* Subheadings 3.3 and 3.4) or standard microscope slide.
8. Image cells with filters for imaging of GFP ($\lambda_{exc}/\lambda_{em}$ 476/480–510 nm, (*see* **Note 14**) or 488/500–530 nm) and Nile Red ($\lambda_{exc}/\lambda_{em}$ 543/550–775 nm (lipid droplets) and 543/600–700 nm (lipid droplets plus internal membranes) [7], (*see* **Note 30**).

3.5.5 Labeling of Vacuoles

1. Cultivate yeast cells to the desired growth stage (*see* Subheading 3.2).
2. Transfer 1 mL of the cell suspension to a 1.5 mL tube.
3. Add 1 μL of FM4-64 stock solution (final concentration 1 μg/mL) to the cell culture.
4. Stain for 15 min (log phase cells) or for 45–60 min (stationary phase cells) (*see* **Note 15**).
5. Centrifuge cells; remove surplus liquid.
6. Mount 1 μL of the cell pellet on an agar sheet or standard microscope slide (*see* **Note 11**).
7. Image cells with filters for GFP ($\lambda_{exc}/\lambda_{em}$ 488/500–535 nm) and FM4-64 ($\lambda_{exc}/\lambda_{em}$ 488/580–650 nm); Fig. 1d (*see* **Note 30**).

3.6 Application of Photobleaching Techniques

3.6.1 Optimization of the Imaging Setup

1. Prepare cells expressing the GFP fusion protein for microscopy (*see* Subheadings 3.1–3.3).
2. Warm-up the microscope lasers for at least 15 min (*see* **Note 16**).
3. Select cells for subsequent analysis using minimum laser intensity and zoom factor to avoid photobleaching.
4. Set up the desired imaging parameters on the microscope (excitation wavelength and laser intensity, emission range, line averaging, zoom factor, raster format).

5. Adjust the focus to the structures of interest.
6. Set the pinhole to Airy 1. Adjust the detector gain to obtain maximum fluorescence signal. Use a lookup-table indicating saturated pixels (*see* **Note 17**).
7. Plan the time-course for your experiment and determine the total number of images to be acquired (prebleach images and post-bleach images).
8. Acquire images in the focal plane using the set parameters.
9. Quantify the loss of fluorescence intensity in the optical sections by using imaging software (*see* **Note 18**).
10. Optimize the imaging setup and adjust the imaging parameters to obtain minimal bleaching but maximum contrast of the fluorescence signal of the structure of interest.

3.6.2 Evaluation of the Lateral and Axial Bleach Patterns

1. Prepare cells expressing the GFP fusion protein for microscopy.
2. Warm-up the microscope lasers for at least 15 min (*see* **Note 16**).
3. Wash cells 2× with 1 mL of 1 M sorbitol (*see* **Note 5**).
4. Add 1 mL of 3.7 % formaldehyde (*see* **Note 19**) in 1 M sorbitol; vortex.
5. Fix cells in a tube rotator for 15 min at RT.
6. Wash cells 2× with 1 mL of 50 mM Tris–HCl pH 7.4.
7. Concentrate cells by centrifugation. Mount 1 μL of the pellet on a standard microscope slide and cover with a large coverslip (*see* **Note 20**).
8. Image the sample by confocal microscopy and select a cell of interest.
9. Acquire a z-stack of the entire cell by using the optimized imaging setup (*see* **Note 21**).
10. Check the GFP intensity distribution by calculating mean intensities of the acquired z-stacks and visually inspect the stacks using lateral and axial software slicing (*see* **Note 18**).
11. Define a region-of-interest (ROI) close to a cell area showing a homogeneous GFP signal. Set the size and shape of the ROI as for the planned experiment later on.
12. Acquire pre-bleach images, apply (repeated) bleach pulse(s) and acquire post-bleach images in the same single optical plane as for the planned experiment later on.
13. Acquire a z-stack of the entire cell(s) (*see* **Note 21**).
14. Check and compare the GFP intensity distribution by calculating mean intensities of the acquired z-stacks and investigate the x/y and the axial bleaching patterns using lateral and axial software slicing (Fig. 2) (*see* **Notes 18** and **25**).

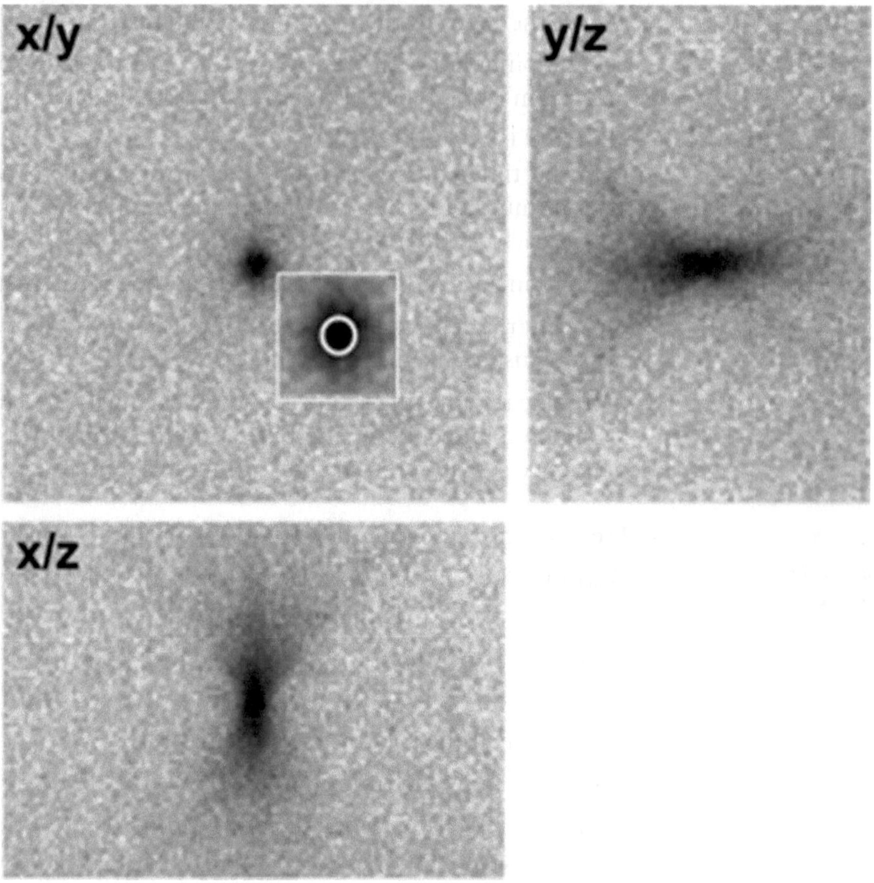

Fig. 2 Nonuniform bleach pattern obtained by repeated bleaching of a fluorescent plastic slide (Chroma Inc.) with a 1 μm circular ROI for 30 s, line average 16 and using a 63× NA1.4 oil objective. Bleaching was performed with higher zoom factor as typically used for yeast imaging. Both, in lateral and axial direction the bleach volume does not correspond to the defined ROI but shows a distribution corresponding to the point spread function (PSF) of the optical system. Note: this effect is stronger with GFP fusion proteins since GFP fluorescence is more sensitive to laser illumination than the fluorescent plastic material. Thus, areas significantly apart from the ROI (x/y and z) may be bleached particularly in FLIP experiments, using GFP

3.6.3 Setup for Fluorescence Recovery After Photobleaching (FRAP)

1. Select the cell of interest on the confocal microscope. Ensure that the field of view contains three or more control cells that will not be bleached (*see* **Note 22**).
2. Set the focus. Acquire an image of the entire cell using the optimized imaging setup (*see* Subheading 3.6.1).
3. Define a ROI for the bleaching procedure.
4. Collect three to five pre-bleach images, photobleach the ROI with intense laser illumination until the fluorescence intensity of the ROI is similar to the background. Acquire post-bleach images in the same single optical plane to record the recovery of the fluorescence until the recovery process has reached a steady state.

5. Collect at least 10 datasets to obtain statistically relevant data (*see* **Notes 23** and **24**).

3.6.4 Setup for Fluorescence Loss in Photobleaching (FLIP)

1. Select the cell of interest under the microscope. Ensure that control cells, which will not be bleached, are in the field of view (*see* **Note 22**).
2. Set the focus. Acquire an image of the entire cell using the optimized imaging setup (*see* Subheadings 3.6.1 and 3.6.2). Define a ROI for the bleaching procedure.
3. Collect three to five pre-bleach images; apply repeated bleach pulses to photobleach the ROI and acquire post-bleach images in the same single optical plane (Fig. 3).
4. Collect at least 10 datasets to obtain statistically relevant data (*see* **Notes 23** and **24**).

3.7 Four-Dimensional Live Cell Imaging of Yeast Cells

1. Cultivate fusion protein-expressing cells or fluorescently labeled cells in appropriate media under vigorous shaking at 30 °C to the desired growth stage (*see* Subheadings 3.1 and 3.2).
2. Mount an objective heater system to the objective; adjust to 30 °C (*see* **Note 26**).
3. Prepare a solid agar medium sheet (*see* Subheadings 3.3 and 3.4).
4. Concentrate cells in 1.5 mL tubes by centrifugation.
5. Immediately mount 1 μL of the cell pellet to the preheated agar sheet (*see* **Notes 3**, **4** and **11**).
6. Find cells for subsequent analysis using minimum laser intensity and zoom factor.
7. Define the least invasive setup for acquisition of z-stacks over extended periods of time (*see* **Note 27**).

3.8 Imaging of Low-Abundant Fusion Proteins

1. Wash cells 2× in SC media without amino acids to reduce background fluorescence that may be caused by components present in rich media (*see* **Note 1**).
2. Mount live cells on an agar sheet (*see* **Note 11**). Alternatively, mount fixed cells on a standard microscope slide.
3. Use a high numerical aperture water immersion lens (NA 1.2) with a correction collar. Alternatively, use a high numerical aperture glycerol objective (NA 1.3) with correction collar to image cells suspended in fixative solution.
4. Activate reflection mode on the microscope.
5. Switch to x/z scan mode on the microscope.
6. Activate the desired laser e.g. the 488 nm laser line and set it to ~5 % of the maximum output.
7. Apply a lookup table to detect areas of saturation within the image (*see* **Note 17**).

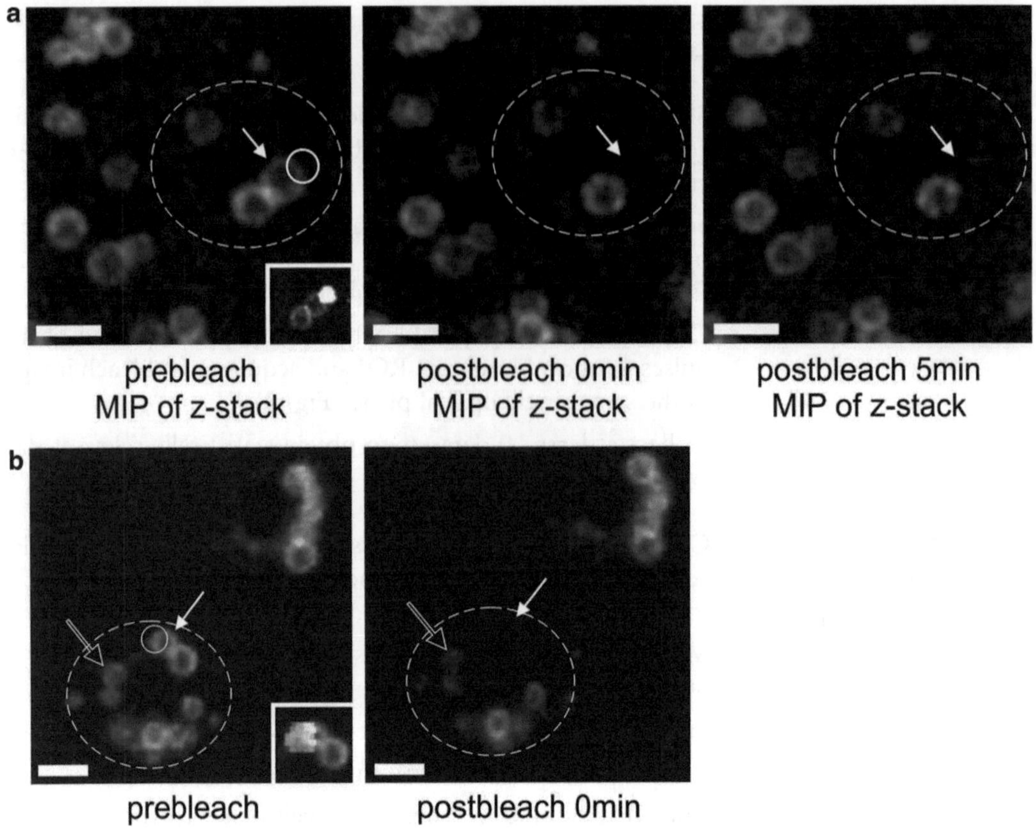

Fig. 3 Visualizing the potential interaction between lipid droplets in living cells by FLIP. Faa4-GFP localizes to the surface of lipid droplets; two closely associated LD were selected in this experiment to address the question whether there was an exchange of protein between adjacent lipid droplets. (**a**) Using an optimized bleaching setup the ROI was set towards the periphery of one LD (smaller image at *bottom right*); bleaching of this individual LD resulted in a total loss of the fluorescence; however, the second closely associated LD was not affected by the bleaching. This experiment indicates that these LD do not exchange their surface proteins during the timeframe of the experiment (5 min post-bleach) and that there is no recovery of Faa4-GFP on the bleached LD from other subcellular compartments or de novo synthesis. Images represent maximum-intensity projections. (**b**), FLIP of Faa4-GFP in living yeast cells using nonoptimized imaging conditions. Again, the ROI was set to the periphery of one of two closely associated LD (smaller image at *bottom right*). Both, the LD within the ROI and the closely associated LD were totally bleached (*closed arrows*). In addition, LDs significantly apart from the ROI show a decreased fluorescence (*open arrows*), most likely caused by lateral and axial bleaching of regions outside the ROI, as shown in Fig. 2. MIP: maximum intensity projection. Bar = 2 μm

8. Start an x/z scan to image the reflection line of the coverslip. Scan within the dynamic range of the photomultiplier.
9. An upper and lower horizontal line become visible. The upper and usually brighter line derives from the light reflection at the interface between the cover slip and the immersion medium. This line is used for optimization of the microscope settings. Consider the orientation of the scan field.

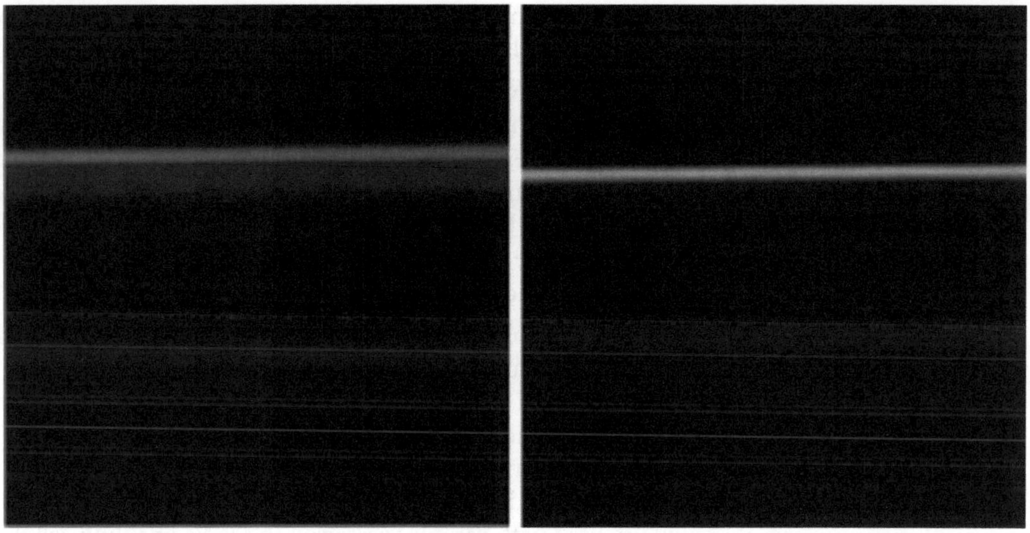

Fig. 4 Improvement of image contrast by correction for coverslide thickness. Nonoptimized reflection image of the coverslide/sample interface (*upper line*). The reflection image of the interface appears blurred (image at *left*). Corrected reflection image of the coverslide/sample interface. Adjustment of the correction collar on the objective lens results in a bright, sharp reflection line (image at *right*)

10. Start a continuous scan on the microscope.
11. While scanning, adjust the correction collar until the maximum intensity of the reflection line is visible and becomes a sharp line (Fig. 4).
12. Switch back to x/y/z fluorescence scanning mode.
13. Set a low zoom factor. Minimize the laser output.
14. Start simultaneous acquisition of fluorescence and transmission images.
15. Search the structure of interest as fast as possible to avoid photobleaching during scanning.
16. Stop the scan.
17. Without scanning, set the desired zoom factor or mark the cell of interest for automated zooming. Start a single scan, if required with line averaging to improve signal to noise ratio (*see* **Note 29**).

4 Notes

1. Yeast media such as complete media are rich in fluorescent metabolites, which may obscure weak GFP fluorescence signals. In contrast, background fluorescence is significantly reduced in minimal media. Thus, if applicable, cells are preferably cultivated or washed in minimal media with or without amino acid and

purine and pyrimidime supplementation. Cellular growth may be somewhat delayed or halted, respectively, but cells remain vital for an extended period of time during microscopy.

2. Growth conditions strongly determine staining efficiency of vital dyes and GFP expression levels. Thus, great care has to be taken to maintain reproducible growth conditions. The use of culture volumes below 2 mL is not recommended since aeration may become limiting in small tubes, and cells tend to sediment in small microtiter plate wells.
3. Agar sheets enable effective immobilization of yeast cells as a monolayer and are particularly useful for long-term observations. Typically, 1–2 μL of the cell suspension is pipetted on the agar sheet. The liquid film that forms underneath the cover slip should not reach the border of the cover slip, which indicates excess liquid, and the cells may start to drift. The agar can be supplemented with various yeast growth media enabling cellular growth and imaging of yeast cells under more physiological conditions (*see* Subheading 3.7). The agar reduces the quality of transmission images, which, however, does not interfere with fluorescence imaging. By cutting slits into the agar multiple samples (strains) can be mounted on a single agar slide and imaged under identical incubation conditions; the slits avoid cross contamination between the strains. This setup also enables high content screens of yeast mutant collections [6, 13].
4. Fluorescence labeling of yeast cells may be quite heterogeneous e.g. due to the activity of efflux pumps [7, 18]. Adding fluorescence dyes to the agar enables continuous uptake of the dye during long-term experiments such as four-dimensional live cell imaging (not applicable for Draq5™).
5. Sorbitol stabilizes GFP fluorescence; the concentration of sorbitol needs to be reduced when using osmo-sensitive strains.
6. Draq5™ is described as a vital stain for different cell types [19]. However, we found that efficient and homogeneous staining of yeast cell populations is only achieved with fixed samples.
7. Logarithmically growing yeast cells are typically more susceptible to aldehyde fixation than cells in stationary growth phase. Thus, cells in the logarithmic growth stage are fixed with a lower concentration of the fixative and incubated for a shorter time period.
8. Draq5™ labels nuclei very fast. Pipetting the concentrated dye directly to the cell suspension frequently causes strong background fluorescence in cells; we thus recommend use of a diluted Draq5™ working solution. Labeling of log-phase cells requires a lower concentration of the dye than stationary phase cells, to reduce background fluorescence.

9. Draq5™ requires only short labeling times; extended labeling may increase background fluorescence.
10. Accumulation of MitoTracker Red CM-H2XRos in mitochondria is dependent on the membrane potential. Thus, effects on cell physiology caused by the preparation have to be considered (*see* **Note 11**). Labeling remains stable after aldehyde fixation.
11. Direct mounting of yeast cells on a standard microscope slide without agar-sheets may limit oxygen and nutrient supply of the cells and may, therefore, rapidly (within seconds!) alter the cellular physiology, which results in insufficient or artificial labeling [6] (*see* Subheading 3.5.2).
12. Concentrations of LD540 above 3 μg/mL (final concentration) in combination with increased laser illumination may induce nonphysiological fusion of lipid droplets (Wolinski, unpublished). Since the dye also emits in the 'green' fluorescence channel, elevated concentrations may lead to fluorescence overlap with GFP fluorescence.
13. Due to the activity of efflux pumps (pleiotropic drug resistance pumps [7, 18]) labeling of growing yeast cell with fluorescence dyes such as Nile Red may be quite heterogeneous; aldehyde fixation significantly improves homogeneous labeling.
14. Excitation of GFP using the 476 nm line (Ar laser) and emission detection between 480 and 510 nm reduces crosstalk between GFP and Nile Red fluorescence. However, this setup limits the excitation efficiency and is not suitable for low-abundance GFP fusion proteins, which are therefore difficult to detect in the presence of Nile Red.
15. Uptake of FM4-64 by endocytosis to label the vacuolar membrane is an energy-dependent process; for labeling cells in stationary growth phase nutrients should be added to the cell suspension (e.g. 1/10 vol YPD) to facilitate efficient vacuolar membrane staining.
16. Older mixed-gas lasers need some time to reach their output maximum. An easy way to estimate the output of the lasers and to determine stable imaging conditions is to record and quantify the mean intensity of a series of transmission images (e.g. acquired in 5 min time-intervals) for 15–30 min, prior to the (quantitative) imaging experiment. Under 'stable' conditions the mean intensity of the images should not vary significantly.
17. Imaging within the dynamic range of the detector is essential for a quantitative analysis in bleaching experiments. Look-up tables indicating saturated pixels (e.g. >255 gray level for an 8-bit image) are included in any confocal software package.
18. Features for image processing such as computation of single optical section intensities, image histograms, or for visualization

of z-stacks are typically included in the microscope software. Alternatively, ImageJ (freeware, [20]) provides all relevant tools for image processing.

19. Fixation of the sample is essential for an evaluation of the axial and lateral bleaching patterns of the microscope setup since under this condition the flux of GFP fusion proteins is blocked.
20. The use of agar sheets for bleaching experiments is not recommended due to significant axial shrinkage of the agar over time, which results in movement of subcellular structures out of the focal plane.
21. To avoid photobleaching z-stacks should be acquired with high scan speed and low laser intensity.
22. For each experiment several unbleached control cells must be in the imaged field of view to estimate the overall loss of fluorescence during image acquisition.
23. A number of protocols and software tools are available for quantitative analysis of bleaching experiments [21]. Several parameters such as intracellular movement of subcellular structures, cell movement, or nonoptimized imaging parameters may bias the results of bleaching experiments. Thus, the acquisition of a larger number of data sets is essential to obtain statistically relevant information.
24. For qualitative bleaching experiments we use longer time intervals (1–3 s) and line-averaging 2–4 for the acquisition of pre- and post-bleach images, to improve the image quality. In FRAP experiments in yeast GFP fusion proteins are typically fully bleached within 1–3 s, using a 25 mW argon laser at full power. After bleaching, the fluorescence intensity of the ROI should be similar to the background. In contrast, for FLIP experiments repeated bleaching of the ROI over a longer time period (up to several min) may be required.
25. Due to the small size of yeast cells bleaching experiments are typically performed with higher zoom factor or raster format, which result in smaller pixel sizes and improved resolution. However, under such conditions the pixel dimensions can be smaller than the diameter of the laser beam causing a mismatch between the ROI and the effectively bleached region. Thus, the actual photobleach profile in the specimen is indeed broader than defined by the ROI. This phenomenon can cause a significant error when evaluating bleaching results of structures of a size comparable or smaller than the laser beam focal spot (~200 nm). The effect is particularly pronounced when intense and repeated bleach pulses are applied in FLIP experiments [22]. Moreover, the bleach profile is also not homogeneous in axial direction. Particularly, by using high numerical aperture objectives an axial bleaching pattern corresponding to

the point-spread-function (PSF) of the optical system is generated, thus limiting long-term bleaching experiments of small subcellular yeast structures [23, 24]. In addition, fluorescence saturation due to high laser illumination may also significantly alter the bleaching volume [25]. Calibration of the ROI and estimation of the bleach pattern dimensions in x/y/z direction is required to avoid imaging artifacts. Under optimized conditions, which can be tested in fixed cells that are devoid of any dynamic movement of fluorescent material, only the ROI and none of the adjacent structures should be bleached.

26. Controlled heating of the sample is recommended to maintain yeast cells on the microscope table under optimal physiological conditions. High numerical aperture immersion lenses act as thermal coupling media and heat sink between the specimen and the objective. Heating of the objective using an objective heater system enables sufficiently stable temperatures.

27. Manual adjustment of the focal plane is required at the different image acquisition time-points since the agar may shrinks over time, particularly in axial direction. Transfer of cells from liquid medium to the agar sheet should be done as quickly as possible to minimize temperature shock and physiological stress. Depending on the doubling time and time course of the experiment the cell density within the field of view needs to be considered, to avoid overgrowth. Yeast cells are sensitive to repeated and long-term laser illumination as used for 4D live cell imaging, and phototoxic effects may lead to reduced or total stop of cellular growth and the formation of large vacuoles and optically dense subcellular structures. We recommend to initially evaluate the sensitivity of the individual yeast strain to light exposure by imaging the cells using a 63× or 100× NA1.4 oil immersion objective (standard water immersion objectives may dry out during the experiment) at ~100 × 100 × 270 nm (undersampled) sampling, 800 Hz scan speed, bidirectional scan mode and 512 × 512 raster point size and in time intervals of 30 min between the z-stacks. Subsequently, the parameters can be optimized for the individual experiment. Typically, image acquisition using optimized settings to avoid phototoxicity in 4D experiments results in noisy 3D images with reduced image quality and contrast. However, image deconvolution typically improves the quality of such images significantly [26, 27]; (*see* **Note 28**). In most cases, the application of methods for noise reduction such as 3D Gaussian or 3D Median filtering are efficient for improving image quality [28, 29]. Such filters are implemented e.g. in ImageJ [20, 30] and in most microscope software packages.

28. Image deconvolution using image restoration algorithms such as the maximum-likelihood estimation (MLE) approach e.g.

implemented in Huygens Deconvolution Pro software (Scientific Volume Imaging, Inc.) requires optimized sampling conditions. Significant undersampling of the z-stacks should be avoided to minimize potential deconvolution artifacts [27]. 3D Gaussian filtering may improve the quality of noisy images in x/y/z direction sufficiently well [29].

29. Correction for coverglass thickness using water (for live cell imaging) or glycerol (preferably for fixed samples) immersion objectives with correction collar improves image quality significantly and allows imaging of fluorescence signals, which may not be detectable at all under nonoptimized conditions.

30. A major challenge in colabeling experiments is to avoid fluorescence crosstalk between two fluorescence dyes. Fluorescence excitation and emission spectra provided by companies are typically recorded in vitro and in organic solvents and may not reflect the 'true' spectral properties in a biological sample. Thus, it is recommended to perform a lambda (wavelength) scan to determine the emission wavelength range of the individual fluorescence dyes in the biological sample. Laser intensity, detector gain, and emission wavelength ranges are set accordingly. Features for lambda scanning are typically integrated in any modern confocal microscope system.

Acknowledgements

Work in the authors' laboratories is supported by grants from the Austrian Science Funds FWF (project SFB Lipotox F3005 and the PhD program Molecular Enzymology W901-B12) and the Austrian Federal Ministry for Science and Research (project GOLD, in the framework of the Austrian Genomics Program, GEN-AU), and by NAWI Graz.

References

1. Hell SW, Wichmann J (1994) Breaking the diffraction resolution limit by stimulated emission: stimulated-emission-depletion fluorescence microscopy. Opt Lett 19:780–782
2. Schermelleh L, Heintzmann R, Leonhardt H (2010) A guide to super-resolution fluorescence microscopy. J Cell Biol 190:165–175
3. Ball G, Parton RM, Hamilton RS et al (2012) A Cell Biologist's Guide to High Resolution Imaging. Methods Enzymol 504:29–55
4. Huang B, Babcock H, Zhuang X (2010) Breaking the diffraction barrier: super-resolution imaging of cells. Cell 143:1047–1058
5. Neumann B, Held M, Liebel U et al (2006) High-throughput RNAi screening by time-lapse imaging of live human cells. Nat Methods 3:385–390
6. Wolinski H, Natter K, Kohlwein SD (2009) The fidgety yeast: focus on high-resolution live yeast cell microscopy. Methods Mol Biol 548:75–99
7. Wolinski H, Kohlwein SD (2008) Microscopic analysis of lipid droplet metabolism and dynamics in yeast. Methods Mol Biol 457:151–163
8. Fei W, Shui G, Gaeta B et al (2008) Fld1p, a functional homologue of human seipin, regulates

the size of lipid droplets in yeast. J Cell Biol 180:473–482
9. Li Z, Vizeacoumar FJ, Bahr S et al (2011) Systematic exploration of essential yeast gene function with temperature-sensitive mutants. Nat Biotechnol 29:361–367
10. Niu W, Hart GT, Marcotte EM (2011) High-throughput immunofluorescence microscopy using yeast spheroplast cell-based microarrays. Methods Mol Biol 706:83–95
11. Saito TL, Ohtani M, Sawai H et al (2004) SCMD: *Saccharomyces cerevisiae* Morphological Database. Nucleic Acids Res 32:D319–D322
12. Vizeacoumar FJ, van Dyk N, Vizeacoumar FS et al (2010) Integrating high-throughput genetic interaction mapping and high-content screening to explore yeast spindle morphogenesis. J Cell Biol 188:69–81
13. Wolinski H, Petrovic U, Mattiazzi M et al (2009) Imaging-based live cell yeast screen identifies novel factors involved in peroxisome assembly. J Proteome Res 8:20–27
14. Bassett DE Jr, Boguski MS, Hieter P (1996) Yeast genes and human disease. Nature 379:589–590
15. Kohlwein SD (2000) The beauty of the yeast: live cell microscopy at the limits of optical resolution. Microsc Res Tech 51:511–529
16. Spandl J, White DJ, Peychl J et al (2009) Live cell multicolor imaging of lipid droplets with a new dye, LD540. Traffic 10:1579–1584
17. Greenspan P, Mayer EP, Fowler SD (1985) Nile red: a selective fluorescent stain for intracellular lipid droplets. J Cell Biol 100: 965–973
18. Ivnitski-Steele I, Holmes AR, Lamping E et al (2009) Identification of Nile red as a fluorescent substrate of the *Candida albicans* ATP-binding cassette transporters Cdr1p and Cdr2p and the major facilitator superfamily transporter Mdr1p. Anal Biochem 394:87–91
19. Martin RM, Leonhardt H, Cardoso MC (2005) DNA labeling in living cells. Cytometry A 67:45–52
20. Rasband WS (2011) Image J, U. S. National Institutes of Health, Bethesda, Maryland, USA, http://imagej.nih.gov/ij/.
21. Snapp EL, Altan N, Lippincott-Schwartz J (2003) Measuring protein mobility by photobleaching GFP chimeras in living cells. Curr Protoc Cell Biol, Chapter 21, Unit 21 21
22. Weiss M (2004) Challenges and artifacts in quantitative photobleaching experiments. Traffic 5:662–671
23. Diaspro A, Mazza D, Krol S et al (2005) Quantitative FRAP by means of diffusion through 3D polyelectrolyte shells using confocal and two-photon excitation approaches. Microsc Microanal 11:786–787
24. Mazza D, Cella F, Vicidomini G et al (2007) Role of three-dimensional bleach distribution in confocal and two-photon fluorescence recovery after photobleaching experiments. Appl Opt 46:7401–7411
25. Braeckmans K, Stubbe BG, Remaut K et al (2006) Anomalous photobleaching in fluorescence recovery after photobleaching measurements due to excitation saturation–a case study for fluorescein. J Biomed Opt 11:044013
26. Wallace W, Schaefer LH, Swedlow JR (2001) A workingperson's guide to deconvolution in light microscopy. Biotechniques 31:1076–1078, 1080, 1082 passim
27. Scientific Volume Imaging. *Huygens Deconvolution Pro User Manual.*
28. Landmann L (2002) Deconvolution improves colocalization analysis of multiple fluorochromes in 3D confocal data sets more than filtering techniques. J Microsc 208:134–147
29. Centonze V, Pawley JB (2006) Tutorial on practical confocal microscopy and use of the confocal test specimen. In: Pawley JB (ed) Handbook of Biological Confocal Microscopy 627–649. Springer, 3rd ed. 2006
30. Papadopulos F, Spinelli M, Valente S et al (2007) Common tasks in microscopic and ultrastructural image analysis using ImageJ. Ultrastruct Pathol 31:401–407

[illegible]

Chapter 8

Microfluidic Platforms for Generating Dynamic Environmental Perturbations to Study the Responses of Single Yeast Cells

Anjali Bisaria, Pascal Hersen, and Megan N. McClean

Abstract

Microfluidic platforms are ideal for generating dynamic temporal and spatial perturbations in extracellular environments. Single cells and organisms can be trapped and maintained in microfluidic platforms for long periods of time while their responses to stimuli are measured using appropriate fluorescence reporters and time-lapse microscopy. Such platforms have been used to study problems as diverse as *C. elegans* olfaction (Chronis et al. Nature Methods 4:727–731, 2007), cancer cell migration (Huang et al. Biomicrofluidics 5:13412, 2011), and *E. coli* chemotaxis (Ahmed et al. Integr Biol 2:604–629, 2010). In this paper we describe how to construct and use a microfluidic chip to study the response of single yeast cells to dynamic perturbations of their fluid environment. The method involves creation of a photoresist master mold followed by subsequent creation of a polydimethylsiloxane (PDMS) microfluidic chip for maintaining live yeast cells in a channel with two inputs for stimulating the cells. We emphasize simplicity and the methods discussed here are accessible to the average biological laboratory. We cover the basic toolbox for making microfluidic lab-on-a-chip devices, and the techniques discussed serve as a starting point for creating sophisticated microfluidic devices capable of implementing more complicated experimental protocols.

Key words Microfluidics, Lab-on-a-chip, *Saccharomyces cerevisiae*, Environmental perturbation, Protocol

1 Introduction

The earliest microfluidic devices adopted technology from the semiconductor industry to etch channels into glass or silicon. This construction process required specialized equipment and expensive materials precluding rapid design prototyping. A breakthrough came with the invention of soft lithography [1], which allows for the repeated casting of polymer substrates such as polydimethylsiloxane (PDMS) from a master mold to generate multiple fluidic chips. Once a master mold is created in a cleanroom, the casting

Jeffrey S. Smith and Daniel J. Burke (eds.), *Yeast Genetics: Methods and Protocols*, Methods in Molecular Biology, vol. 1205, DOI 10.1007/978-1-4939-1363-3_8,

process can be carried out in a standard lab space, making the creation of basic microfluidic chips accessible to experimentalists without cleanroom fabrication experience.

Microfluidic devices operate at low Reynold's number with volumes in the nanoliter and picoliter range. Flow is laminar (fluids mix only by diffusion) and this physical property combined with small volumes allows for the precise spatial and rapid temporal control of the fluid environment. Devices can be designed to trap living single cells or small populations of cells for hours or days. When combined with the appropriate fluorescence reporters and time-lapse microscopy microfluidic platforms are ideal for studying the responses of single cells to perturbations of their extracellular environment [1–4]. Single-cell resolution allows for the analysis of response variability between large numbers of isogenic cells [5]. Recent studies in *Saccharomyces cerevisiae* have taken advantage of microfluidic technologies to dissect in vivo signaling kinetics in the hyperosmotic glycerol response pathway [6, 7], to discover new network connections in the galactose utilization network [8], and to understand the role of Msn2 transcription factor dynamics in stress-induced gene expression [9].

In this review we describe the process for creating a simple microfluidic device designed to expose yeast cells to dynamic perturbations of their extracellular environment. The device is designed to allow observation of the cells on a standard inverted epifluorescence microscope. We describe the process starting with the design and creation of the master mold in a cleanroom. Not all labs will have access to cleanroom facilities, and the notes section contains suggestions for how to outsource creation of the master mold or find commercially available alternatives. Subsequent steps in the process do not require any specialized facilities. This simple device will be adequate for most experiments involving media exchange. For experimenters who wish to go further and design their own sophisticated microfluidic chips, the techniques described in this chapter provide a practical basis.

2 Materials

2.1 Transparency Mask and Master Mold Creation

1. Appropriate design software (Auto Cad, Layout Editor, Adobe Illustrator, or similar).
2. Photolithography Transparency Mask (25400 dpi, Negative Polarity, Emulsion Down, CAD/Art Services, Bandon, OR).
3. Borosilicate Type Glass Plates, 5" × 5" × 1/16" Thick, Beveled Edges (Chemglass Life Sciences, Vineland, NJ, USA).
4. 4" Silicon Wafers (P(100) 0–100 ohm-cm SSP 500um Test 100 crystal orientation, one-side polished) (University Wafers, South Boston, MA).

5. Negative Photoresist SU8 2050 (Microchem Corp, Newton, MA).
6. SU8 Developer Propylene Glycol Methyl Ether Acetate (PGMEA) (Microchem Corp, Newton, MA).
7. HTG III mask aligner (Hybrid Technology Group, Silicon Valley, CA, USA).
8. SCS G3 spin coater (Specialty Coating Systems, Indianapolis, Indiana, USA).
9. KLA Tencor D-120 surface profiler (KLA-Tencor, Milpitas, CA, USA) (*see* **Note 1**).
10. Glass 4" petri dish with lid.
11. 3M Scotch tape, Matte Finish, Magic Tape (3M, St. Paul, MN).
12. Hexamethyldisilazane, HDMS (Sigma, 379212) (*see* **Note 2**).
13. Disposable plastic dropper.
14. Aluminum foil.
15. Small disposable aluminum trays.
16. 3-aminopropyl-triethoxysilane, 99 % (Sigma-Aldrich, St Louis, MO).
17. Hot plate at 65 °C.
18. Hot plate at 95 °C.
19. Wafer tweezers (Unifit).
20. 150 mm × 15 mm polystyrene petri dishes.
21. Large glass container for developing wafer (at least 4" deep and 6" in diameter).
22. Isopropyl alcohol in a squeeze bottle.

2.2 Soft Lithography

1. Dow Corning SYLGARD 184 Elastomer kit containing Sylgard 184 base, Sylgard184 curing agent (Polydimethylsiloxane-PDMS) (Ellsworth Adhesive Systems).
2. Cover Glass 23 × 60 mm No. 1.5 Cat No 48393 (VWR International Inc).
3. 3M Scotch tape, Matte Finish, Magic Tape (3M).
4. Plasma-Preen Cleaner/Etcher (Terra Universal).
5. VWR Gravity Convection Oven Model 1300U (VWR International Inc).
6. Harris Uni-Core 0.75 mm Punches (Ted Pella Inc).
7. 21G 1½" PrecisionGlide Needles #305198 and #305167 (Becton-Dickinson).
8. 150 mL Plastic disposable beaker (Fisher Scientific).
9. Plastic fork.
10. 190 mm vacuum desiccator (Bel-Art) and ¼" I.D. vacuum tubing.

2.3 Manual Fluid Control

1. Tygon Tubing, 0.38 mm I.D. 2.31 mm O.D. (Cole-Parmer).
2. VWR Talon Regular Clamp Holder (VWR International Inc).
3. 50 mL Sterile Plastic Conical Tubes (Corning).
4. O-Ring Stand (VWR International Inc).
5. Tube Clamps (VWR International Inc).
6. Type 304 SS Dispensing Needle 23 G, 0.017" I.D., 0.025" O.D., 1/2" Long (McMaster-Carr).
7. Metal pins, NE-1310-02, 0.025 O.D. × 0.013 I.D. × 0.500 long, Type 304, WD, Full Hard (New England Small Tube).
8. Male Luer Slip to 200 Series Barb, 1/16" (1.6 mm) I.D. Tubing, Natural Polypropylene (Value Plastics).
9. Three Way Stopcock, Capped Female Luer Thread Style & Capped Male Luer Lock, Polycarbonate Body w/Polyethylene Handle (Value Plastics).
10. Tygon 2275 High Purity Tubing I.D. 1/16" O.D. 3/16" (Cole-Parmer).
11. 16G 1½" PrecisionGlide Needles #305198 and #305167 (Becton-Dickinson).

2.4 Yeast Culture Conditions and Cell Loading

1. Low Fluorescence Minimal (LFM) Yeast Media: 2 % glucose, 5 g/L $(NH_4)_2SO_4$, 1 g/L KH_2PO_4, 0.5 g/L $MgSO_4$, 0.1 g/L NaCl, 0.1 g/L Ca_2Cl, 0.5 mg/L H_3BO_4, 0.04 mg/L $CuSO_4$, 0.1 mg/L KI, 0.2 mg/L $FeCl_3$, 0.4 mg/L $MnSO_4$, 0.2 mg/L Na_2MoO_4, 0.4 mg/L $ZnSO_4$, 2 μg/L biotin, 0.4 mg/L calcium pantothenate, 2 mg/L inositol, 0.4 mg/L niacin, 0.2 mg/L PABA, 0.4 mg/L pyridoxine HCl, 0.4 mg/L thiamine (*see* **Note 3**).
2. Concanavalin A Solution: 2 mg/mL Concanavalin A, 5 mM $MnSO_4$, 5 mM $CaCl_2$, dissolved in sterile water. Solution should be around pH 6.5 (*see* **Note 4**).
3. Intramedic Tubing: 0.86 mm Inner Diameter, 1.27 mm Outer Diameter (Becton-Dickinson).
4. 1 mL syringe with male luer-lock.
5. 21G 1½" PrecisionGlide Needles (Becton-Dickinson Franklin Lakes, NJ).

2.5 Microscopy

1. Slide Holder (1 mm thick aluminum, cut 1.25" × 3" with 0.9" × 0.9" inner square hole) (*see* **Note 5**).
2. Lab Labeling Tape (VWR international).
3. Inverted microscope equipped for epi-fluorescence (*see* **Note 6**).

3 Methods

3.1 Design of Transparency Mask

The pattern of your master mold will be created from photoresist, a light sensitive resist that changes its susceptibility to attack by solvent when exposed to UV illumination. The transparency mask is used to define what regions of the photoresist are exposed to UV illumination. The resist we will use, SU8, is a negative resist meaning that the resist exposed to UV light is more resistant to the developer solution and remains on the silicon wafer to form the raised pattern that will be printed into the PDMS microfluidic chip.

1. Create the pattern for you mask in your design software of choice (Auto Cad, Layout Editor, L-Edit, Adobe Illustrator, or similar). The y-shaped design used in this paper with relevant dimensions can be seen in Fig. 1.
2. Send your design to a company capable of printing the transparency at high resolution in a cleanroom environment (to avoid particle contamination). To ensure you have an appropriate transparency mask, order it at the highest resolution possible (e.g., 254,000 dpi) as a negative mask with emulsion down (*see* **Note 7**).

3.2 Fabrication of the Master Mold

To avoid defects associated with particle contamination, construction of the microfluidic mask should be done in a Class 1000 or better cleanroom. A Class 1000 cleanroom is designed to prevent more than 1000 particles (>0.5 μm) per cubic foot of air (*see* **Note 8**). The fabrication process is illustrated in Fig. 2.

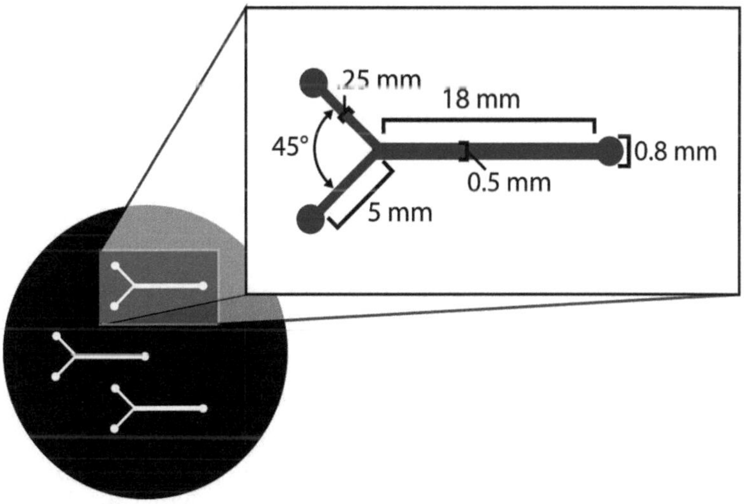

Fig. 1 Diagram of a y-shaped channel used for stimulating yeast cells with two different types of media

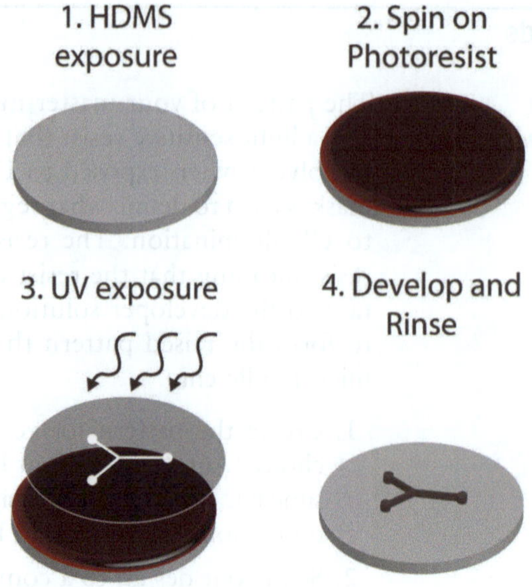

Fig. 2 Photolithography: (1) A clean silicon wafer is first treated with hexamethyldisilazane (HDMS) to increase surface adhesion of photoresist. (2) A spin coater is used to evenly coat the wafer with SU8 photoresist. The spin speed and resist viscosity determine the thickness of the photoresist layer. (3) The photoresist is exposed to UV light after covering it with the transparency mask. (4) Areas of the photoresist that are exposed to the illumination become cross-linked and remain after the wafer is developed

1. Remove a glass plate from its packaging. It will be used to hold your transparency mask in place over your resist-coated wafer when exposing it to UV light. Rinse the glass with deionized water and clean it with an air gun. Make sure that the plate is completely dry.
2. Remove the transparency mask from its packaging. The transparency mask was prepared in a cleanroom to minimize particle contamination, so make sure to open the package inside of a cleanroom. Cut your design from the mask and tape onto a glass plate using scotch tape. Make sure the emulsion (dark) side is facing away from the glass.
3. If your SU8 is stored in a refrigerator (as is the case in many facilities) remove it and allow it to reach room temperature before continuing.
4. Expose your silicon wafer to HDMS. Put a small foil pan inside of the 4" glass petri dish and add a few drops of HDMS with the disposable plastic dropper. Place your wafer in the same dish and incubate the two together for 2 min. These steps must be carried out in a fume hood. The same HDMS can be used to treat several wafers before it needs to be replaced (*see* **Note 9**).

5. While your wafer is being treated with HDMS line the spin coater with aluminum foil to protect it from SU8 resist. Place an appropriately sized spin chuck onto the motor shaft in the center of the spin coater. The spin chuck is a circular metal plate that will be used to hold your wafer. It has holes in it that are used to establish a vacuum between the wafer and the chuck to prevent the wafer flying off during the spin coating process.
6. Program the spin coater with the desired spin program. The spin speed sets the height of the resist and therefore the eventual height of the PDMS channel (*see* **Note 10**). For SU8 2050 and a height of 100 μm we use the following program:

   Step 1: Speed 500 rpm/s, Ramp 100 rpm/s, Duration 10 s

   Step 2: Speed 1500 rpm/s, Ramp 300 rpm/s, Duration 30 s

   Step 3: Speed 0 rpm/s, Ramp 500 rpm/s, Duration 0 s
7. Try a practice run with the spin coater and an old wafer to make sure that the system can establish a vacuum and that the program runs correctly.
8. Use wafer tweezers to move your wafer from the glass dish to the chuck of the spin coater.
9. Press "Start" to make sure that a seal is formed between the wafer and the chuck and that the wafer is centered. Stop the spin after a few seconds of observation. If the wafer is askew, readjust it and try again.
10. Pour enough resist onto the wafer to cover ~1/3 of the surface area. Be careful to avoid bubbles as these will cause defects when the resist is spun. If you get bubbles in the resist, they can be gently removed by pulling them to the side of the wafer with a needle.
11. Press "Start" to begin the spin program.
12. Once the program is finished move the wafer to the 65 °C hot plate and bake for 10 min.
13. Place the wafer at 95 °C for 50 min.
14. Allow wafer to cool to 65 °C by changing the hot plate temperature to 65 °C and waiting for the temperature to adjust (approximately 15 min, varies by hot plate).
15. Remove the screws and remove the alignment frame from the mask aligner. Position your glass plate into the holder, transparency side down, and reinstall the frame (*see* **Note 11**).
16. Expose wafer for 30 s to UV light (this time is specific for our machine and light source, the desired exposure is 215–240 mJ/cm^2) (*see* **Note 12**).

17. Put wafers at 65 °C (1 min), allow hot plate temperature to ramp to 95 °C (approximately 7 min), keep wafers at 95 °C for 10 min. Allow temperature to ramp back down to 45 °C (approximately 20 min). Move wafer to the bench and allow it to cool at room temperature for 5–10 min. At this point you should be able to see the outlines of your pattern in the SU8.
18. Put wafer in SU8 developer (PGMEA) in a deep glass dish and develop by gently swirling the container for about 10 min. Develop the wafer in a fume hood. You will begin to see your pattern and the shininess of the wafer showing through as the development proceeds and the dull unexposed SU8 lifts off. Rinse with isopropyl alcohol when the unexposed SU8 has been removed. Dry the wafer with a nitrogen or air gun (*see* **Note 13**).
19. Inspect the master mold under a stereoscope. Make sure that the features are fully formed and clean. If you see particles of undeveloped SU8 or SU8 remaining in corners of the design, develop the master mold for another 30 s or so as in **Step 17**.
20. Measure the features using a contact profilometer. This will tell you the height and width of the master mold features. These dimensions will carry over to your microfluidic chip, so if your master is not the right size you will need to remake it (*see* **Note 14**).
21. Incubate the master mold for 5–10 min in a glass petri dish with 2–3 drops of 3-aminopropyl-triethoxysilane in a separate disposable aluminum tray (*see* **Note 15**).
22. Place the completed master mold into a polystyrene petri dish and cover it to keep it particle-free.

3.3 Soft Lithography

The soft lithography steps can be carried out in any standard laboratory space, though some universities and research institutions will have special Class 1000 cleanrooms for this purpose. The soft-lithography process is illustrated in Fig. 3.

1. Master molds patterned with SU8 are constructed on 4" silicon wafers using standard photolithography techniques for microfluidics as described in the previous section. The master mold can be reused many times to make multiple flow cells.
2. Place the master mold in a plastic petri dish if you have not already done so.
3. Prepare PDMS by mixing curing agent and polymer in a 1:9 ratio by weight in the plastic disposable beaker using a plastic fork (*see* **Note 16**). Use about 60 total grams of PDMS the first time, this will cover the entire wafer. The PDMS should be very well mixed, with fine bubbles throughout (*see* **Note 17**).
4. Remove bubbles from the PDMS by degassing it in a vacuum desiccator hooked up to the house vacuum or similar source. Be cautious that your mixing container is not overfull, or the PDMS will bubble over during degassing. Periodically release the vacuum to avoid the PDMS bubbling over.

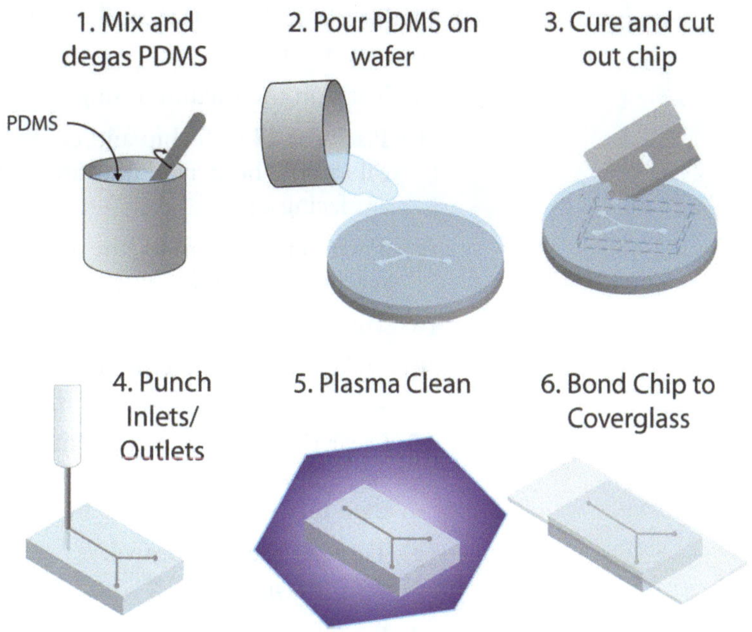

Fig. 3 Soft lithography: (1) Curing agent and polymer are mixed in a 1:9 ratio and degassed. (2) Uncured PDMS is poured over the master mold. (3) Once the PDMS has cured, the chip is cut out (4) and shaped into a chip and the inlet and outlet ports are punched. (5) Plasma cleaning exposes silanol groups on the PDMS allowing it to (6) covalently bond to a glass coverslip forming the completed microfluidic chip

5. Pour the PDMS into the petri dish being careful not to introduce new bubbles. If bubbles form, remove them carefully with a needle (*see* **Note 18**).
6. Cure the PDMS for 1–2 h at 65 °C until the surface of the PDMS is no longer tacky. Test the surface of the PDMS at an edge (away from the features) with tweezers or nitrile gloves. If the PDMS is not sticky, it is cured.
7. Using a razor blade, cut around the flow cell design. Cut gently without pressure so as not to break the silicon master mold. Do not allow the razor blade to make contact with the master mold. When the PDMS separates from the underlying mask remove the PDMS from the mold. Cut to the desired shape. The mold can be reused in the future by repeating **steps 3–4** and filling the hole made in the cured PDMS.
8. Punch holes for the inlets and outlets using the Harris Uni-Core 0.75 mm punch. Be careful to make your punches as vertical as possible.
9. Clean the PDMS block with scotch tape. With the feature side face up on the bench, apply tape being careful not to touch the feature-side of the block with your gloves. Repeat this step three times (*see* **Note 19**).

10. Plasma clean the PDMS block and a 22×60 mm coverslip in the Plasma Preen:
    (a) Turn on the vacuum pump.
    (b) Place the PDMS chip and coverslip into the glass cylindrical sample holder. Make sure that the sides to be bonded are facing up.
    (c) Turn on vacuum to the chamber, making sure that it reaches 30 mmHg.
    (d) Turn on the gas control.
    (e) Make sure that the air flow meter registers ~3 standard cubic feet per minute (SCFM), adjust if necessary.
    (f) Clear the timer, and then hit 1. The microwave will start. As soon as you see purple plasma formation, wait 5 s, and then hit "Cancel" on the microwave control panel.
    (g) Turn off the gas control.
    (h) Turn off the vacuum.
    (i) Remove the chip and coverslip from the glass cylindrical sample holder (*see* **Note 20**).
11. Apply the PDMS to the coverslip feature-side down to create the finished microfluidic flow cell. If the plasma cleaning worked effectively you will see instant bonding between the PDMS and the glass (*see* **Note 21**).
12. Cure the flow cell in the convection oven at 65°C for several hours or overnight. This increases the strength of the bond between the PDMS and the glass (*see* **Note 22**).

3.4 Fluid Handing

In this section we describe how to set up manual fluid handling to allow rapid switching between two liquids, A and B, within the microfluidic chip. Here media B is marked with a tracer dye that does not conflict with the fluorescent probes used to measure cellular response. More complicated experimental designs are possible with more sophisticated microfluidic chips. We describe manual switching because it requires no specialized skills to set-up. It is also possible to automate the fluid handling (*see* **Note 23**). The manual switching configuration is illustrated in Fig. 4a.

1. Fill two 50 mL conicals with liquid A and one 50 mL conical with liquid B. Label the two conicals filled with liquid A "LOW" and "HIGH." Punch holes in the plastic tops of the conicals using a 16 G needle and insert the Tygon 2275 tubing. Provide a container for effluent (*see* **Note 24**).
2. Place the conical on the ring stand in the order High-B-Low as shown in Fig. 4a.
3. Connect the two conicals filled with liquid A to the female luer connections of the stopcock using the Tygon 2275 1/16"

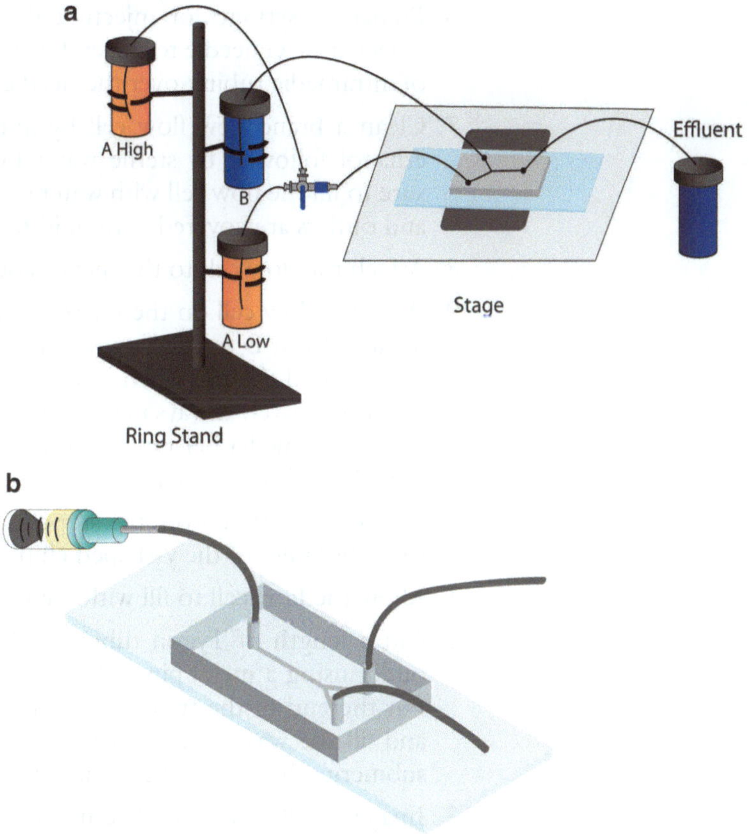

Fig. 4 Fluid handling: (**a**) The microfluidic chip is positioned on the microscope stage. One inlet is fed by media A and the other by media B. By using the stopcock to switch between media A "HIGH" and media A "LOW" either media A or B will fill the channel. (**b**) Concanavalin A and cells are loaded from the outlet port using a syringe. Care should be taken to avoid filling the flow cell with bubbles during the loading process

inner diameter tubing and the male luer Slip to 200 Series Barb. Connect the Tygon Tubing, 0.15" I.D. to the male luer of the stopcock by attaching the blunt-end needle and sliding the tubing over the needle. This is the tubing that will connect to your PDMS chip. Make sure you use enough tubing so that the chip can sit easily on the stage of the microscope with a small amount of slack to allow the stage to move.

4. Insert a length of Tygon tubing into the conical with fluid B (long enough so that you can connect to the chip).
5. Allow liquid to fill all tubing and the stopcock. Allow liquid to continue to flow into a waste container while you proceed with the remaining sets. If you have trouble getting liquid to flow through the tubing you can use a syringe with a 21 G needle to start flow by drawing fluid out with the syringe.

6. Prepare a syringe for injecting fluid through the flow cell. Attach a 21 G needle to the end of a 1 ml syringe, and slip 4 in. of intramedic tubing over the needle (*see* Fig. 4b).
7. Clean a brand new flow cell by injecting the cell with 70 % ethanol followed by sterile water by syringe injection. Make sure to fill the flow cell with water completely, so that the inlets and outlets are covered with fluid to prevent air bubbles.
8. Attach the flow cell to the metal slide holder with lab tape.
9. With the flow cell on the microscope stage, attach the tubing feeding from B to the flow cell using a metal pin. Be careful not to bend the pin when inserting it into the chip, and slide the tubing over. Always make sure that all tubing is completely filled with media before you attach it to the chip, or you will introduce air bubbles into the chamber (*see* **Note 25**).
10. Attach the A tube from the stopcock as you did the B tube to the other inlet of the y-shaped chip.
11. Allow the flow cell to fill with media.
12. Cut a length of Tygon tubing and attach it to the flow cell outlet using a metal pin. Allow this tubing to fill with media. Put the end of the tubing into the waste collection container and fill the waste container with enough water to completely submerge the end of the effluent tube.
13. Image the flow cell in "live mode" on the microscope in the channel of your tracer dye. Change the relative heights of the TOP and BOTTOM tubes to adjust the position of the ON and OFF states in the chamber.
14. Test that you are able to achieve the desired switching dynamics (*see* **Note 26**) as demonstrated in Fig. 5.
15. Once the line is set you are ready to load cells into the flow cell. Remove the tubing and place it in an effluent container to avoid spills. Remove the holder and flow cell from the microscope and place on a flat surface.

3.5 Preparation of Yeast Samples and Cell Loading

1. Prepare two more syringes as in **Step 6** above. One will be for loading the Concanavalin A solution and one will be for loading cells.
2. Grow yeast cells to saturation overnight in low fluorescence media (LFM) at 30 °C with shaking (OD_{600} ~2).
3. Reinoculate at a low density into low fluorescence media in a flask such that cells are in mid-log growth before being loaded into the flow cell (*see* **Note 27**).
4. Clean flow cell with 70 % ethanol by injection. Then inject Milli-Q water. Use 1–2 mL of ethanol and water (*see* **Note 28**).
5. Load the 2 mg/mL Concanavalin A solution into the flow cell using a syringe. It is best to load from the effluent port as

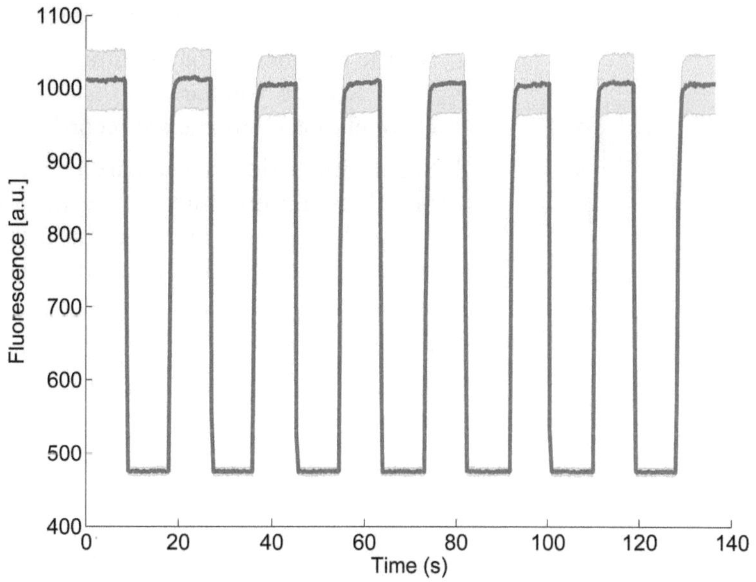

Fig. 5 The switching dynamics of the device are tested using fluorescence microscopy. Here the stopcock was manually switched every 10 s between media A (water) and media B (100 μM fluorescein). *Shaded regions* indicate standard deviation from the mean fluorescence

illustrated in Fig. 4b. Allow the solution to incubate in the flow cell at room temperature for 2 min.

6. When the Concanavalin A is done incubating, load cells by injecting them into the flow cell with the syringe. Dilute or concentrate the cells into LFM as necessary before loading to ensure an even and dense coverage of the channel's surface. A final OD_{600} of 0.5 is usually sufficient.
7. Allow cells to sit in the flow cell for 5 min at room temperature on the bench.
8. Check that cells have stuck at 40× magnification. Stuck cells will not move when the flow cell is gently tapped.
9. Put the flow cell back on the microscope and replace the appropriate tubing being cautious not to introduce air bubbles.
10. Unclamp all tubing and check for leaks. Make sure that the desired initial media is flowing in the chamber.

3.6 Microscopy Time course

The microscopy protocol will necessarily vary with the available equipment and software. The only real requirements are an inverted microscope (because this allows visualization from the bottom of the chip and room for fluid handling above the chip) and appropriate automation of the microscope to acquire a time course (*see* **Note 29**). Figure 5 illustrates integration of the fluidic set-up with the microscope. The outline of a typical experiment is:

1. Switch the microscope objective to the desired magnification.
2. Set the parameters for the time course on the microscope software. Set the software so that the stage position, time, and channel of each image are recorded in a log file.
3. Set stage positions to acquire in the middle of the flow channel. This is important as the middle of the channel is not exposed to the diffusion region near the media interface and therefore is exposed to the appropriate input signal.
4. Once cells have equilibrated start the acquisition and then start fluid switching.
5. Acquire images at time points spaced to capture the desired phenomena.
6. Periodically check the experiment for leaks.

3.7 Cleaning the Setup

1. Once the experiment is finished remove all tubing and place in an effluent container to avoid drips. Clean the chip with 70 % ethanol by injection from a syringe.
2. Clean all tubing and the stopcock with 70 % ethanol and then sterile water.
3. The flow cell can be reused but cell adhesion goes down with repeated use of the chip (*see* **Note 30**). It is also possible to reuse the tubing and stopcock if they are thoroughly rinsed and sterilized between experiments.

4 Notes

1. The manufacturer details of the mask aligner, spin coater, and profilometer are given for completeness. Most cleanrooms will have appropriate equipment, but the manufacturers and details of operation will vary. Be sure to get appropriate training before attempting these protocols on the equipment available to you.
2. Many of the chemicals used in this protocol, including hexamethyldisilazane, pose serious health hazards. Be sure to read the Material Safety Data Sheet for all chemicals and follow appropriate safety precautions. Follow your university and local regulations when disposing of hazardous chemicals.
3. This low fluorescence yeast media is as described in [10]. Note that the media contains no amino acids, which must be supplemented for auxotrophic strains. We recommend this media because the lack of riboflavin and folic acid reduces cellular autofluorescence. However, it is not necessary to use low fluorescence media and most synthetic yeast media (synthetic complete,

synthetic drop-out) work well [11]. YPD is not an acceptable media for imaging as it is intrinsically very fluorescent.

4. Do not adjust the pH of the solution with NaOH as this will cause the manganese to precipitate out. We store this solution aliquoted at –20 °C and thaw just before use.
5. We find that the holder makes it easier to manipulate the flow cell without breaking the delicate coverslip. It is possible to perform experiments without such a holder. However, most university machine shops can easily and cheaply make a holder to your desired specifications from scrap metal leftover from larger projects.
6. Most imaging modalities are compatible with microfluidics (confocal, TIRF, 2-photon, etc). We discuss epifluorescence here because it is a widely used technique accessible to most laboratories.
7. A negative mask will have the design areas clear (to allow penetration of the UV light) and all other areas black. Emulsion down means that the pattern is printed on the bottom side of the page when looking at the page with the patterns in the correct orientation. This ensures that you can place the mask directly against the wafer with the pattern as close as possible to the surface giving the highest feature resolution in your design.
8. Cleanroom facilities can often be found in the Electrical Engineering or Materials Science department of most universities. Additionally, many universities are creating designated "foundries" for making microfluidic devices for biological applications. These facilities contain the necessary equipment for making the master mold. If you cannot or do not want to make the master mold there are several companies which offer this service. See for example, the foundry at Stanford (http://www.stanford.edu/group/foundry/), which will make both master molds and completed microfluidic chips for a fee. In addition, some companies have started to make premade microfluidic chips (see for example dolomite-microfluidics.com or www.ibidi.com).
9. Hexamethyldisilazane $[(CH_3)_2Si]_2NH$ is used to promote adhesion of photoresist to the silicon wafer. This compound reacts with surface hydroxyl groups to form a new siloxane end product, i.e., $Si\text{-}O\text{-}Si(CH_3)_3$, which leaves the surface more hydrophobic leading to improved wetting by photoresist.
10. SU8 resists are available in several standard viscosities. The combination of the resist viscosity and the spin speed sets the height of the coating on the wafer and therefore the height of the microfluidic channels. The datasheets for the various resists (available from the manufacturer Web page) contain estimates

for the desired spin speeds needed to get specific heights. However, it is best to calibrate this yourself for your spin coater before proceeding with manufacture of your master mold.

11. You may add a band-pass filter at this point if desired. Short-wavelength (<350 nm) radiation can cause negative sidewall profiles or "T-topping." To reduce the occurrence of these features a band-pass filter can be used to filter the short wavelength radiation. The PL360-LP filter from Omega Optical or filters V-42 plus UV-D35 from Asahi Technoglass are both appropriate. If you choose to use a band-pass filter you will need to increase exposure time by approximately 40 % to reach the optimum exposure dose. The filter can be placed immediately above the glass plate that contains the photo-transparency mask.
12. The exposure energy (mJ/cm^2) equals the product of the field intensity (mW/cm^2) multiplied by the exposure time. The field intensity can be measured using a radiometer or similar instrument. In any case, the exposure doses given in the technical data sheet for SU8 are benchmarks for resist of a certain thickness but need to be calibrated for your particular mask aligner and UV lamp. It is necessary, when perfecting a process, to do an exposure-series. This can be done on the same wafer by shielding different parts of the design for different amounts of exposure time.
13. Problems in the fabrication process often become apparent during development. If the developed mask pattern does not remain in contact with the wafer or there is excessive cracking this indicates an under cross-linking condition and can be corrected by increasing the exposure time or increasing the length of the post-exposure bake.
14. It is necessary the first time you develop a new master mask to optimize the photolithography process. The SU8 datasheet contains details on what steps to modify to optimize the process for your particular set of equipment. They include:
    (a) Optimizing spin speed or adjusting resist viscosity to optimize the height of spun resist
    (b) Optimizing pre-exposure bake times
    (c) Adjusting UV exposure time
    (d) Adjusting post-exposure bake, in particular the amount of time at 65 °C and the ramp speed from 65 °C to 95 °C
15. This silanization step allows for easy removal of the polymerized PDMS that will be cast onto the master mold. This in turn decreases wear-and-tear on the mold features and increases the lifetime of the master mold. This step is optional, however, it is highly recommended if you want to make multiple PDMS chips from a single master mold.

16. The ratio of curing agent to monomer determines the firmness of the cured PDMS. For a firmer chip, use more curing agent, for a softer chip use less. Care should be taken to avoid contaminating surfaces with uncured PDMS as it is difficult to clean. All soft lithography steps should be carried out while wearing particle-free nitrile gloves. Oils in your skin can prevent PDMS from curing correctly.
17. Some facilities may also be equipped with a mixer (for example, the Thinky ARE-310 planetary centrifugal mixer) that can simultaneously mix and degas viscous mixtures such as PDMS. This is not necessary, but convenient if you have access to one.
18. If you create a significant number of bubbles when pouring your PDMS onto the mold, it is a good idea to degas again. The entire petri dish can be placed into the vacuum desiccator and degassed.
19. If desired, the flow cell can be stored in this state (with tape covering the attachment side) for several days before plasma cleaning and bonding to the coverslip.
20. These instructions are specific to the Plasma Preen plasma cleaner, however, any available plasma cleaner will work. A handheld corona treater can be used to produce a similar effect (see [12]). If no plasma treater is available it is also possible to bond the chip to glass by placing it on the cover glass and baking at 65 °C for about a day.
21. Surface oxidation is believed to expose silanol groups (OH) at the surface of the PDMS that when brought together with the cover glass form covalent siloxane bonds (Si–O–Si) [13].
22. In theory the chip should be functional immediately after plasma-cleaning and bonding. However, we have greater success with allowing the chip to bake at 65 °C for at least an hour. The bake step increases the strength of the bond between the PDMS and the glass.
23. There are several ways to automate the fluid handling. For this particular chip, a simple method is described in [14]. Code for controlling the switching and a parts list are freely available upon request.
24. The switching employs gravity flow to create the alternating media profiles. The higher the media container, the faster the flow in the flow cell. When there is a difference in flow speed between the flows coming into the inlets of the y-junction, the faster flow will dominate the main channel. The media switch alternates between media "A" at the highest point and media "A" at the lowest point. When A "HIGH" flows through the switch, it dominates the main channel, when A "LOW" flows

through, media B now has the faster flow and dominates the channel. This allows for the quick alternation between media A and B.

25. Bubbles will remove cells and clog the microfluidic device. Take care to avoid them at all costs. Make sure all fluid lines are full of liquid and that there is no air between the pin and the inlet when you plug tubing into the chip. If you accidentally feed air bubbles into the chip you can remove them by pulling media from the outlet using a syringe. If you do this too vigorously you may detach cells and need to start over.
26. To set the switching dynamics we used the set-up illustrated in Fig. 4 with water in the "A" conical and 100 μM fluorescein in water in the "B" conicals. All imaging was done on a Nikon Eclipse-TI inverted microscope using a Nikon 40×/0.95 NA DIC Plan Apo objective. The microscope is equipped with a PerfectFocus system (Nikon) for maintaining the correct focal plane, and a Clara CCD Camera (Andor) for recording fluorescence emission. Fluorescein emission was visualized at 525 nm (50-nm bandwidth) upon excitation at 470 nm (40-nm band-width, Chroma 49002_Nikon ETGFP filter cube). Image acquisition was automatically controlled using NIS Elements software (Nikon). Once the heights of the conical tubes were set as described in **step 13** we started continuous image acquisition using the NIS Elements software. Continuous acquisition corresponds to images every 0.5 s on our system. Images were quantified in ImageJ (http://rsbweb.nih.gov/ij/) by defining a region-of-interest (ROI) corresponding to the entire image frame and using the ROI Manager "multi-measure" utility to find the average fluorescence in the region throughout the time course. Media switching was done by hand, using a timer to keep track of 10 s intervals.
27. Growth conditions will vary depending on the experiment. These serve as general guidelines only; none of the techniques discussed here require that the cells be in mid-log growth.
28. Safety glasses should be worn when injecting liquids into the chip.
29. An inverted microscope is not strictly necessary for this simple microfluidic device. It is possible to punch L-shaped inlets and outlets (using the Harris Uni-Core at 90° angles) and then run the flow cell "upside-down" so that the cover glass is beneath the objective on an upright scope. However, this is painful even for a simple device and next to impossible for more complicated microfluidic chips.
30. We generally do not find it profitable to reuse chips as the cell adhesion decreases significantly. We make a new PDMS chip for each experiment.

References

1. Qin D, Xia Y, Whitesides G (1996) Rapid prototyping of complex structures with feature sizes larger than 20 μm. Adv Mater 8: 917–919
2. Chronis N, Zimmer M, Bargmann C (2007) Microfluidics for in vivo imaging of neuronal and behavioral activity in Caenorhabditis elegans. Nat Method 4:727–731
3. Huang Y, Agrawal B, Sun D et al (2011) Microfluidics-based devices: new tools for studying cancer and cancer stem cell migration. Biomicrofluidics 5:13412
4. Ahmed T, Shimizu T, Stocker R (2010) Microfluidics for bacterial chemotaxis. Integr Biol 2:604–629
5. Ryley J, Pereira-Smith O (2006) Microfluidics device for single-cell gene expression analysis in *Saccharomyces cerevisiae*. Yeast 23:1065–1073
6. Hersen P, McClean MN, Mahadevan L et al (2008) Signal processing by the HOG MAP kinase pathway. Proc Natl Acad Sci U S A 105:7165–7170
7. Mettetal JT, Muzzey D, Gomez-Uribe C et al (2008) The frequency dependence of osmo-adaptation in *Saccharomyces cerevisiae*. Science 319:482–484
8. Bennett MR, Pang WL, Ostroff NA et al (2008) Metabolic gene regulation in a dynamically changing environment. Nature 454: 1119–1122
9. Hao N, O'Shea EK (2011) Signal-dependent dynamics of transcription factor translocation controls gene expression. Nat Struct Mol Biol 19:31–39
10. Sheff M, Thorn K (2004) Optimized cassettes for fluorescent protein tagging in *Saccharomyces cerevisiae*. Yeast 21:661–670
11. Burke D, Dawson D, Stearns T (2000) Methods in yeast genetics: a cold spring harbor laboratory course manual, 2000th edn. Cold Spring Harbor Laboratory Press, Woodbury, New York
12. Haubert K, Drier T, Beebe D (2006) PDMS bonding by means of a portable, low-cost corona system. Lab Chip 6:1548–1549
13. McDonald JC, Whitesides GM (2002) Poly (dimethylsiloxane) as a material for fabricating microfluidic devices. Acc Chem Res 35:491–499
14. McClean MN, Hersen P, Ramanathan S (2011) Measuring in vivo signaling kinetics in a mitogen-activated kinase pathway using dynamic input stimulation. Methods Mol Biol 734:101–119

Chapter 9

Using Two-Hybrid Interactions to Identify Separation-of-Function Mutations

Brian Haarer and David C. Amberg

Abstract

Protein functions within cells frequently require they interact physically with a number of partner proteins to coordinate the appropriate biochemical processes. Mutational analysis has been quite useful for analyzing how the loss of a gene or protein impacts cell function or more specifically particular pathways. However, the genetics approach to studying gene function can be limited by not having the right mutations; for example because the mutant allele ablates all function, as is the case for deletion (null) alleles and most temperature-sensitive alleles. To dissect the relative contributions of a protein's interactions, the researcher needs mutations that specifically affect one but not all of the protein's interactions. In genetics parlance such mutations are called separation-of-function mutations. The yeast two-hybrid system has been exploited for two decades to identify protein-binding partners. Here we describe a fairly simple protocol, within reach of laboratories with molecular biology experience, for using the two-hybrid system to identify separation-of-function mutations.

Key words Two-hybrid, Mutagenesis, Separation of function, Protein–protein interactions

1 Introduction

In 1989, Fields and Song first introduced the yeast two-hybrid system in a proof-of-principal paper showing the system could detect biologically relevant protein–protein interactions [1]. Subsequently, they showed the system could be harnessed to screen libraries for binding partners of a protein of interest [2]. The latter application has had a huge impact on biology by giving researchers a powerful tool for unraveling the spectrum of protein interactions governing biological pathways. A number of variations and applications of this basic methodology have been developed over the years. One is the use of the system, in conjunction with mutants, to map sites of protein–protein interaction [3, 4]. This application suggested that perhaps the system could be reversed to screen for mutations that disrupt specific protein–proteins interactions to facilitate structure/function studies of protein complexes.

Jeffrey S. Smith and Daniel J. Burke (eds.), *Yeast Genetics: Methods and Protocols*, Methods in Molecular Biology, vol. 1205, DOI 10.1007/978-1-4939-1363-3_9, © Springer Science+Business Media New York 2014

Application of a reverse two-hybrid screening approach was described in 1996 [5].

A problem with the early versions of reverse two-hybrid strategies was that a very large percentage of the mutants isolated were complete null alleles, most usually frameshift mutations. We describe here a constrained reverse two-hybrid screening procedure in which we ask that mutants bind at least one binding partner while losing an interaction with a different binding partner. We refer to this procedure as a differential interaction two-hybrid screen. This procedure was first applied to the double ß-propeller, actin regulatory protein Aip1p which cooperates with the protein cofilin to disassemble actin filaments [6]. Briefly, *AIP1* as a fusion to the *GAL4* DNA binding domain in a two-hybrid expression vector was mutagenized by PCR and the resulting mutant library was screened for loss of the actin or cofilin interactions while discarding mutants that had lost both interactions. The resulting separation-of-function mutants were extremely informative in that they allowed us to map the actin binding site to a specific location on the N-terminal ß-propeller of Aip1p and the cofilin binding site to the C-terminal propeller of Aip1p [7]. Here we describe the details of how this procedure can be applied to any protein that has multiple binding partners that are amenable to two-hybrid analysis.

2 Materials

1. Yeast Strain Y187: *MATα gal4 gal80 his3 trp1-901 ade2-101 ura3-52 leu2-3,112 cyh^r^* P_{GAL}*-lacZ* [8].
2. Yeast Strain Y190: *MAT***a** *gal4 gal80 his3 trp1-901 ade2-101 ura3-52 leu2-3,112 URA3::* P_{GAL}*-lacZ LYS2::* P_{GAL}*-HIS3 cyh^r^* [8].
3. Plasmid pDAb1: *GAL4* DNA binding domain fusion vector, *CEN, TRP1* [3].
4. Plasmid pACTII: *GAL4* activation domain fusion vector, 2 μ, *LEU2* [8].
5. Plasmid pSE1111: *GAL4-SNF4* activation domain fusion vector, 2 μ, *LEU2* [8].
6. Plasmid pSE1112: *GAL4-SNF1* DNA binding domain fusion vector, 2 μ, *TRP1* [8].
7. Yeast Strain JTY32: *MAT***a** *ade2 ade8Δ::kanR ura3 leu2 his3 trp1*.
8. Plasmid pJT10: 2 μ *URA3 ADE8*.
9. Primer BHo-JT10-1: 5′-gccactcctcaattggattag-3′.
10. Primer BHo-JT10-2: 5′-gctaaggtagagggtgaacg-3′.
11. *GAL4* Primer #1: 5′-gactggaacagctatttctactg-3′ Sense strand primer to *GAL4* DBD in pDAb1, upstream of the polylinker.

12. *GAL4* Primer #2: 5′-gcagctggcacgacaggtttccc-3′ Antisense strand primer to the terminator in pDAb1, downstream of the polylinker.
13. YPD liquid medium [9].
14. SC-leu liquid medium [9].
15. SC-trp plates [9].
16. SC-leu plates [9].
17. SD + 100 μg/mL adenine + 25 mM 3-aminotriazole plates [9].
18. Sterile velvets: http://www.corastyles.com/.
19. Replica plating block: http://www.corastyles.com/.
20. Lithium Acetate (1 M).
21. 10× TE pH 7.5 (100 mM Tris, 10 mM EDTA).
22. PEG 4000 (50 % w/v).
23. LiOAc mix: 0.1 M lithium acetate, 1× TE; made fresh from the above stocks.
24. PEG mix: 0.1 M lithium acetate 1× TE, 40 % PEG 4000; made fresh from the above stocks.
25. ssDNA carrier (10 mg/mL): sheared and boiled salmon sperm, calf thymus or some similar source of DNA.
26. STES Buffer:

	Stocks	mL
0.5 M NaCl	5 M	1.0
0.2 M Tris pH 7.5	1 M	2.0
10 mM EDTA	0.5 M	0.2
1 % SDS	10 %	1.0
H_2O		5.8
		10 mL

3 Methods

3.1 Cloning of "Bait" and "Prey" Constructs into Two-Hybrid Vectors

The application of this method requires the construction of a plasmid fusing the *GAL4* DNA binding domain of plasmid pDAb1 to a bait that is the target for mutagenesis. In addition, at least two "prey" binding partner constructs must be cloned fusing their open reading frames to that of the *GAL4* activation domain in plasmid pACTII. Ideally, you should know that the bait and prey proteins do in fact bind to each other, although you may not already know that the interaction can be detected as two-hybrid fusions; if so you will need to confirm this. Shown below, in Fig. 1,

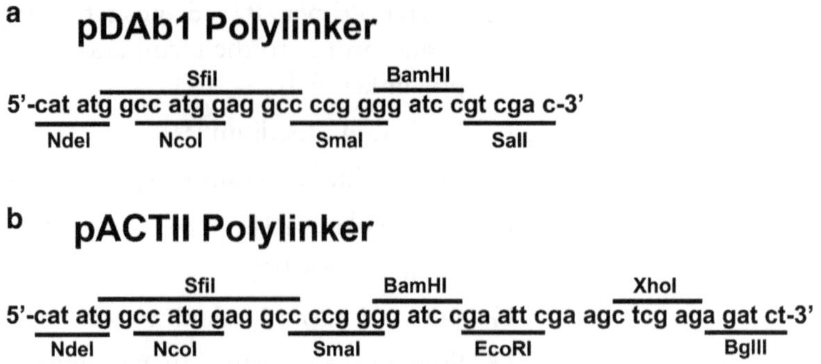

Fig. 1 Polylinker restriction enzyme cut sites for two-hybrid vectors. The restriction enzyme cut sites are shown for plasmids pDAb1 (**a**) and pACTII (**b**). Reading frame is indicated by groupings of three bases

are the polylinker sequences and restriction enzyme cut sites for plasmids pDAb1 and pACTII. Note the reading frame is indicated by groups of three nucleotides. Standard cloning methods can be used to clone in frame fusions of the bait and prey protein-reading frames. We usually provide for some conformational flexibility by inserting 3–5 codons for alanine between the *GAL4* and bait or prey coding sequences.

3.2 Testing the Bait and Prey Constructs for Binding and Two-Hybrid Activation

Given the ease with which plasmids can be sequenced, we recommend that the bait and prey construct be confirmed by sequencing, particularly since you are likely to have used a PCR-based method for the construction of the plasmids.

The bait and prey constructs can be co-introduced into a two-hybrid reporter strain by co-transformation, but for ease we prefer to transform the bait constructs into a two-hybrid reporter strain of one mating type (Y187) and the prey constructs into a reporter of the opposite mating type (Y190). We use pSE1112 transformed into Y187 and pSE1111 transformed into Y190 as a positive control; these constructs express Gal4 fusions to Snf1 and Snf4, respectively, and give very strong activation in the two-hybrid system. The empty pDAb1 and pACTII plasmids can be used as negative controls. The Y187 and Y190 transformants can be patched together on YPD medium plates and allowed to mate for 1 day at 30 °C. Diploids that contain both plasmids are then selected on SC-trp-leu medium plates either by replica plating with sterile velvets or by streaking for single colonies. At 30 °C it will take approximately 3 days for the diploids to grow from streaks, after which they can be replica plated or streaked onto SD + Adenine + 25 mM 3-aminotriazole plate medium and incubated at 30 °C. Note that one of the two-hybrid reporters is *HIS3* and the amino-triazole (3-AT) selects for cells that make higher levels of the His3 enzyme. Growth on the 3-AT medium, for positive interactions, can take 3 days to a week. The Snf4–Snf1 interaction will be evident within 3 days. Although we have found that 25 mM 3-AT works well for most interactions, this may need to be adjusted to reduce backgrounds; we have used concentrations as high as 200 mM. We do not recommend using less than 25 mM because such low levels of activation are unlikely to work well for this procedure.

3.3 Adjusting the Level of PCR Mutagenesis; Knowing the Fidelity of Your Enzyme

Mutagenesis of the bait construct insert will be performed by using PCR as opposed to chemical mutagenesis of the entire plasmid; this will allow for targeting only the bait protein sequences so as to avoid mutations in the *GAL4* DNA binding domain insert portion of the vector. In addition, PCR is not heavily biased in the types of mutations it creates and the extent of mutagenesis can be adjusted, most readily by the choice of enzyme used.

The level of mutagenesis will be most heavily influenced by the choice of enzyme used as their fidelity can vary greatly. This will be a critical choice for each project with the desired outcome being the isolation of many single mutants of interest without a large background of irrelevant base pair substitutions. Many PCR mutagenesis protocols call for skewing nucleotide ratios or including salts (such as manganese) that lower enzyme fidelity, but we have found that these steps are unnecessary; Taq has pretty low fidelity all on its own, producing mutations every few hundred base pairs. While there are many reportedly high-fidelity DNA polymerases available for use in polymerase chain reaction (PCR) technology, we have found that many suffer from higher-than-expected frequencies of mutations upon sequencing of the resulting products. Because of the critical nature of this step, we are including a protocol below for determining enzyme fidelity. You may be tempted to believe the fidelities reported for enzymes by the companies that sell them but as you will see below, we found little relationship between advertised fidelities and how those enzymes behave in our hands.

To get a better idea of the practical and relative mutation frequencies of such enzymes, we devised an in vivo assay for the comparison of thermostabile DNA polymerases used in standard PCR reactions. Our assay makes use of the red or white colony colors exhibited by yeast strains carrying various mutations in the adenine biosynthetic pathway. For example, yeast that carry *ade2* mutations turn red due to the buildup of a biosynthetic intermediate (P-ribosylamino-imidazole), but if they also carry an epistatic ("upstream") mutation in another adenine biosynthetic gene (such as in *ADE8*), they remain white. Our assay starts with an *ade2 ade8* (white) yeast strain into which we introduce PCR-generated *ADE8*; if this PCR product is essentially mutation-free (note that silent, conservative, and non loss-of-function mutations will remain undetected by this method), the resulting yeast colonies will turn red. Thus, the frequency of white colonies is an indicator of the *relative* mutation frequencies when comparing different polymerases.

In practice, we use co-transformation of a cut vector and PCR product to introduce *ADE8* products. Recipient yeast are able to "gap-repair" the cut vector [10] to generate a replicating *ADE8-bearing* plasmid. In a proof-of-principle experiment, we utilized several common commercially available high-fidelity (Primestar HS, Pfx50, Deep Vent, EHF Plus, and a mix of LA-Taq + PfuTurbo) and low-fidelity (Taq) polymerases to generate PCR-based *ADE8* product using plasmid pJT10 (2 μ *URA3 ADE8*) as a template and the primers BHo-JT10-1 and BHo-JT10-2 described above. Manufacturer's buffers and nucleotides were used when provided; in the LA-Taq/Pfu Turbo mix, LA-Taq buffer and nucleotide mix were used. Resulting PCR products were co-transformed with *Msc*I-*Nru*I-digested and gel purified pJT10 (i.e. lacking the *ADE8* gene) into yeast strain JTY32 (*MATa ade2 ade8Δ100::kanR ura3Δ leu2Δ1 his3Δ200 trp1Δ63*) and plated on low-adenine media (promotes color development) lacking uracil. After growth at 30°C, resulting red and white colonies were quantified. Since strains lacking normal mitochondrial function ("petite" strains) fail to develop red color, we also replica plated colonies to media containing the nonfermentable carbon source glycerol; on these plates, white colonies should arise only if cells fail to acquire a functional copy of *ADE8*.

We were quite surprised to find that one enzyme, Primestar HS (TaKaRa), performed markedly better than all others in this particular pilot experiment (*see* Fig. 2). Equally surprising was the observation that some of the other "high-fidelity" enzymes performed no better than Taq polymerase, which lacks any proofreading function. While we understand that this assay is an indirect measure of enzyme fidelity and that it does not give absolute mutation frequencies, it did in fact point us to an enzyme that has since performed remarkably well in our cloning experiments (unlike the several previous "high-fidelity" enzymes that we had been using).

As high throughput sequencing continues to improve and becomes more affordable, it may become the method of choice for labs to directly measure mutation frequencies of PCR products. However, such technology is still relatively expensive and requires significant troubleshooting. Thus, for the "common" laboratory, our method provides a reasonably quick and inexpensive measure of the relative effectiveness of DNA polymerases as used to generate PCR products.

3.4 Mutagenesis of Bait Coding Sequence

In our experience we have found that using regular Taq polymerase in a standard PCR reaction can be quite effective as a mutagen and so the procedure described below follows such a protocol. However, you may find that this level of mutagenesis is too high, particular if your bait is a large protein. Given that "jackpot" mutants can accumulate in an amplified reaction, thus reducing the complexity of the mutant library, we highly recommend performing several separate PCR reactions, keeping the resulting libraries separate (do not pool

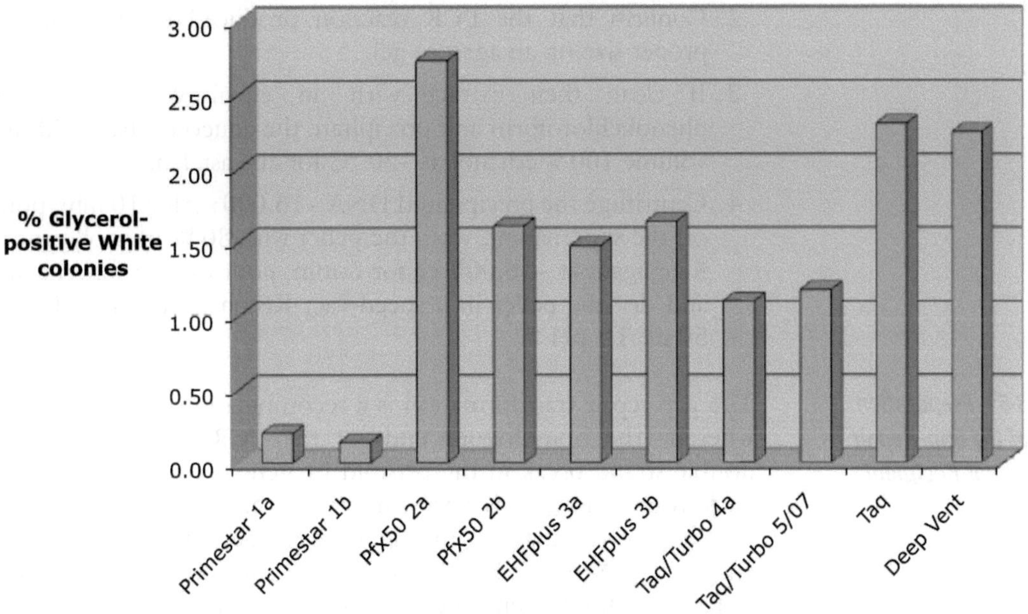

Fig. 2 Measurement of relative mutation rates of thermocycle DNA polymerases. The *ADE8* gene was amplified using the indicated enzymes and the products were gap-repair transformed to reconstitute plasmid pJT10 in strain JTY32. The percentages of *ade8* inactivating mutations were quantified by determining the percentages of nonpetite (glycerol positive), white colonies. *N* = 500–2,000

the PCR reactions) and isolating only a few mutants from each mutagenic PCR reaction.

1. Mutagenic PCR Reaction:

   1 μL 10 ng/μL of bait plasmid DNA (pDAb1-based).

   5 μLl 5 μM *GAL4* Primer #1.

   5 μL 5 μM *GAL4* Primer #1.

   5 μL 10× Taq buffer (supplied by manufacturer).

   5 μL 2 mM dNTP's.

   0.5 μL 10 mg/mL acetylated BSA.

   1 μL Taq polymerase.

   26.5 μL H_2O.

   Run a standard PCR reaction:

   94 °C × 4 min.

   25 Cycles of:

   94 °C × 1 min.

   55 °C × 1 min.

   72 °C × 1 min for every 1 kb of insert.

   72 °C × 20 min.

   4 °C soak.

2. Confirm that the PCR reaction produced product of the proper size on an agarose gel.
3. If clean then extract with an equal volume of 1:1 phenol:chloroform and precipitate the aqueous phase with 3× volume 100 % ethanol at −20 °C for at least 1 h.
4. Centrifuge the precipitated DNA ~16,000 × *g* for 10 min, pour off the supernatant, wash the pellet with 80 % ethanol, centrifuge again at ~16,000 × *g* for 5 min, pour off the supernatant and dry the pellet in a speed-vac. Reconstitute the pellet in 50 μL TE pH 8.

3.5 Preparation of the Gap-Repair Vector Recipient

The gap repair transformation is a recombination-mediated event whereby the homologous ends of the PCR fragment heals a double-strand break in the plasmid by gene conversion. To keep backgrounds low, it is very important that all, or as much as possible, of the plasmid molecules are digested with restriction enzyme (*see* **Note 1**). For this reason, we cut the plasmid with two different restriction enzymes, well known for having high activity, and we use a fairly large concentration of these enzymes.

1. Digest of vector pDAb1 with *Sal*I and *Bam*HI:

   77.8 μL H_2O.

   10 μg pDAb1.

   10 μL 10× *Bam*HI buffer.

   4 μL (80 units) *Sal*I.

   4 μL (80 units) *Bam*HI.

   1 μL 10 mg/mL acetylated BSA.

   Digest at 37 °C for 5 h.
2. Extract with an equal volume of 1:1 Phenol-chloroform and ethanol precipitate the aqueous phase overnight at −20 °C by the addition of three volumes 100 % ethanol.

Next Day

3. Centrifuge at ~16,000 × *g* for 10 min.
4. Wash the pellet with 80 % ethanol.
5. Dry pellet and suspend in 80 μL TE pH 8.0.

3.6 Gap Repair Transformation and Mutant GAL4-DBD-Query Library Construction

1. Start a 5 mL overnight of strain Y190 in YPD and incubate at 30 °C in preparation for the gap repair transformation the next morning.

Next Day

2. Inoculate 100 mL YPD with the overnight culture of strain Y190 such that after at least three doublings the final concentration will be 2×10^7 cells/mL. The doubling time of this

strain is ~90 min and one can assume ~2–2.5 × 10^8 cells/mL in the overnight culture if it has reached stationary phase. In addition, cells require ~30 min to recover from stationary phase.

3. Spin down cells 5 min at ~3,000 × *g* in a tabletop centrifuge and discard supernatant.
4. Wash cells with 5 mL LiOAc mix, spin again, and discard supernatant.
5. Resuspend cells in 1 mL LiOAc mix, and incubate on a roller drum for 1 h at 30 °C.
6. Boil the ssDNA carrier for 5 min then place on ice. Set up the transformations as follows:

   10 μL 10 mg/mL ssDNA.

   1 μL *Bam*HI/*Sal*I cut pDAb1.

   5 μL Bait PCR.

   200 μL Y190 cells in LiOAc mix.
7. Vortex well to mix and incubate on a roller drum at 30 °C for 30 min.
8. Set up one transformation per PCR reaction and be sure to include a control with no DNA and a control with *Bam*HI/*Eco*RI cut pDAb1 alone.
9. Add 1 mL PEG mix to each tube, vortex and incubate on a roller drum for 30 min at 30 °C.
10. Heat shock each tube at 42 °C for 15 min, pellet the cells 5 min at ~2,500 × *g* in a micro-centrifuge, and reconstitute the pellet in 200 μL TE pH 7.5.
11. Plate all of the no DNA control and pDAb1 alone transformations onto two SC-trp plates. Plate 20 μL of the gap repair transformations onto SC-trp plates. Incubate at 30 °C. Note that this step may need to be repeated to adjust the plating density of the gap-repair transformations. The goal is to obtain ~200–500 colonies per plate.

3.7 Screening for Differential Interacting Mutants

A procedure called replica plating using sterile squares of velvet and a replica plating block will be used to mate the mutant query libraries transformed into strain Y190 with strain Y187 that has been transformed with at least two different prey constructs previously shown to interact by two hybrid with the bait. The diploid colonies containing both two-hybrid plasmids are selected again by replica plating, then finally replica plated to medium that selects for cells expressing the two-hybrid reporter *HIS3*. The goal is to identify bait mutants that activate with one prey but not the other and vice versa (*see* **Notes 2–5**). Since colonies increase in size at every replica plating step, the plating density must be sparse enough that there is sufficient space between the colonies but dense enough so that many colonies can be screened per plate (usually 200–500 colonies per plate).

1. Start 5 mL overnights of the Y187 prey transformants in SC-leu. Incubate at 30 °C.
2. The next day plate 200 μL of the overnight cultures of the Y187 prey transformants on separate SC-leu plates to create lawns of cells. Make enough lawns plates for the number of gap repair transformation plates. Incubate at 30 °C.
3. The next day replica plate each plate of the gap repair transformants onto two YPD plates. Onto one, replica plate a lawn of Y187 + prey construct #1 and onto the other replica plate a lawn of strain Y187 + prey construct #2. Be sure to mark the orientation of the plates. Incubate at 30 °C. Save the original plates of the gap repair transformants at 4 °C.
4. The next day replica plate the diploid cells to SC-trp-leu plates. Incubate at 30 °C.
5. Two days later replica plate the mated cells to SD + adenine + 25 mM 3-AT and incubate at 30 °C.
6. Approximately 5 days later, identify the colonies that grew on the 3-AT plate with one mating partner but not the other. Go back to the original gap repair transformation plate and recover the mutants of interest by streaking for single colonies onto fresh SC-trp plates.
7. Retest the bait mutants to confirm the differential interaction phenotypes by a similar procedure to that used to identify the mutants as described above. We prefer to do this by making lines on plates of the transformants, 6 lines per plate, and using replica plating to create a mating grid with the prey transformants.

3.8 Recovery and Identification of the Differential Interaction Mutants

In order to identify the mutations in the bait protein that result in the separation of function/differential interactions with the prey proteins, the plasmids that encode these alleles need to be retrieved from the yeast strains carrying the *GAL-bait* mutant plasmids and transferred into *E. coli* so they can be amplified and sequenced.

1. Start 2 mL overnight cultures of the mutants of interest in Y190 in SC-Trp and incubate overnight at 30 °C on a roller drum.
2. Centrifuge 1.5 mL of stationary phase yeast culture in a microfuge tube, 5 s at ~16,000 × *g*, decant the supernatant and resuspend the pellet in 60 μL STES buffer.
3. Add acid washed glass beads to the level of the liquid in the tube and vortex on high for 1–2 min, then add 50 μL phenol:chloroform (1:1) and vortex 1 min on high.
4. Centrifuge 5 min at ~16,000 × *g* in a microcentrifuge.

5. Remove the supernatant to a fresh tube and use 6 μL to transform *E. coli* by standard methods.
6. Use whatever preferred method for DNA sequencing and to mini-prep the plasmid DNA from the *E. coli* transformants.
7. *GAL4* Primers #1 and #2 can be used to amplify the inserts for DNA sequencing.

4 Notes

There are a number of problems that can occur that will either limit the success of the method or lead it to fail entirely. These are listed below:

1. The pDAb1 vector recipient must be well digested with enzyme or most of the transformants from the gap repair transformation will be pDAb1 lacking insert. This will be indicated by the transformation control using the digested pDAb1 vector without the PCR fragment present. From this control you should be able to estimate the percentage of colonies on the gap repair transformations that contain empty vector. If this percentage is high you will be reducing the number of mutants you are able to screen and it would be in your interest to go back and digest the vector more thoroughly.
2. The level of PCR mutagenesis may be too low such that no mutants of interest are isolated. This will be obvious if none of the gap repair transformation libraries give colonies that show two-hybrid activation failure. If faced with this situation (not one we have experienced) you should go back to the mutagenic PCR step and try a lower fidelity enzyme or try using some of the conditions in the literature suggested for lowering PCR fidelity.
3. The level of PCR mutagenesis is too high. This will most likely be evident only after sequencing some mutants of interest. If the mutants have base pair changes at several locations, it is likely to be time consuming to ascertain which of these mutations are responsible for the loss of the interaction. Given the ease of the procedure, it will be less work to repeat the PCR using an enzyme or conditions with increased fidelity such that the isolated mutants are less complex, preferably with a single site mutation.
4. All of the mutants identified are defective for both interactions. This result is informative in that it suggests the interactions are not separable and there are a limited number of molecular explanations for such a result. For example, they may share overlapping binding sites. However, the mutants retrieved from such a screen are likely to be null alleles resulting from frame-shifts

or premature stop codons and it would be difficult to identify mutants that specifically disrupt the common binding site(s).

5. The mutants may show specificity for only one prey protein, that is 2 classes of mutants are isolated. One class fails to interact with prey A but can interact with prey B, and a second class of mutants fails to interact with both A and B. This result could be informative in that it may suggest interesting influences on cooperative or co-dependent interactions within a trimeric complex. This was the result we obtained in our screen for Aip1p mutants; we could get cofilin-specific mutants but not actin-specific mutants. This result agreed with the biochemistry where we observed that the cofilin interaction requires that Aip1p be able to interact with actin.

References

1. Fields S, Song O (1989) A novel genetic system to detect protein–protein interactions. Nature 340:245–246
2. Chien CT, Bartel PL, Sternglanz R et al (1991) The two hybrid system: a method to identify and clone genes for proteins that interact with a protein of interest. Proc Natl Acad Sci U S A 88:9578–9582
3. Amberg DC, Basart E, Botstein D (1995) Defining protein interactions with yeast actin in vivo. Nat Struct Biol 2:28–35
4. Amberg DC, Botstein D (1997) Obtaining structural information about protein complexes with the two-hybrid system. In: Bartel PL, Fields S (eds) The yeast two-hybrid system. Oxford University Press, New York
5. Leanna CA, Hannink M (1996) The reverse two-hybrid system: a genetic scheme for selection against specific protein/protein interactions. Nucleic Acids Res 24:3341–3347
6. Rodal AA, Tetreault JW, Lappalainen P et al (1999) Aip1p interacts with cofilin to disassemble actin filaments. J Cell Biol 145: 1251–1264
7. Clark MG, Teply J, Haarer BK et al (2006) A genetic dissection of Aip1p's interactions leads to a model for Aip1p-cofilin cooperative activities. Mol Biol Cell 17:1971–1984
8. Durfee T, Becherer K, Chen PL et al (1993) The retinoblastoma protein associates with the protein phosphatase type 1 catalytic subunit. Genes Dev 7:555–569
9. Amberg DC, Burke DJ, Strathern JN (2005) Methods in yeast genetics: a cold spring harbor laboratory course manual. Cold Spring Harbor Laboratory Press, Cold Spring Harbor, NY
10. Ma H, Kunes S, Schatz PJ et al (1987) Plasmid construction by homologous recombination in yeast. Gene 58:201–216

Chapter 10

Synthetic Genetic Array Analysis for Global Mapping of Genetic Networks in Yeast

Elena Kuzmin, Sara Sharifpoor, Anastasia Baryshnikova, Michael Costanzo, Chad L. Myers, Brenda J. Andrews, and Charles Boone

Abstract

Genetic interactions occur when mutant alleles of two or more genes collaborate to generate an unusual composite phenotype, one that would not be predicted based on the expected combined effects of the individual mutant alleles. Synthetic Genetic Array (SGA) methodology was developed to automate yeast genetic analysis and enable systematic genetic interaction studies. In its simplest form, SGA consists of a series of replica pinning steps, which enable the construction of haploid double mutants through mating and meiotic recombination. For example, a strain carrying a query mutation, such as a deletion allele of a nonessential gene or a conditional temperature sensitive allele of an essential gene, could be crossed to an input array of yeast mutants, such as the complete set of ~5,000 viable deletion mutants, to generate an output array of double mutants, that can be scored for genetic interactions based on estimates of cellular fitness derived from colony-size measurements. A simple quantitative measure of genetic interactions can be derived from colony size, which serves as a proxy for fitness. Furthermore, SGA can be applied in a variety of other contexts, such as Synthetic Dosage Lethality (SDL), in which a query mutation is crossed into an array of yeast strains, each of which overexpresses a different gene, thus making use of SGA to probe for gain-of-function phenotypes in specific genetic backgrounds. High Content Screening (HCS) also integrates SGA to perform genome-wide screens for quantitative analysis of morphological phenotypes or pathway activity based upon fluorescent markers, extending genetic interaction analysis beyond fitness-based measurements. Genetic interaction studies offer insight into gene function, pathway structure, and buffering, and thus a complete genetic interaction network of yeast will generate a global functional wiring diagram for a eukaryotic cell.

Key words Yeast, Genetics, Genetic interactions, Synthetic genetic array, Synthetic dosage lethality, High-content screening

1 Introduction

A complex interplay of genetic interactions may underlie the relationship between an organism's genotype and phenotype [1–3]. However, genetic interactions based upon natural variation remain elusive and thus while we suspect that genetic networks may

Jeffrey S. Smith and Daniel J. Burke (eds.), *Yeast Genetics: Methods and Protocols*, Methods in Molecular Biology, vol. 1205, DOI 10.1007/978-1-4939-1363-3_10, © Springer Science+Business Media New York 2014

underlie the genotype to phenotype relationship, the topology of such networks remains to be elucidated. To begin to understand the general principles of genetic networks, we have undertaken a systematic and unbiased study of how different genes interact with one another to modulate essential biological functions in the budding yeast model system [4–6]. A digenic interaction occurs when a double mutant shows a phenotype that could not be predicted based on the combined effects associated with the individual mutants [7]. For example, if two different yeast deletion mutants, each of which fails to alter cellular fitness as a single mutant, combine to lead to a lethal double mutant, then this is referred to as a synthetic lethal genetic interaction. Indeed, researchers have made use of genetically tractable model organisms, such as yeast and worms, as well as isogenic cultured cell lines, derived from fruit flies and mammals, as platforms to map genetic interactions and further our understanding of phenotypes associated with complex genetics [8].

Genome-wide genetic interaction mapping was pioneered in the budding yeast, *Saccharomyces cerevisiae* [4, 6] and was first made possible by the yeast gene deletion collection of ~5,000 nonessential genes [9, 10]. This collection has been constructed by replacing every yeast open reading frame (ORF) with kanMX4 antibiotic selectable marker [9, 10]. Each deletion strain also contains two unique 20 bp molecular barcodes flanked by common PCR-amplification primer sites, which can be used to identify the mutant in highly parallel assays that contain a pool composed of thousands of different mutants [9, 10]. Additional mutant strain collections have been assembled to enable large-scale genetic interaction studies of the set of ~1,000 essential genes. A gene is termed essential when its deletion in a haploid cell causes lethality [10]. The essential gene mutant collections include conditional temperature-sensitive [11, 12], tetracycline-repressible [13] and hypomorphic DAmP alleles [14]. Thus, there are tools in yeast to examine digenic interactions of all the genes within the genome, a set representing ~18 million gene–gene combinations.

Synthetic Genetic Array (SGA) methodology was developed to automate genetic interaction studies [6]. SGA is used to cross any haploid query strain of interest (carrying mutant alleles of one or more genes) to an input array composed of a set of different yeast strains (each carrying a mutant allele of a different gene, such as the set of ~5,000 nonessential gene deletion mutants). A series of replica-pinning steps is used to first generate heterozygous diploid cells and ultimately select an output array in which each colony is composed of haploid double mutant cells containing the query and array strain mutations. The resulting mutants can be scored for a variety of phenotypes including colony-size and used to measure genetic interactions amongst the mutant alleles.

Our first large-scale SGA genetic interaction study qualitatively scored synthetic sick and lethal phenotypes to map ~4,000 genetic interactions involving ~1,000 genes, to reveal functional connections between genes and pathways and gain insight into the extent of buffering within the cell [4]. A synthetic lethal genetic interaction often occurs between parallel pathways that impinge on an essential function [8]. Global genetic interaction studies can be utilized to gain knowledge of the roles of previously uncharacterized genes, pathway composition, and the connections between these pathways, to begin to map a functional wiring diagram of the cell [15].

We also developed a scoring system to quantitatively measure genetic interactions [16]. We use colony size as a proxy for cellular fitness, which means that double mutant fitness can be compared to a model that combines the single mutant phenotypes to estimate the expected fitness of the double mutant. If the measured double mutant fitness differs significantly from the expected fitness, then we score it as either a negative (synthetic sick or lethal) or a positive (e.g. suppression) genetic interaction. In our most recent quantitative analysis of ~5.4 million SGA-derived double mutants we identified a total of ~170,000 genetic interactions (~2/3 negative and ~1/3 positive) providing a global view of the functional organization of a eukaryotic cell [16, 17].

Since SGA is a methodology for combining different alleles of genes in a high-throughput fashion, it is amenable for an application in a variety of contexts and can be used to study various types of genetic interactions using a multitude of biological read-outs. For instance, Synthetic Dosage Lethality (SDL) genome-wide screens make use of SGA with gene overexpression arrays to probe for gain-of-function phenotypes within specific genetic contexts [18, 19]. Systematic analysis of ectopic gene overexpression in wild-type cells demonstrated that increased gene dosage, driven constitutively by *GAL1* promoter, was toxic for cell growth for ~15 % of the genes within the yeast genome [19]. Comparing the gene overexpression to the corresponding gene deletion phenotypes revealed that they rarely resembled one another, suggesting that overexpression phenotypes can often be driven by a neomorphic gain-of-function effect rather than inhibitory hyperactivation [19]. In some cases overexpression phenotypes do resemble loss-of-function phenotypes, which occurs due to a disruption of protein complex stoichiometry [19]. In particular, increased gene dosage effects can also be studied further in different genetic backgrounds sensitized for regulatory enzymes, such as kinases, histone deacetylases, or ubiquitin ligases, which often keep pathways in check through negative regulation [18, 19]. Combining the overexpression of a given gene with the deletion of another gene to compromise cellular fitness is referred to as an SDL interaction.

SGA is also applicable to dosage suppression experiments, in which overproduction of one gene rescues the mutant phenotype of another [20]. In this case, the most productive approach involves the overexpression of genes on a high-copy plasmid, rather than the constitutive *GAL1* promoter, such that the timing of gene expression is largely preserved [20]. Dosage suppression genetic interactions often connect genes of similar functionality and while they significantly overlap with protein–protein or synthetic lethal genetic interactions, they often identify novel connections between different genes and their corresponding pathways [20].

In addition, High-Content Screening (HCS) can be coupled with SGA to introduce fluorescently marked alleles into the deletion collection to perform genome-wide screens of morphological phenotypes [21]. In a proof-of-principle study, an integrated SGA-HCS pipeline was used to quantitatively assess spindle morphology defects, which identified novel genes involved in spindle function, and characterize a set of genes involved in the dynamic process of mitotic spindle disassembly [21]. In theory, virtually any pathway can be analogously examined within the context of genetic and environmental perturbations.

Other applications of SGA include SGA Mapping (SGAM) to identify random mutations that suppress the growth defect associated with either the query or the array strain mutations [22]; to combine more than two mutant alleles to study higher-order genetic interactions [4, 23]; and to integrate SGA with chemical genomic approaches to elucidate drug targets of bioactive compounds by comparing their chemical-sensitivity profiles to genetic interaction profiles [17].

In this chapter we describe in detail the procedure of SGA, tools for genetic interaction data processing and analysis, as well as the manner in which SGA can be applied to SDL and coupled with HCS to study genetic interactions. We encourage researchers from diverse fields to apply SGA to suit their particular interests.

2 Materials

2.1 Media and Stock Solutions

1. *G418* (Geneticin, Invitrogen): dissolve in water to a concentration of 200 mg/mL, filter-sterilize, and store aliquots at 4 °C (*see* **Note 1**).
2. *clonNAT* (Nourseothricin, Werner BioAgents, Jena, Germany): dissolve in water to a concentration of 100 mg/mL, filter-sterilize, and store aliquots at 4 °C.
3. *Canavanine* (L-canavanine sulfate salt; Sigma): dissolve in water to a concentration of 100 mg/mL, filter-sterilize, and store aliquots at 4 °C.

4. *Thialysine* (S-[2-aminoethyl]-L-cysteine hydrochloride; Sigma): dissolve in water to a concentration of 100 mg/mL, filter-sterilize, and store aliquots at 4 °C.
5. *Complete synthetic medium amino acid supplement powder*: 3 g adenine, 2 g uracil, 2 g inositol, 0.2 g *p*-aminobenzoic acid, 2 g alanine, 2 g arginine, 2 g asparagine, 2 g aspartic acid, 2 g cysteine, 2 g glutamic acid, 2 g glutamine, 2 g glycine, 2 g histidine, 2 g isoleucine, 10 g leucine, 2 g lysine, 2 g methionine, 2 g phenylalanine, 2 g proline, 2 g serine, 2 g threonine, 2 g tryptophan, 2 g tyrosine and 2 g valine. To prepare a drop-out (DO) mixture, the desired amino acid is excluded from the powder. To make 1 L of synthetic medium use 2 g of DO mixture (*see* **Note 2**).
6. *Sporulation amino acid supplement powder*: 2 g histidine, 10 g leucine, 2 g lysine and 2 g uracil. To make 1 L of sporulation medium use 0.1 g of sporulation amino acid supplement powder (*see* **Note 2**).
7. *Glucose* (Dextrose): Prepare 40 % w/v stock solution in water, autoclave, and store at room temperature.
8. *Galactose*: Prepare 40 % w/v stock solution in water, autoclave, and store at room temperature.
9. *YEPD* (Yeast Extract Peptone Dextrose): Add 120 mg adenine, 10 g yeast extract, 20 g peptone and 20 g bacto agar to a 2-L Erlenmeyer flask with 950 mL water. Autoclave. Add 50 mL of 40 % glucose solution and stir to mix. Pour plates once the medium has cooled to ~65 °C.
10. *YEPD+G418*: Cool YEPD medium to ~65 °C, add 1 mL of G418 stock solution (200 mg/mL) to a final concentration of 200 µg/mL, stir to mix, and pour plates.
11. *YEPD+clonNAT*: Cool YEPD medium to ~65 °C, add 1 mL of clonNAT stock solution (100 mg/mL) to a final concentration of 100 µg/mL, stir to mix, and pour plates.
12. *YEPD+G418/clonNAT*: Cool YEPD medium to ~65 °C, add 1 mL of G418 stock solution (200 mg/mL) to a final concentration of 200 µg/mL, 1 mL of clonNAT stock solution (100 mg/mL) to a final concentration of 100 µg/mL, stir to mix, and pour plates.
13. *Enriched sporulation*: Add 10 g potassium acetate, 1 g yeast extract, 0.5 g glucose, 0.1 g of sporulation amino acid supplement powder, 20 g bacto agar to 1 L of water in a 2-L Erlenmeyer flask. Autoclave. Cool the medium to ~65 °C, add 250 µL G418 stock solution (200 mg/mL) to a final concentration of 50 µg/mL. Stir to mix and pour plates (*see* **Note 4**).
14. *SD—His/Arg/Lys+canavanine/thialysine*: Add 6.7 g yeast nitrogen base without amino acids, 2 g (DO—His/Arg/Lys)

amino acid supplement powder mixture, and 200 mL water in a 250-mL Erlenmeyer flask. Separately, add 20 g bacto agar to 750 mL water in a 2-L Erlenmeyer flask. Autoclave the two flasks. Cool the medium to 65 °C, add 50 mL of 40 % glucose, 0.5 mL canavanine stock solution (100 mg/mL) to a final concentration of 50 μg/mL, 0.5 mL thialysine stock solution (100 mg/mL) to a final concentration of 50 μg/mL, stir to mix, and pour plates (*see* **Note 5**).

15. *SD—Leu/Arg/Lys+canavanine/thialysine*: Add 6.7 g yeast nitrogen base without amino acids, 2 g (DO—Leu/Arg/Lys) amino acid supplement powder mixture, and 200 mL water in a 250-mL Erlenmeyer flask. Separately, add 20 g bacto agar to 750 mL water in a 2-L Erlenmeyer flask. Autoclave the two flasks. Cool the medium to 65 °C, add 50 mL of 40 % glucose, 0.5 mL canavanine stock solution (100 mg/mL) to a final concentration of 50 μg/mL, 0.5 mL thialysine stock solution (100 mg/mL) to a final concentration of 50 μg/mL, stir to mix, and pour plates (*see* **Note 5**).
16. *SD_{MSG}—His/Arg/Lys+canavanine/thialysine/G418*: Add 1.7 g yeast nitrogen base without amino acids or ammonium sulfate, 1 g monosodium glutamic acid (MSG), 2 g complete synthetic medium amino acid supplement powder to 200 mL water in a 250-mL Erlenmeyer flask. Separately, add 20 g bacto agar to 750 mL water in a 2-L Erlenmeyer flask. Autoclave the two flasks. Cool the medium to 65 °C, add 50 mL of 40 % glucose, 0.5 mL canavanine stock solution (100 mg/mL) to a final concentration of 50 μg/mL, 0.5 mL thialysine stock solution (100 mg/mL) to a final concentration of 50 μg/mL, 1 mL of G418 stock solution (200 mg/mL) to a final concentration of 200 μg/mL, stir to mix, and pour plates (*see* **Note 6**).
17. *SD_{MSG}—His/Arg/Lys+canavanine/thialysine/G418/clonNAT*: Add 1.7 g yeast nitrogen base without amino acids or ammonium sulfate, 1 g monosodium glutamic acid (MSG), 2 g complete synthetic medium amino acid supplement powder to 200 mL water in a 250-mL Erlenmeyer flask. Separately, add 20 g bacto agar to 750 mL water in a 2-L Erlenmeyer flask. Autoclave the two flasks. Cool the medium to 65 °C, add 50 mL of 40 % glucose, 0.5 mL canavanine stock solution (100 mg/mL) to a final concentration of 50 μg/mL, 0.5 mL thialysine stock solution (100 mg/mL) to a final concentration of 50 μg/mL, 1 mL of G418 stock solution (200 mg/mL) to a final concentration of 200 μg/mL, 1 mL of clonNAT stock solution (100 mg/mL) to a final concentration of 100 μg/mL, stir to mix, and pour plates (*see* **Note 6**).
18. *SD_{MSG}—Ura+clonNAT*: Add 1.7 g yeast nitrogen base without amino acids or ammonium sulfate, 1 g monosodium glutamic acid (MSG), 2 g complete synthetic medium amino acid

supplement powder to 200 mL water in a 250-mL Erlenmeyer flask. Separately, add 20 g bacto agar to 750 mL water in a 2-L Erlenmeyer flask. Autoclave the two flasks. Cool the medium to 65 °C, add 50 mL of 40 % glucose, 1 mL of clonNAT stock solution (100 mg/mL) to a final concentration of 100 μg/mL, stir to mix, and pour plates.

19. *SD_{MSG}—Ura/His/Arg/Lys+ canavanine/thialysine/clonNAT*: Add 1.7 g yeast nitrogen base without amino acids or ammonium sulfate, 1 g monosodium glutamic acid (MSG), 2 g complete synthetic medium amino acid supplement powder to 200 mL water in a 250-mL Erlenmeyer flask. Separately, add 20 g bacto agar to 750 mL water in a 2-L Erlenmeyer flask. Autoclave the two flasks. Cool the medium to 65 °C, add 50 mL of 40 % glucose, 0.5 mL canavanine stock solution (100 mg/mL) to a final concentration of 50 μg/mL, 0.5 mL thialysine stock solution (100 mg/mL) to a final concentration of 50 μg/mL, 1 mL of clonNAT stock solution (100 mg/mL) to a final concentration of 100 μg/mL, stir to mix, and pour plates.
20. *SG_{MSG}—Ura/His/Arg/Lys+ canavanine/thialysine/clonNAT*: Add 1.7 g yeast nitrogen base without amino acids or ammonium sulfate, 1 g monosodium glutamic acid (MSG), 2 g complete synthetic medium amino acid supplement powder to 200 mL water in a 250-mL Erlenmeyer flask. Separately, add 20 g bacto agar to 750 mL water in a 2-L Erlenmeyer flask. Autoclave the two flasks. Cool the medium to 65 °C, add 50 mL of 40 % galactose, 0.5 mL canavanine stock solution (100 mg/mL) to a final concentration of 50 μg/mL, 0.5 mL thialysine stock solution (100 mg/mL) to a final concentration of 50 μg/mL, 1 mL of clonNAT stock solution (100 mg/mL) to a final concentration of 100 μg/mL, stir to mix, and pour plates.

2.2 Strains and Plasmids

For a list of strains and plasmids *see* Table 1.

2.3 Accessories

OmniTrays (Nunc) are used for all the replica pinning steps of SGA.

2.4 Manual Pin Tools: Low-Throughput SGA Analysis

Manual pin tools are available for purchase from V&P Scientific, Inc. (San Diego, CA).

1. 96 floating pin E-clip style manual replicator.
2. 384 floating pin E-clip style manual replicator.
3. Extra floating pins (FP): 1.58 mm diameter with chambered tip (*see* **Note 7**).
4. Registration accessories: Library Copier, Colony Copier.
5. Pin cleaning accessories: Plastic bleach or water reservoirs pyrex alcohol reservoir with lid pin cleaning brush (*see* **Note 8**).

Table 1
Strains and plasmids

Strain	Genotype (S288c)	Comments
Y7092	*MATα can1Δ::STE2pr-Sp_his5 lyp1Δ ura3Δ0 leu2Δ0 his3Δ1 met15Δ0*	Background strain for SGA query strain construction through PCR-mediated gene deletion.
SGA Query Strains (Y7092 background)	*MATα yfgΔ::natMX4 can1Δ::STE2pr-Sp_his5 lyp1Δ ura3Δ0 leu2Δ0 his3Δ1 met15Δ0*	Genotype of the majority of SGA query strains; *yfg* is your favorite gene.
Y8205	*MATα can1Δ::STE2pr-*Sp*_his5 lyp1Δ::STE3pr_LEU2 ura3Δ0 leu2Δ0 his3Δ1 met15Δ0*	Background strain for SGA query strain construction through the switching method.
SGA Query Strains (Y8205 background)	*MATα yfgΔ::natMX4 can1Δ::STE2pr-Sp_his5 lyp1Δ::STE3pr_LEU2 ura3Δ0 leu2Δ0 his3Δ1 met15Δ0*	Genotype of some SGA query strains; *yfg* is your favorite gene.
Y8835	*MATα can1Δ::STE2pr-Sp_his5 lyp1Δ ura3Δ0::natMX4 leu2Δ0 his3Δ1 met15Δ0*	Wild-type control strain for *natMX4*-marked query strains. Provides information about single mutant colony size when crossed to the *MAT***a** *kanMX4* Deletion Array.
hoΔ from the Deletion Array	*MAT***a** *hoΔ::kanMX4 his3Δ1 leu2Δ0 met15Δ0 ura3Δ0*	Wild-type control strain for *kanMX4*-marked array strains. Provides information about single mutant colony size when crossed to an array of *MATα natMX4* query strains.
Y9687	*MAT***a***/α can1Δ::STE2pr-Sp_his5/+; lyp1Δ::STE3pr-LEU2/+; his3Δ1/his3Δ1 leu2Δ0/leu2Δ0 ura3Δ0/ura3Δ0met15Δ0/met15Δ0 LYS2+/+*	The background strain for SGA query strain construction through PCR-mediated gene integration of a conditional allele.
p4339	*pCRII-TOPO::natMX4*	Plasmid used to amplify the *natMX4* cassette.
Yeast Deletion Collection	*MAT***a** *his3Δ1 leu2Δ0 met15Δ0 ura3Δ0*	The collection of *MAT***a** deletion strains is available for purchase as 96-well agar plates from Invitrogen, American Type Culture Collection, EUROSCARF; or 96-well agar plates and 96-well plate of frozen stock from Open Biosystems.
GST Overexpression Array	*MAT***a** *his3Δ1 leu2Δ0 met15Δ0 ura3Δ0+pEGH (pGAL1/10-6xHIS-GST-ORF-URA3)*	5,380 ORFs are on the array; 96-well plates as frozen stock can be purchased from Open Biosystems.
SGA-FLEX Array	*MATα can1Δ::STE2pr-Sp_his5, lyp1Δ::STE3pr-LEU2, his3Δ1, leu2Δ0, ura3Δ0, met15Δ0 + pGAL-ORF-URA3*	5,336 ORFs are on the array [31].
b-FLEX Array	*MATα can1Δ::STE2pr-Sp_his5, lyp1Δ::STE3pr-LEU, his3Δ1, leu2Δ0, ura3Δ0, met15Δ0; hoΔ::kanMX4+*(barcoded) *pGAL-ORF-URA3*	5,111 ORFs are on the array; every ORF is barcoded and thus can be used for pooled experiments [31].

2.5 Robotic Pinning System: High-Throughput SGA Analysis

Singer RoTor benchtop robot can be purchased from Singer Instruments, Somerset, UK (www.singerinst.co.uk). It uses plastic replicators (RePads, Singer Instruments, UK), which are sterile and disposable (*see* **Note 9**).

2.6 Robotic Pinning System: Ultra High-Throughput SGA Analysis

BioMatrix colony arrayer robot can be purchased from S&P Robotics, Toronto, ON (www.sprobotics.com). This machine uses metal pinning systems rather than plastic replicators. The pins are washed between replica pinning procedures and thus the system is fully automated.

3 Methods

3.1 Query Strain Construction

3.1.1 Nonessential Genes: PCR-Mediated Gene Deletion

1. Synthesize a pair of gene-deletion primers: upstream gene deletion primer consists of 55 bp immediately upstream (including the start codon) of the gene of interest and 22 bp of 5′ end of *natMX4* cassette (*see* MX4-F in Table 2); downstream gene deletion primer consists of 55 bp immediately downstream (including the stop codon) of the gene of interest and 22 bp of 3′ end of *natMX4* cassette (*see* MX4-R in Table 2).
2. PCR amplify *natMX4* cassette with your gene-deletion primers. Set up 100 μL PCR (74.2 μL H_2O, 10 μL of 10× buffer, 2 μL of 10 mM dNTPs, 4 μL of 50 μM forward primer, 4 μL of 50 μM reverse primer, 0.1 μg p4339 DNA template in 0.5 μL H_2O, 5 μL DMSO, 0.3 μL 5 U/μL Taq DNA Polymerase) (*see* **Note 10**). Initiate the amplification of the cassette with a 5 min denaturation at 95 °C, followed by 30 cycles of: 95 °C 30 s, 55 °C 30 s, 68 °C 2 min; terminate the reaction with a 10 min extension at 68 °C and hold at 4 °C, if necessary. Store PCR products at −20 °C.
3. Transform 20 μL of PCR product into the SGA background strain, Y7092 using standard LiAc procedure [9] and select for transformants on YEPD + clonNAT agar plates, 3–4 days, 30 °C (*see* **Note 11**).
4. PCR confirm the integration using primers that anneal inside the marker and either 200 bp upstream or 200 bp downstream of the deleted gene of interest. Use mating type test to confirm

Table 2
Primer sequences for *natMX4* amplification

Primer	Sequence (5′–3′)	Comments
MX4-F	ACATGGAGGCCCAGAATACCCT	Forward amplification primer of MX4 series cassettes [32]
MX4-R	CAGTATAGCGACCAGCATTCAC	Reverse amplification primer of MX4 series cassettes [32]

MATα mating type. Perform fluorescence-activated cell sorting (FACS) analysis to ensure a haploid query strain.

5. Freeze down in 20 % glycerol and use the query strain for SGA screens.

3.1.2 Nonessential Genes: Marker Switching Method

1. Mate a deletion strain of interest from the *MAT***a** deletion collection (*yfgΔ*::*kanMX4*) with Y8205 and isolate diploid zygotes by micromanipulation (*see* **Note 25**).
2. To switch *kanMX4* to *natMX4*, PCR amplify *natMX4* from p4339 using primers from Table 2. Transform 20 μL of PCR product into the resulting diploid and select for transformants on YEPD + clonNAT agar plates, 3–4 days, 30 °C.
3. Transfer a NAT-resistant colony to the enriched sporulation medium and incubate at 22 °C for 7 days.
4. Resuspend a small amount of spores (tip of the yellow pipette tip) in 1 mL of sterile water and plate 100 μL on SD—Leu/Arg/Lys + canavanine/thialysine. Incubate at 30 °C for 2–3 days to select for *MATα* meiotic progeny.
5. To select for NAT-resistant *MATα* meiotic progeny, replica plate the resulting plate from **step 4** on YEPD + clonNAT and incubate for 1 day, 30 °C.
6. Freeze down in 20 % glycerol and use the query strain for SGA screens.

3.1.3 Essential Genes: Two-Step PCR-Mediated Integration of Conditional Allele

1. Step 1: Synthesize a pair of gene-deletion primers to amplify the conditional allele of interest and 200 bp downstream of its stop codon to create a fragment that overlaps with the 5′ region of the *natMX4* cassette. Separately, synthesize another pair of primers that amplify *natMX4* cassette with 3′ containing the sequence immediately downstream of the gene of interest.
2. Step 2: Combine PCR products and co-transform into the SGA background strain, Y9687, as previously described.
3. Select transformants on YEPD + clonNAT and incubate at the permissive temperature ~22 °C, 3–5 days.
4. Verify the integration of the conditional allele by replica plating on YEPD + clonNAT and incubating at the restrictive temperature (*see* **Note 12**) for 1–2 days.
5. Freeze down in 20 % glycerol and use the query strain for SGA screens.

3.2 Sterilization Procedure for Pin Tools

3.2.1 Manual Pin Tools

1. Set up five sterile reservoirs as follows: (1) 30 mL sterile water, (2) 50 mL sterile water, (3) 70 mL sterile water, (4) 40 mL 10 % v/v bleach, and (5) 90 mL 95 % ethanol (*see* **Note 13**).
2. Soak the replicator pins for 1 min in 30 mL sterile water reservoir to remove the cells from the pins.

3. Immerse the replicator pins in 10 % bleach for ~20 s.
4. Transfer the replicator to the 50 mL sterile water reservoir and then to the 70 mL sterile water reservoir to rinse the bleach from the pins.
5. Immerse the replicator pins in 95 % ethanol for 30 s.
6. Shake off the ethanol and flame the replicator pins.
7. Allow the replicator pins to cool before use.

3.2.2 Robotic Pin Tools: BioMatrix Colony Arrayer

Prior to the use of the robot, clean and sterilize the replicator pins:

1. Fill the sonicator bath with 550 mL sterile distilled water.
2. Clean the replicator pins in the sonicator bath for 5 min.
3. Replace the water with 550 mL 70 % ethanol.
4. Sterilize the replicator pins in the sonicator bath for 20 s. Repeat.
5. Dry the replicator over a fan for 22 s.

At the end of each replica pinning step, clean and sterilize the replicator pins:

1. Program the robot to do the following:
2. Automatically fill the waste bath with sterile distilled water from a bottle supply source. Manually, fill the brush station with 320 mL sterile distilled water. Fill the sonicator with 550 mL 70 % ethanol.
3. Soak the replicator pins in the water bath for 10 s. Repeat for a total of four times to remove the residual cells from the pins.
4. Transfer the replicator to the brush station. Clean the pins for a total of three cycles.
5. Sterilize the replicator in the 70 % ethanol sonicator bath twice for 20 s.
6. Dry the replicator over a fan for 22 s.

3.3 Deletion Mutant Array Construction: 1536-Format

1. Slowly peel off the aluminum sealing tape from the 96-well plates with frozen glycerol stock of deletion mutants. Do not cross-contaminate the wells.
2. Thaw the plates.
3. Using Singer RoTor benchtop robot mix the stock by stirring with a 96 pin replicator and replicate on YEPD + G418 agar plates.
4. Reseal the glycerol stock 96-well microtiter plates with a fresh aluminum sealing tape and immediately return to −80 °C.
5. Let cells grow at room temperature for 2 days.

6. Using the BioMatrix Colony Arrayer condense four plates of 96-format to one plate of 384-format. Incubate at room temperature for 2 days (*see* **Note 14**).
7. Using the BioMatrix Colony Arrayer replicate each 384-format deletion mutant array (DMA) in quadruplicate onto a single plate to generate 1536-format DMA. Incubate at room temperature for 1 day.
8. Duplicate the 1536-format DMA, to generate a working copy and a master copy (*see* **Note 15**).

3.4 SGA Procedure

A schematic of the SGA procedure is illustrated in Fig. 1.

All the steps can be performed using manual pin tools, Singer RoTor benchtop robot or the BioMatrix Colony Arrayer and the accompanying software.

1. Prepare query strain lawns.
   (a) Streak out the query strains of interest on YEPD + clonNAT and incubate at 30 °C, 2–3 days.
   (b) Inoculate 5 mL YEPD with a single colony and incubate at 30 °C, 2 days, at 250 rpm shaking incubator.
   (c) Spread 800 μL of the query strain saturated culture on a YEPD OmniTray. Prepare four lawns for one genome-wide screen with 14 DMA plates. Allow the lawns to dry, then incubate them at 30 °C, 2 days (*see* **Note 3**).
2. Prepare a fresh nonessential gene deletion array. *See* Subheading 3.3 **step 8**.
3. To mate the query strain with the DMA, pin the query strain from the lawn to fresh YEPD plates and in a second step pin the DMA on top of the query. Incubate at room temperature, 1 day.
4. To select for diploids, pin the resulting *MAT***a**/α diploid zygotes to YEPD + G418/clonNAT. Incubate at 30 °C, 2 days.
5. To sporulate the diploids, pin the resulting KAN^R/NAT^R *MAT***a**/α diploids onto the enriched sporulation agar plates. Incubate at 22 °C, 7 days.

Fig. 1 (continued) Sporulated cells are, then, transferred to a synthetic medium lacking histidine, but containing canavanine and thialysine to allow for the selective germination of *MAT***a** haploid meiotic progeny. The *MAT***a** haploids are then transferred to medium containing geneticin to select for array mutants and, then, to the medium containing geneticin and nourseothricin to select for double mutants. The illustration was adapted from [24]. (**b**) A sample portion of a plate image following the final SGA selection step. The deletion of either a query gene (*query*Δ) or an array gene (*xxx*Δ) does not result in any observable fitness defects, but the deletion in both genes is lethal. Shown is an example of an extreme negative genetic interaction termed synthetic lethality. The mutant is represented four times on the array and is highlighted with a *white box* (color figure online)

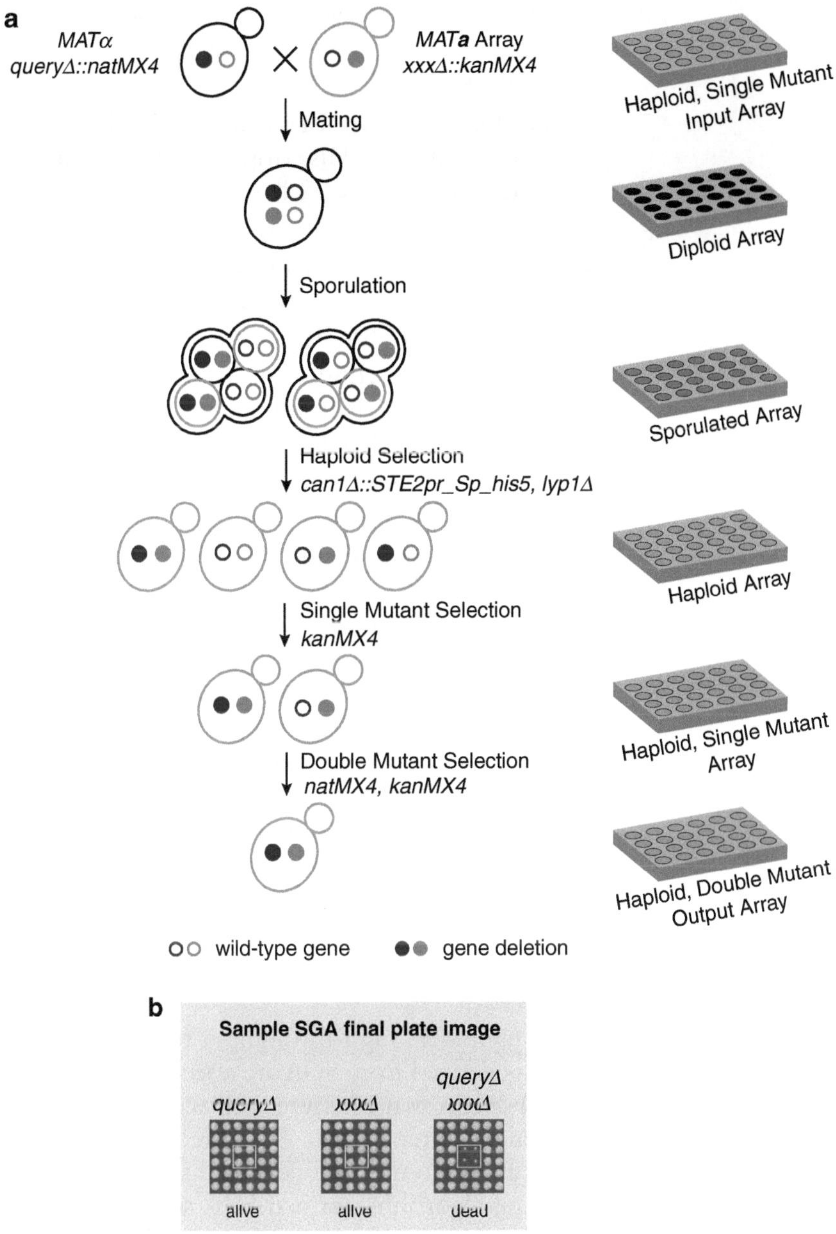

Fig. 1 Synthetic Genetic Array (SGA) methodology. (**a**) A *MATα* query mutant strain carries a deletion in the gene of interest linked to a dominant selectable marker (*natMX4*), which confers resistance to the antibiotic nourseothricin (*filled blue circle*). The query strain also carries deletions of the arginine and lysine permease genes, *CAN1* and *LYP1*, respectively. *CAN1* is replaced by the SGA reporter, *STE2pr_Sp_his5*, in which the *STE2 MAT***a**-specific promoter (*STE2pr*) controls the expression of the *Schizosaccharomyces pombe his5* gene. Deletions of *CAN1* and *LYP1* loci confer sensitivity to canavanine and thialysine, respectively, and are used to select against diploids following the sporulation step. In a typical SGA screen the query strain is crossed to an ordered array of *MAT***a** nonessential gene deletion strains ("array" mutants) that are marked by a dominant selectable marker, *kanMX4*, which confers geneticin resistance (*filled gray circle*). The resulting heterozygous diploids are transferred by replica-pinning to a medium containing reduced carbon and nitrogen sources to induce sporulation.

6. To select for *MAT***a** meiotic haploid progeny, pin the spores onto SD—His/Arg/Lys + canavanine/thialysine. Incubate at 30 °C, 2 days.
7. To select for KAN^R *MAT***a** meiotic haploid progeny, pin the spores onto SD_{MSG}—His/Arg/Lys + canavanine/thialysine/G418. Incubate at 30 °C, 2 days.
8. To select for KAN^R/NAT^R *MAT***a** meiotic haploid progeny, pin the single mutant haploids onto SD_{MSG}—His/Arg/Lys + canavanine/thialysine/G418/clonNAT. Incubate at 30 °C, 2 days (*see* **Note 16**).
9. The resulting array consists of a collection of yeast double mutant strains in which each nonessential KAN^R-marked deletion mutant is also deleted for the query gene of interest, which is marked with NAT^R (*see* **Note 24**).
10. Image the double mutant array using a high-resolution digital imaging system, such as the one developed by S&P Robotics, Inc. (Toronto, ON).
11. Identify genetic interactions qualitatively by comparing double mutant colony size to a single mutant control screen (*see* **Note 19**). Alternatively, identify genetic interactions quantitatively. Process the images using custom-developed image-processing software that distinguishes colonies from the background and measures their area in pixels [4]. Correct the raw colony sizes for systematic effects and quantify genetic interactions using the single mutant control screens as reference [16]. *See* next section for details.
12. Confirm genetic interactions using random spore or tetrad analysis [24] (*see* **Note 26**).

3.5 Quantitative Scoring of Genetic Interactions: Colony Size—Proxy for Fitness

Like most array-based, high-throughput technologies, colony size measurements derived from SGA are affected by several systematic experimental effects that interfere with accurate estimations of fitness and genetic interactions. To correct for these effects and enhance our ability to measure genetic interactions we have developed a scoring algorithm, termed SGA score [16]. It provides a several fold increase in genetic interactions data quality and functional prediction capacity compared to other available methodologies. Briefly, the SGA score implements the following systematic experimental effects normalizations:

1. Plate effect normalization: corrects for plate-to-plate variance in colony growth, which is caused by differences in individual plate incubation times (*see* **Note 21**).
2. Row/column effect normalization: corrects for within-plate variance in colony growth; due to their arrangement on the plate, colonies have a different access to nutrients (*see* **Note 22**).

3. Spatial effect normalization: corrects for within-plate gradients in colony growth; uneven spreading or drying of the query lawn, media thickness, or uneven distribution of heat in the incubator that cause one side of the plate to have larger colonies than another.
4. Competition effect normalization: corrects for local competition for nutrients between neighboring colonies that can cause differences in colony sizes. For instance, if a colony size is immediately next to a sick colony, it will grow slightly larger not because it is more fit, but because it has a larger access to nutrients.
5. Batch effect normalization: corrects for a common nonbiological signature of colony sizes in screens performed on the same day, by the same person, using the same instruments; such colony sizes can be mistaken for true genetic interactions and introduce a major source of error, if not corrected (*see* **Note 20** for corrections of systematic effects for low-throughput experiments).

These normalized fitness estimates for single and double mutants are used to calculate genetic interaction scores using the multiplicative model, whereby the double mutant fitness is expected to be equal to the product of the corresponding single mutant fitnesses [25]. A genetic interaction score is, then, computed as the difference between the observed (f_{ab}) and the expected double mutant fitnesses ($f_a \times f_b$, where f_a and f_b represent the *gene A* and *gene B* single mutant fitnesses, respectively): SGA score $= f_{ab} - f_a \times f_b$ [16]. If the observed double mutant colony size is smaller than expected ($f_{ab} < f_a \times f_b$), then it is considered a negative genetic interaction; if it is larger than expected ($f_{ab} > f_a \times f_b$), then it is a positive genetic interaction. The strength of the interaction is reflected in the magnitude of the SGA score, while our confidence in a given score is reflected in the associated *p*-value, which incorporates the variance of each double mutant across multiple replicates and the variance of its constituent query and array mutants across all control screens available to date. We store our genetic interaction data in a publically accessible Web database, DRYGIN [26].

We also developed a Web-based analysis system called "SGAtools" to aid researchers in scoring their own SGA-based screens [27]. It will enable one to quantify colonies on agar plates, normalize systematic effects, and calculate fitness scores relative to a control experiment. Additional analysis tools are available, such as viewing scored colonies on plates, as well as the shape of the distribution of the genetic interaction scores. The Web site is, also, linked to external sources to aid with Gene Ontology (GO) enrichment analysis. SGAtools can be accessed at the following Web site: http://sgatools.ccbr.utoronto.ca/.

3.6 Analysis Tools for Genetic Interactions Data

Genetic interaction studies have offered much insight into the properties of genetic networks [17]. For instance, genes with similar biological roles tend to share many genetic interactions [4]. Therefore, clustering genes according to the similarity of their genetic interaction profiles (all the interactions of a particular gene) is a powerful tool to predict gene function of unknown genes by "guilt-by-association." Cluster 3.0 is an example of an open source clustering software (http://bonsai.hgc.jp/~mdehoon/software/cluster/software.htm). Once the data is clustered, it can be visualized with an open source application, such as Java TreeView (http://jtreeview.sourceforge.net/). Another approach to find genes with similar genetic interaction profiles is to compute pair-wise Pearson correlation coefficients between all genes in the genome and use a network visualization tool, such as Cytoscape, to organize the network [17, 28]. Applying force-directed network layout will pull highly correlated genes towards each other and away from less correlated genes, shedding light on the genetic architecture of the cell and genes with unknown biological roles [17].

See Fig. 2 for a simplified illustration of the analysis tools of genetic interactions data.

3.7 Applications of SGA: SDL Procedure

Follow the steps of SGA as described above, except where noted (Fig. 3).

1. Query strain lawns *see*Subheading 3.4 **step 1**.
2. Prepare a fresh GST overexpression array in the 1536-format array on SD—Ura agar plates (*see* **Note 17**).
3. Mating as in Subheading 3.4 **step 3**.
4. Diploid selection as in Subheading 3.4 **step 4**, except use SD_{MSG}—Ura + clonNAT.
5. Sporulation as in Subheading 3.4 **step 5**.
6. To select *MAT***a** meiotic haploid progeny with NAT^R and the overexpression plasmid, pin the spores onto SD_{MSG}—Ura/His/Arg/Lys + canavanine/thialysine/clonNAT. Incubate at 30 °C, 2 days.

Fig. 2 (continued) same essential protein complex. Positive genetic interactions often connect members of the same nonessential pathway/complex but also occur between different protein complexes. The position of an unknown gene in the network can be informative of its biological role through "guilt-by-association," because genes belonging to the same pathway or complex tend to share a similar pattern of negative and positive genetic interactions. (**b**) Genetic interactions can also be visualized as a clustered matrix, in which genes sharing similar genetic interaction patterns are grouped together. Similar to the network analysis, groups of genes exhibiting similar genetic interaction profiles tend to be functionally related and, thus, genetic interaction profiles are an effective way to define pathway and complex membership. The figure was adapted from [5]

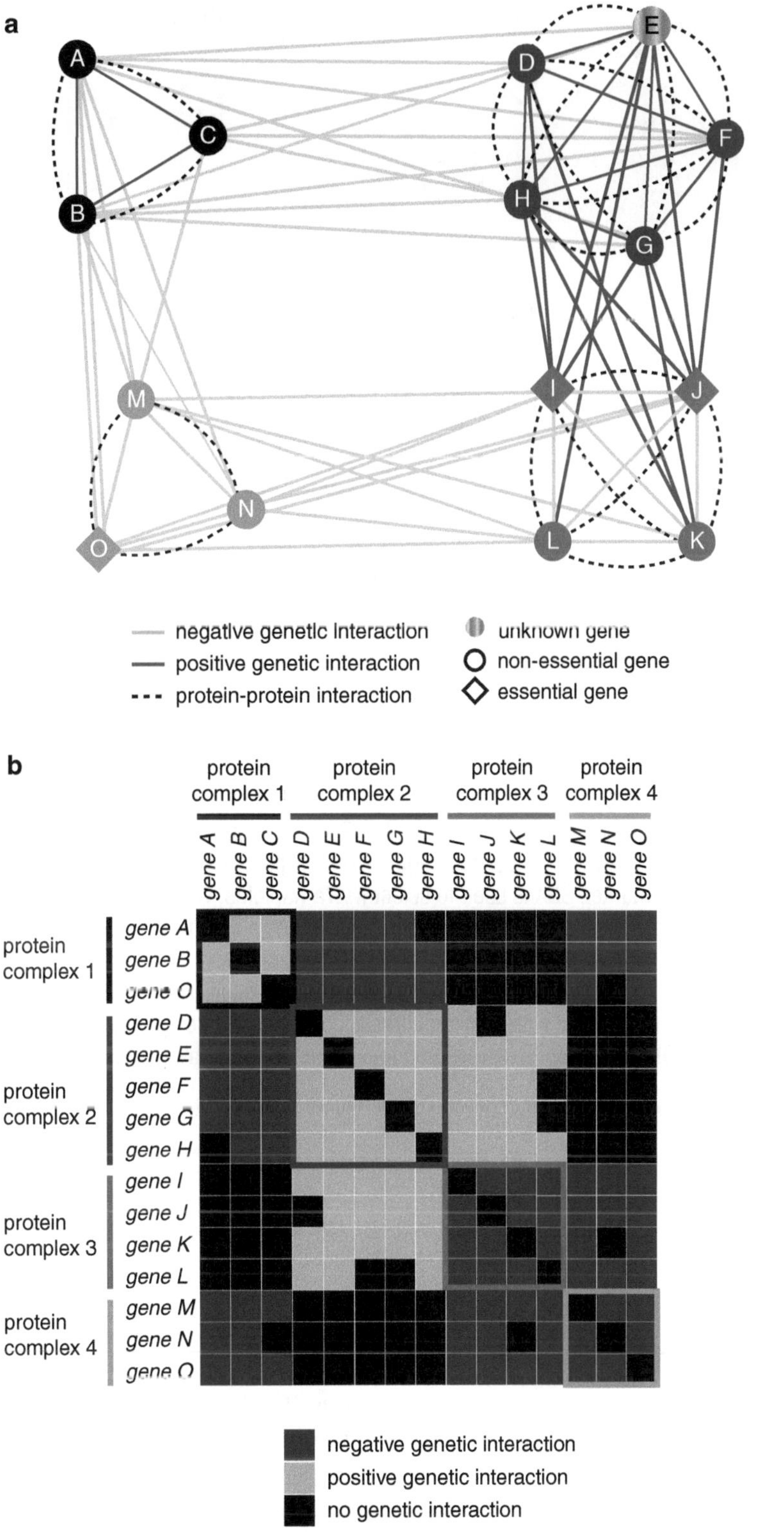

Fig. 2 Analysis of high-throughput genetic interactions data. (**a**) Genetic interactions can be visualized using a network layout. Negative genetic interactions tend to connect different nonessential pathways/complexes, and members of the

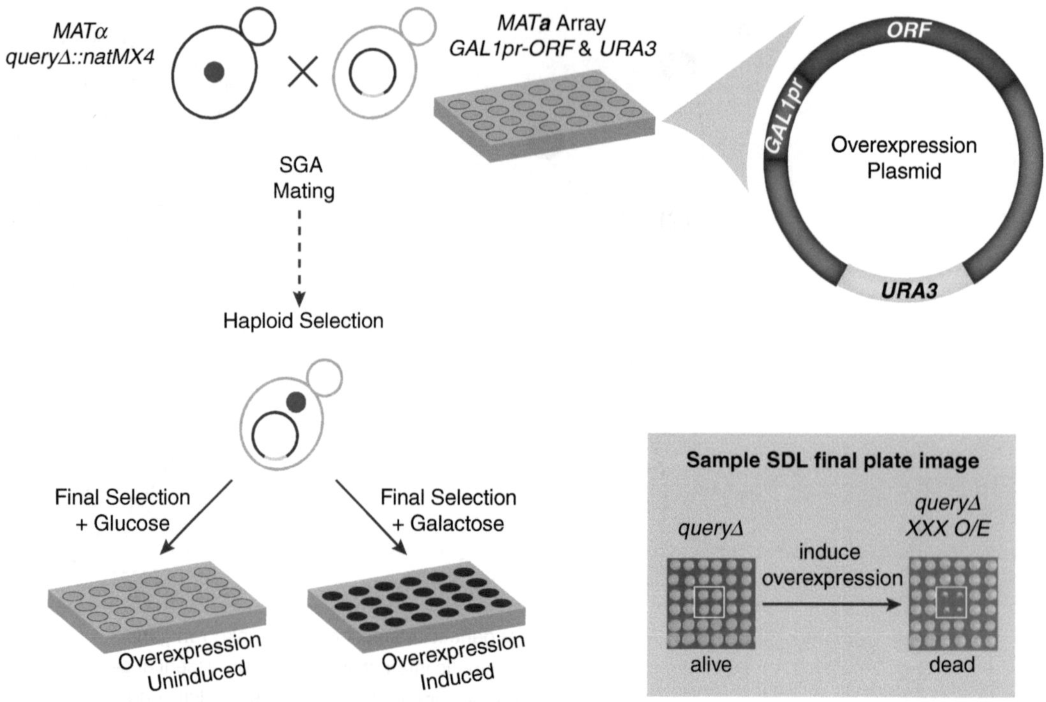

Fig. 3 Synthetic dosage lethality (SDL) methodology. (**a**) The SDL procedure is identical to the SGA methodology outlined in Fig. 1a with one exception. Instead of an array of *kanMX4*-marked deletion mutants, the query mutant strain is crossed to an ordered array of *MAT***a** strains that carry an overexpression plasmid (*burgundy circle*). A diagram of the plasmid from the GST overexpression collection is shown at the *top right*. Each plasmid contains an ORF under the regulation of a galactose inducible promoter, *GAL1pr*, and carries a *URA3* selectable marker. SGA is used to select *MAT***a** haploids, which contain the query gene deletion and the plasmid, and are then plated on glucose and galactose containing media. Galactose induces plasmid expression and glucose serves as an uninduced control. A sample portion of a plate image at the final step of SDL is shown at the *bottom right*. The deletion of a query gene (*query*Δ) does not result in any observable fitness defect, but combined with an overexpression (O/E) in another gene (*XXX*) results in lethality. The mutant showing an SDL interaction is represented four times on the array and is highlighted with a *white box*

7. Repeat for a total of two rounds of haploid selection.
8. To induce overexpression, pin *MAT*a haploids onto galactose medium, SG_{MSG}—Ura/His/Arg/Lys + canavanine/thialysine/clonNAT. For an uninduced control pin *MAT*a haploids onto glucose containing medium SD_{MSG}—Ura/His/Arg/Lys + canavanine/thialysine/clonNAT. Incubate the former at 30 °C, 2 days and the latter at 30 °C, 3 days.
9. Image the uninduced and induced final plates using a high-resolution digital camera.
10. Process the images using image-processing software that distinguishes colonies from the background and measures their area in pixels [4].

11. Use colony size measurements to derive SDL genetic interaction scores quantitatively [18] or qualitatively (*see* **Note 18**).
12. Confirm SDL genetic interactions using serial spot dilutions (*see* **Note 23**).

3.8 SGA-High Content Screening: Profiling of Morphological Phenotypes

Follow the steps of SGA as described above except where noted (Fig. 4).

1. Mate the query strain (marked with a cell biological reporter) with DMA and follow all the SGA **steps** from **1** to 7.
2. To select for KAN^R/NAT^R *MAT***a** meiotic haploid progeny, transfer the KAN^R *MAT***a** meiotic haploid progeny from solid agar plates to deep 96-well plates with 200 μL liquid SD_{MSG}—His/Arg/Lys + canavanine/thialysine/G418/clonNAT. Use 3 mm glass beads in each well for effective mixing of cells and seal the plates with breathable adhesive tape (Corning). Remember to include the auxotrophic selection of the fluorescent reporter throughout the procedure.
3. Incubate liquid cultures overnight at 30 °C.
4. Depending on the optical density of the cell cultures, with the aid of a liquid handling robot (Biomek FX) dispense the appropriate volume of the samples onto 96-well filter plates (Whatmann).
5. To wash the cells, vacuum the filter plates (NucleoVac96 vacuum manifold, Macherey-Nagel) and wash the cells in sterile distilled water using a plate washer (Multidrop384 Liquid Dispenser System, Thermo Electron Corporation). Vacuum the plates again such that only the washed cells remain in the filter plates.
6. Resuspend the washed cells in 600 μL of low-fluorescent medium [29] with the appropriate dug selection.
7. Using the liquid handler, transfer cells from the filter plates to 96-well optical plates (Matrical) and let the cells settle down. Seal the plates with aluminum foil to prevent the samples from drying.
8. Place the plates in an automated incubator (Cytomat, Thermo Fisher Scientific, Inc) and store the plates at 4 °C before imaging to prevent overgrowth. Prior to imaging program the platform to incubate the plates at 30 °C for 30–45 min.
9. To image the cells, use a robotic arm (CRS Catalyst Express, Thermo Electron Corporation) to transfer the plates to the wide-field HCS imager (ImageXpress5000A, MDS Analytical, Inc.). Capture 4–6 images per plate using a 60× air-objective to obtain ~200 cells of each mutant.
10. Store the images and analyze them with an image-analysis software, such as MetaXpress v1.6 (MDS Analytical, Inc.) or CellProfiler [30].
11. Perform statistical analysis and data mining [21].

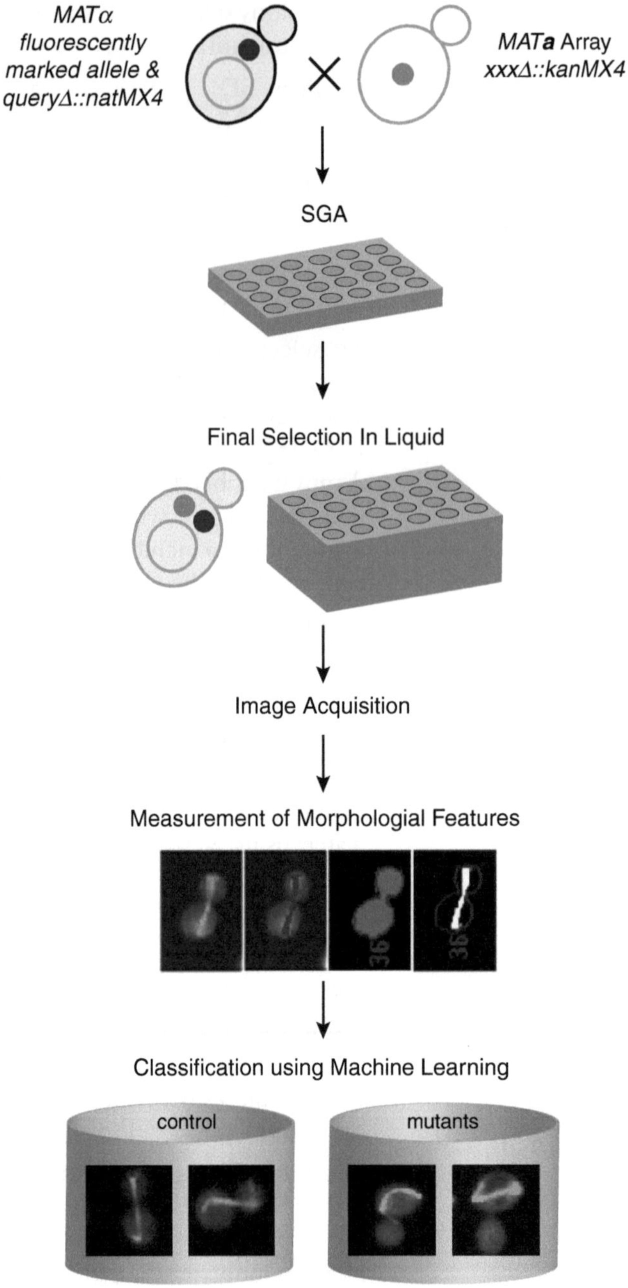

Fig. 4 SGA and high-content screening (HCS) methodology. (**a**) A *MAT*α query strain carries (1) a plasmid with a gene of interest fused to a fluorescent reporter (*green circle*; note that these markers can also be integrated into the genome) and (2) a query gene deletion linked to a dominant selectable marker (*natMX4*) (*filled blue circle*) along with SGA reporters. The query strain is crossed to an ordered array of 4,294 *MAT***a** nonessential gene deletion strains (filled gray circle) and SGA is used to select *MAT***a** haploids, which contain the plasmid and the array gene deletions. The final selection step is performed in a liquid medium to select for *MAT***a** haploids, which contain the fluorescent reporter, the query gene

4 Notes

1. Geneticin is very stable and is able to withstand autoclaving (Yeast Deletion Webpages). We observed no loss of its activity after it has been added to agar plates and stored for at least 2 months at 4 °C.
2. To make a stock of amino acid supplement powder mixture, combine all the ingredients in a container and invert it end-to-end for 15 min for a thorough mixing. Store the contents at room temperature in a dark bottle.
3. For every step of SGA OmniTrays are filled with 40 mL of medium per plate. For making a query lawn, use a pipette to spread the lawn on agar plates that have been dried for 3 days after pouring. Dry agar plates will ensure that your lawns set evenly. For all other pinning steps dry the agar plates for 2 days after they were poured. Dry plates will ensure that the surface of the agar is not too wet and colonies will not mix with one another and create smears.
4. Enriched sporulation medium ensures that cells sporulate, but also grow slowly so that there are enough cells for the next replica pinning step onto the haploid medium. G418 (50 μg/mL) minimizes the risk of contamination, but is insufficient for G418 resistance selection.
5. This medium does not contain antibiotics, G418 and clonNAT and thus yeast nitrogen base (which contains ammonium sulfate) can be used as a nitrogen source.
6. Ammonium sulfate cannot be used as the nitrogen source in this medium, because it interferes with the activity of antibiotics, G418 or clonNAT. Thus, MSG is used instead of ammonium sulfate as an alternative nitrogen source.
7. FP6 pins can be used as an alternative, but they transfer more cells than FP pins and may not be suitable for the construction of high-density arrays with 1,536 colonies/plate.
8. Empty tip boxes can be used instead of the reservoirs for bleach and water.
9. RePads can be reused, if desired. They should be washed in bleach and sterilized by UV exposure.
10. The addition of 5 % dimethyl sulfoxide (DMSO) to the PCR enhances the product yield of the *natMX4* cassette.

Fig. 4 (continued) deletion and the array gene deletions and imaged using the HCS imaging system. Morphological features are measured and used for classification by a machine-learning algorithm. For example, wild-type cells have normal spindle morphology, but some mutants with or without fitness defects exhibit a defect in spindle morphology [21]. The spindle is labeled with Tub1p fused to the green fluorescent protein (GFP) (color figure online)

11. The background strain, Y7092, used to make query strains has important features: *MAT***a**-specific reporter, *STE2pr_*Sp*_his5*, in which *MAT***a**-specific promoter drives the expression of the *Schizosaccharomyces pombe his5* gene, which complements *S. cerevisiae his3Δ1*, but has a sufficiently dissimilar sequence as to not cause gene conversion. This *MAT***a**-specific reporter replaces the *CAN1* locus. *LYP1* is also deleted in this strain. The deletions of *CAN1* and *LYP1* loci are used to select against diploid cells that have been carried over to haploid selection plates from the sporulation plates [24]. Diploid cells are heterozygous for *CAN1* and *LYP1* transporters and are thus sensitive to canavanine and thialysine (toxic analogs of arginine and lysine, respectively), whereas the desired haploids cells are *can1Δ* and *lyp1Δ* and therefore survive this selection step.
12. This type of a query strain will have a conditional allele inserted in place of the wild-type allele followed by 200 bp of the sequence downstream of the stop codon followed by the antibiotic resistance marker *natMX4*.
13. The pins are submerged in increasingly larger volumes of water for more effective cleaning. Initially, only the tips of the pins are submerged but by the end of the washing procedure half of the total pin length is in the washing solution.
14. The 384 colonies/plate density format DMA has a border of *his3Δ1::kanMX4* (*see* **Note 22**). It also excludes 432 slow growing strains, because they have a slower growth rate than other strains and should be examined in an unbiased manner on an array exclusively consisting of slow growing mutants. In total the array has 4,294 unique strains on 14 plates. After assembling the array perform some quality control steps. For example, to ensure that known auxotrophs are in their correct positions replica-pin the array on minimal media lacking the corresponding amino acids.
15. We recommend passaging both 384-format and 1536-format DMA master copies every month and assemble a fresh 1536-format DMA master copy every 2 months. DMA can be reused for up to four times for mating.
16. For temperature-sensitive mutant query or array strains, conduct all the steps at 22 °C and add an extra day of incubation to each step. In the last step, pin the single mutant haploids on two final selection plates: incubate one at 26 °C and the other at 30 °C or other appropriate semi-permissive temperature.
17. The GST overexpression array has been traditionally used for SDL screens [19]. It contains 5,380 of BY4741 strains each with a unique high-copy plasmid, in which each ORF is tagged at the N-terminus with a 30 kDa glutathione *S*-transferase (GST) protein tag. In theory, any plasmid collection that

harbors compatible markers can be introduced into the yeast gene deletion collection (or other available collections) using SGA methodology. SDL can also be performed with the SGA-FLEX overexpression array [31]. It is a collection of yeast strains with fully sequence-verified untagged overexpression plasmids and integrated SGA markers. Thus, this array is not compatible with SGA query strains, but it is compatible with other strains that do not carry SGA markers, such as those from the yeast deletion collection. Another version of this array, the b-FLEX overexpression array, can also be used for SDL [31]. In this array each strain is barcoded enabling one to conduct pooled fitness assays for competitive strain analysis in liquid cultures. The b-FLEX overexpression collection can be applied to study overexpression lethality under various environmental conditions to reveal condition-specific SDL interactions.

18. SDL interactions can be qualitatively scored. We recommend using three replicate screens, such that the final putative interaction list would consist of colonies that appear smaller in the induced versus the uninduced media in two of the three replicates.

19. SGA control screens are used to obtain single mutant fitnesses and should be conducted together with the screens of interest. Y8835, with *natMX4* at a neutral locus (*URA3*), is crossed to the DMA to generate a single mutant array, in which every strain carries a different gene deletion marked with *kanMX4* and *ura3Δ* marked with *natMX4*. One can then find single mutant colonies of interest and compare them to the double mutant to see if the latter is more or less sick than expected by the multiplicative model. If the double mutant colony size is smaller than expected from the product of the corresponding single mutant colony sizes, then the two mutant alleles exhibit a negative genetic interaction: synthetic sick or lethal phenotype. However, if the colony size of the double mutant is larger than expected, then there is a positive genetic interaction between the mutant alleles [8].

20. The batch-effect is very difficult to distinguish in small scale SGA studies in which only a few screens are conducted at a time. The following two experimental strategies will ensure that their genetic interaction results are of the highest possible quality [16]. First, each double mutant screen should be paired with a control SGA screen, in which the query strain carries the *natMX4* selectable marker at a neutral genomic locus (e.g., *URA3*). Any genetic interaction displayed by the control screen is likely to be a false positive caused by systematic experimental effects and thus should be filtered out from all the double mutant screens performed at the same time. Second, each SGA screen should be repeated multiple times (three or more)

varying any potential source of systematic biases as much as possible, for instance, the day of the experiment, media batches, instruments, and plate positions of individual array mutants. Averaging the results obtained from such replicates will minimize the contribution of systematic effects and provide the best estimate of true genetic interactions.

21. It takes slightly more than an hour to replica-pin the entire nonessential gene deletion collection during each SGA step. Thus, depending on the number of screens being done there is going to be a substantial difference in growth time between the first plate that was pinned and the last one.

22. Colonies grow larger along the edges of the plates than in the middle, because of a greater nutrient availability. However, their large size does not reflect an increased fitness or a positive genetic interaction and so this border effect should be normalized, which can be very difficult to do. To avoid false positives a control strain is grown along the border of the DMA and these colonies are excluded from the analysis. The border strain is *his3Δ1::kanMX4* from the yeast deletion collection; it survives because the SGA reporter, *STE2pr*_Sp_*his5*, in the query strain complements its histidine auxotrophy (*see* **Note 11**). The top two and the bottom two rows and columns in the 1536-format DMA exclusively consist of the border strain.

23. Transform overexpression plasmids into control and query yeast strains. Grow 200 μL of overnight cultures in the appropriate liquid synthetic medium (SD-Ura) for 2 days at 30 °C in sterile 96-well plates. Make 15-fold dilutions of saturated cultures and spot 5 μL of serially diluted culture onto noninducing (SD_{MSG}—Ura + clonNAT) and inducing (SG_{MSG}—U + clonNAT) synthetic solid media; incubate them at 30 °C for 2 and 3 days, respectively.

24. *STE2* and *STE3* promoters were the most reliable in SGA experiments and result in the lowest level of inappropriate expression of Sp_*his5* in *MAT*α and *MAT***a**/α cells compared to other *MAT***a**-specific reporters [24]. Also, it is very unlikely for three recombination events to occur simultaneously near the *MAT*, *can1Δ*, and *lyp1Δ* loci to generate *MAT***a**/**a** *can1Δ/can1Δ* and *lyp1Δ/lyp1Δ* and thus decreasing the chance of *MAT***a**/**a** diploids contributing to the false negative SGA scores [24]. However, depending on the assay that an investigator prefers to use, the mutants generated through SGA may have to be subjected to colony purification. A simple and effective approach is to streak them out for single colonies and confirm their ploidy by FACS.

25. A *MAT*α-specific reporter (*STE3pr-LEU2*) was integrated at the *LYP1* locus in Y8205 to provide a convenient method for selecting *MAT*α meiotic progeny.

26. For random spore analysis, a small amount of spores (approximately the size of a pinprick) is resuspended in 1 mL of sterile water and mixed well. Then, 20 μL of the suspension is plated on SD—His/Arg/Lys + canavanine/thialysine; 40 μL on SD_{MSG}—His/Arg/Lys + canavanine/thialysine/G418; 40 μL on SD_{MSG}—His/Arg/Lys + canavanine/thialysine/clonNAT; and 80 μL on SD_{MSG}—His/Arg/Lys + canavanine/thialysine/G418/clonNAT [24]. Incubated at 30 °C for 2 days and scored by comparing the single to the double drug selection. Note that the expected number of meiotic progeny on each medium should be equal. Petri dishes which are 6 cm in diameter filled with ~ 8 mL of media are suitable for this assay. For tetrad analysis, we recommend using SD_{MSG} complete medium for tetrad dissections because it most closely resembles the final double mutant selection conditions of SGA screens and thus is more sensitive than rich YEPD medium in detecting subtle growth defects [24].

References

1. Waddington CH (1957) The strategy of the gene. Allen and Unwin, London
2. Hartman JL IV, Garvik B, Hartwell L (2001) Principles for the buffering of genetic variation. Science 291:1001–1004
3. Zuk O, Hechter E, Sunyaev SR et al (2012) The mystery of missing heritability: genetic interactions create phantom heritability. Proc Natl Acad Sci U S A 109:1193–1198
4. Tong AH, Lesage G, Bader GD et al (2004) Global mapping of the yeast genetic interaction network. Science 303:808–813
5. Costanzo M, Baryshnikova A, Myers CL et al (2011) Charting the genetic interaction map of a cell. Curr Opin Biotechnol 22:66–74
6. Tong AH, Evangelista M, Parsons AB et al (2001) Systematic genetic analysis with ordered arrays of yeast deletion mutants. Science 294: 2364–2368
7. Bateson WRSE, Punnett RC, Hurst CC (1905) Reports to the Evolution Committee of the Royal Society, report II. Harrison and Sons, London
8. Dixon SJ, Costanzo M, Baryshnikova A et al (2009) Systematic mapping of genetic interaction networks. Annu Rev Genet 43: 601–625
9. Winzeler EA, Shoemaker DD, Astromoff A et al (1999) Functional characterization of the *S. cerevisiae* genome by gene deletion and parallel analysis. Science 285:901–906
10. Giaever G, Chu AM, Ni L et al (2002) Functional profiling of the *Saccharomyces cerevisiae* genome. Nature 418:387–391
11. Li Z, Vizeacoumar FJ, Bahr S et al (2011) Systematic exploration of essential yeast gene function with temperature-sensitive mutants. Nat Biotechnol 29:361–367
12. Ben-Aroya S, Coombes C, Kwok T et al (2008) Toward a comprehensive temperature-sensitive mutant repository of the essential genes of *Saccharomyces cerevisiae*. Mol Cell 30:248–258
13. Mnaimneh S, Davierwala AP, Haynes J et al (2004) Exploration of essential gene functions via titratable promoter alleles. Cell 118:31–44
14. Schuldiner M, Collins SR, Thompson NJ et al (2005) Exploration of the function and organization of the yeast early secretory pathway through an epistatic miniarray profile. Cell 123:507–519
15. Boone C, Bussey H, Andrews BJ (2007) Exploring genetic interactions and networks with yeast. Nat Rev Genet 8:437–449
16. Baryshnikova A, Costanzo M, Kim Y et al (2010) Quantitative analysis of fitness and genetic interactions in yeast on a genome scale. Nat Methods 7:1017–1024
17. Costanzo M, Baryshnikova A, Bellay J et al (2010) The genetic landscape of a cell. Science 327:425–431
18. Sharifpoor S, van Dyk D, Costanzo M et al (2012) Functional wiring of the yeast kinome

revealed by global analysis of genetic network motifs. Genome Res 22:791–801

19. Sopko R, Huang D, Preston N et al (2006) Mapping pathways and phenotypes by systematic gene overexpression. Mol Cell 21:319–330
20. Magtanong L, Ho CH, Barker SL et al (2011) Dosage suppression genetic interaction networks enhance functional wiring diagrams of the cell. Nat Biotechnol 29:505–511
21. Vizeacoumar FJ, van Dyk N, Vizeacoumar FS et al (2010) Integrating high-throughput genetic interaction mapping and high-content screening to explore yeast spindle morphogenesis. J Cell Biol 188:69–81
22. Costanzo M, Boone C (2009) SGAM: an array-based approach for high-resolution genetic mapping in *Saccharomyces cerevisiae*. Methods Mol Biol 548:37–53
23. Zou J, Friesen H, Larson J et al (2009) Regulation of cell polarity through phosphorylation of Bni4 by Pho85 G1 cyclin-dependent kinases in *Saccharomyces cerevisiae*. Mol Biol Cell 20:3239–3250
24. Tong AH, Boone C (2007) High-throughput strain construction and systematic synthetic lethal screening in *Saccharomyces cerevisiae*. In: Stansfield I, Stark MJR (Series Editor) Methods in microbiology, vol. 36. Elsevier, Amsterdam. pp. 369–386
25. Mani R, St Onge RP, Hartman JL IV et al (2008) Defining genetic interaction. Proc Natl Acad Sci U S A 105:3461–3466
26. Koh JL, Ding H, Costanzo M et al (2007) DRYGIN: a database of quantitative genetic interaction networks in yeast. Nucleic Acids Res 38:D502–D507
27. Wagih O, Usaj M, Baryshnikova A et al (2013) SGAtools: one-stop analysis and visualization of array-based genetic interaction screens. Nucleic Acids Res 41(Web Server issue): W591–W596
28. Shannon P, Markiel A, Ozier O et al (2003) Cytoscape: a software environment for integrated models of biomolecular interaction networks. Genome Res 13:2498–2504
29. Sheff MA, Thorn KS (2004) Optimized cassettes for fluorescent protein tagging in *Saccharomyces cerevisiae*. Yeast 21:661–670
30. Carpenter AE (2007) Image-based chemical screening. Nat Chem Biol 3:461–465
31. Douglas AC, Smith AM, Sharifpoor S et al (2012) Functional analysis with a barcoder yeast gene overexpression system. G3 (Bethesda) 2:1279–1289
32. Goldstein AL, McCusker JH (1999) Three new dominant drug resistance cassettes for gene disruption in *Saccharomyces cerevisiae*. Yeast 15:1541–1553

Chapter 11

Chemical Genetic and Chemogenomic Analysis in Yeast

Namal V.C. Coorey, Liam D.P. Sampson, Jacqueline M. Barber, and David S. Bellows

Abstract

Chemogenomics is the systematic genome-wide study of the cellular response to small molecule agents. Modern high-throughput genetic techniques allow massively parallel examination of the genetic effects of such biologically active small molecules (BASM). Here we present methodology for the identification and characterization of potentially bioactive compounds using the budding yeast *Saccharomyces cerevisiae* as a model organism. First, we present a method for screening libraries of compounds for growth inhibition in solid or liquid phase, followed by techniques for potency determination using a half-log dose response. Then the Deletion Mutant Array (DMA), a genome-wide library of single gene deletion strains, is used to probe the chemical genetic interactions of individual BASMs on genetic networks—a process that can be achieved with a solid phase pinning assay or a pooled liquid assay utilizing barcode microarray techniques. Finally, we offer some considerations for optimizing these protocols.

Key words *Saccharomyces cerevisiae*, Chemical genetics, Chemogenomics, Deletion mutant array, Synthetic lethality, Chemical genetic interaction profiling, Biologically active small molecules, Barcode microarray, Dose response, Halo assay

1 Introduction

Chemical genetics derives from the principle that biologically active small molecules (BASMs) bind specifically to cellular components and alter their activity, creating a unique phenotype that functionally mimics a mutation at the corresponding gene locus [1]. A corollary of this concept is that the modern armory of genetic techniques can be applied to the dissection of small molecule mechanism and, conversely, that small molecules may be used as aides to genetic dissection. Building on this definition, chemogenomics is the genome-wide analysis of cellular response to small molecule perturbation. Specifically, the concept of genetic enhancement, in which a second-site mutation enhances the phenotype of an initial mutation, is applied in a genome-wide approach to the study of BASM mechanism. Chemogenomic analysis relies on synthetic

Jeffrey S. Smith and Daniel J. Burke (eds.), *Yeast Genetics: Methods and Protocols*, Methods in Molecular Biology, vol. 1205, DOI 10.1007/978-1-4939-1363-3_11,

lethal (SL) interactions, an extreme form of enhancement in which the initial phenotype is enhanced to the point of death by the second mutation [2]. In this case, the BASM acts as a functional surrogate for the initiating genetic lesion. SL interactions identify genes that are required when a specific gene function is disrupted. Thus, chemical genetic interactions identify genetic networks that buffer the phenotype induced by the small molecule. General mechanistic inferences can be drawn in a "guilt-by-association" manner [3].

Traditional approaches for the identification of BASMs normally begin with screening of crude biological extracts or libraries of pure compounds for a desired biological activity. This is analogous to a forward genetic screen where the small molecule induces the phenotype of interest in lieu of a mutation in the genome [4]. Once a candidate compound has been identified, a genome-wide chemical genetic interaction profile is used to infer the mechanism of action and identify a putative target.

Discovery of novel inhibitors within compound libraries requires robust, low-cost and high-throughput screening methods. A growth inhibition assay is the simplest assessment of bioactivity, indicating compound interactions that interfere with essential cellular processes. Additional insight can be obtained by screening mutants defective in the process of interest as sensitized strains to identify compounds that further perturb the process. This simple assay is the point of entry for most yeast laboratories, as it only requires a spectrophotometer to measure culture growth or a camera to photograph colonies. Nearly any phenotype that can be measured in high-throughput is amenable to this approach. Although beyond the scope of this chapter, high-throughput high-content screening holds great promise for identification of phenotypes at the molecular level [5]. This chapter describes two high-throughput yeast growth inhibition assays, which allow cost-effective identification of bioactive molecules, and two complementary genome-wide profiling techniques with the potential to characterize a given molecule's mechanism of action.

2 Materials

2.1 General Materials

1. Methanol.
2. Ethanol (96 %).
3. Ethanol (Absolute).
4. Dimethyl Sulfoxide (DMSO).
5. DNA purification kit: MasterPure Yeast DNA prep kit (Epicenter Biotechnologies).
6. Amino acids supplement powder mixture for SC media: 3 g adenine, 2 g uracil, 2 g inositol, 0.2 g para-aminobenzoic acid,

2 g alanine, 2 g asparagines, 2 g aspartic acid, 2 g cysteine, 2 g glutamic acid (MSG), 2 g glutamine, 2 g glycine, 2 g histidine, 2 g isoleucine, 10 g leucine, 2 g lysine, 2 g methionine, 2 g phenylalanine, 2 g proline, 2 g serine, 2 g threonine, 2 g tryptophan, 2 g tyrosine, 2 g valine. Combine in light-proof container with 3 sterile marbles. Mix by shaking vigorously for several minutes. Store at room temperature away from light.

2.2 Media

1. Agarose gel (4 %): 1.6 g agarose in 40 mL 1× TBE buffer. Add 1 mg/mL ethidium bromide after melting.
2. YPD (Agar): distilled water 760 mL, yeast extract 8 g, peptone 16 g, adenine 0.096 g, agar 16 g, glucose (40 %) 40 mL (Add sterile glucose after autoclaving media to prevent caramelization).
3. Synthetic Complete (SC): distilled water 760 mL, yeast nitrogen base w/o amino acids or ammonium sulfate 1.36 g, MSG 0.8 g, amino acid mixture to suit 1.6 g, glucose (40 %) 40 mL (Add sterile glucose after autoclaving media to prevent caramelization).
4. HEPES-buffered SC broth (H-SC): distilled water 740 mL, yeast nitrogen base w/o amino acids or ammonium sulfate 1.36 g, MSG 0.8 g, amino acid mixture to suit 1.6 g, glucose (40 %) 40 mL, HEPES (1 M, pH 7.2) 20 mL (Add sterile HEPES and sterile glucose after autoclaving media).
5. SC (2×): distilled water 360 mL, yeast nitrogen base w/o amino acids or ammonium sulfate 1.36 g, MSG 0.8 g, amino acid mixture to suit 1.6 g, glucose (40 %) 40 mL (Add sterile glucose after autoclaving media to prevent caramelization).
6. H-SC (2×): distilled water 340 mL, yeast nitrogen base w/o amino acids or ammonium sulfate 1.36 g, MSG 0.8 g, amino acid mixture to suit 1.6 g, glucose (40 %) 40 mL, HEPES (1 M, pH 7.2) 20 mL (Add sterile HEPES and sterile glucose after autoclaving media).
7. Agar (2×): agar 16 g, q.s. to 400 mL with distilled water.

2.3 Antibiotics

1. G418 (Geneticin, Invitrogen): 200 mg/mL in H_2O.

2.4 Buffers

1. Gel electrophoresis buffer: Tris 89 mM, boric acid 89 mM, EDTA 2 mM, pH 8.3.
2. MOPS buffer: MOPS 1 M, pH 6.5, autoclaved.
3. HEPES buffer: HEPES 1 M, pH 7.2, autoclaved.
4. Hybridization buffer: 2 M NaCl, 20 mM Tris HCl pH 7.5, 1 % Triton X-100. Filter-sterilize and add DTT (1 mM final concentration) immediately prior to use.

2.5 PCR Reagents

1. 10× Platinum Taq buffer (Invitrogen).
2. MgCl, 50 mM.
3. dNTPs, 5 mM each.
4. Platinum Taq, 5 units/μL (Invitrogen).
5. Primers:

D1	5′-CGGTGTCGGTCTCGTAG
U1	5′-GATGTCCACGAGGTCTCT
D2comp-Cy3 or -Cy5	5′-Cy [3/5]-CGAGCTCGAATTCATCGAT
U2comp-Cy3 or -Cy5	5′-Cy [3/5]-GTCGACCTGCAGCGTACG

6. Blocking primers:

D2block	5′-ATCGATGAATTCGAGCTCG
U2block	5′-CGTACGCTGCAGGTCGAC

3 Methods

3.1 Identification of Bioactive Compounds

3.1.1 Solid Phase (Halo Assay)

The Yeast Halo assay is a solid-phase growth inhibition assay in which small volumes of candidate small molecules are diffused into nutrient agar plates seeded with yeast cells. Cell growth occurs in areas not occupied by an inhibitory compound, resulting in a visible "halo" around the site of a bioactive molecule. Compound potency is proportional to halo diameter while growth is measured by absorbance, allowing rapid discrimination of candidate potency and toxicity in a single assay [6].

Protocol 1: Yeast Halo Assay

1. Grow 20 mL overnight culture of selected yeast strain (*see* **Note 1**) in buffered SC broth media (*see* **Note 2**).
2. Autoclave the desired quantity of SC agar media, buffered to ~pH 6.5 with MOPS buffer with a final concentration of 25 mM.
3. Cool media to ~50 °C, ensuring it is not so hot as to kill cells (<55 °C).
4. Inoculate yeast cells with stirring from overnight culture to a final concentration of ~2×10^5 cells/mL (~20 mL culture in 1 L media, *see* **Note 3**).
5. Pour 40 mL liquid media into rectangular polystyrene plates such as omnitrays (Nunc) and allow to set.
6. Pin 1–2 μL of compound onto an identifiable location on the plate (*see* **Note 4**). In high throughput, this is best accomplished using a pinning robot, but could be achieved with a liquid

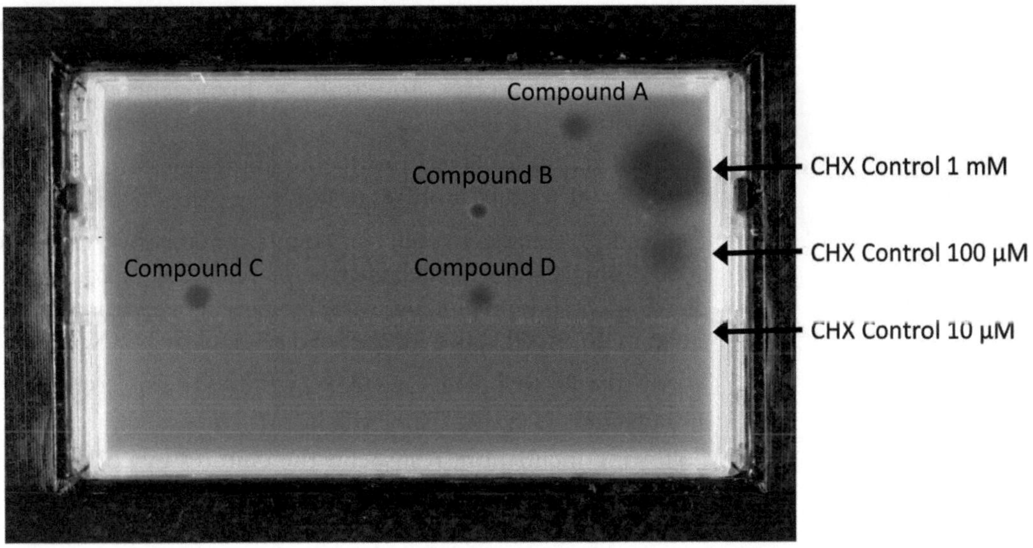

Fig. 1 Halo assay hits. This figure shows zone of inhibition for Cycloheximide (CHX) controls and small molecule compound hits. Note that halo diameter is nonlinearly proportional to concentration and that additional information is encoded in the halo

handling robot, multichannel pipettor or hand pinning tool. If the pinning robot or hand pinning tool is used, pins should be dipped in DMSO (3×), followed by MeOH (3×) between each library plate to wash them and minimize cross-contamination (*see* **Note 5**).

7. Allow compounds to diffuse into agar for 5 min then incubate plates at 30 °C for at least 24 h.
8. Measure halos produced. Measure absorbance at 590 nm directly in a spectrophotometer. Refer to Scott Lokey's method [6] for a rigorous spectrophotometric scoring method. For small libraries, or when using plates that do not fit conveniently in a spectrophotometer (e.g., Singer plates), scoring can be performed manually; either by measuring the diameter of the halos with a micrometer or similar device, or subjectively scoring them on a 1–10 scale where the score is proportional to the observed halo diameter (Fig. 1).

3.2 Liquid Phase (Turbidity Assay)

An alternative to the solid phase "halo" assay is the liquid-phase assay in which small molecules of interest are added to yeast cells in broth at a single concentration. The bioactivity of each compound is then determined by measuring turbidity of the resulting culture following treatment.

Protocol 2: Identifying Bioactives in Liquid Media

1. Inoculate a single yeast colony into 3 mL of HEPES-buffered SC broth (H-SC, 1 M HEPES, pH 7.2 diluted to 25 mM) and incubate at 30 °C on a 40 rpm rotary shaker overnight.

2. Determine cell density of the overnight culture and inoculate H-SC broth to a final concentration of 5×10^5 cells/mL (*see* **Note 6**). Vortex to create a homogeneous culture suspension.
3. Transfer 100 μL of H-SC broth containing cells into each well of a clear 96 well flat-bottom polystyrene plate.
4. Deliver 1 μL volumes of each compound from the small molecule library using a multichannel pipettor, pinning tool, or liquid handler. For compound pinning, wash the pins between each inoculation as described above for the solid phase assay (*see* **Note 4**).
5. Vortex the 96 well plates at $100 \times g$ for 30 s to mix completely and incubate at 30 °C for 15–18 h.
6. Following incubation, resuspend the cultures by vortexing at $100 \times g$ for 30 s. Measure absorbance at 590 nm using a spectrophotometer (*see* **Note 7**).
7. Determine the residual growth (%) by comparing the turbidity in each condition against the mean absorbance for DMSO control (Abs_{590}Exp/Abs_{590} DMSO control) $\times$ 100 for each unique condition (*see* **Note 8**).

3.3 Identifying the IC50 and Minimum Inhibitory Concentration (MIC)

Following identification of bioactive compounds, potency needs to be determined. Each BASM is serially diluted and tested in a dose–response assay against the query yeast strain.

Protocol 3: Solid-Phase Dose–Response Assay (MIC)

1. Construct a half-log dilution series (*see* **Note 9**). Starting from the highest available concentration, create 11 serial dilutions in half-log ($\sqrt{10} \approx 3.162$-fold) decrements by dissolving in the carrier DMSO. To dilute by half logs, start with an initial volume of diluent in all tubes except for the first (for example, add 100 μL DMSO to tubes 2–11). Add 1.462 vol to the first tube (146.2 μL in our example), vortex and transfer 0.462 vol (46.2 μL) to the next tube and repeat.
2. Aliquot 3 μL of compound dilutions into 1.5 mL microcentrifuge tubes.
3. Carefully pipette 297 μL molten SC agar into tubes, one at a time. Gently pipette to mix, avoiding bubble formation.
4. Pipette the mixture into a flat-bottom 48 well plate, again avoiding bubbles. It is advisable to practise this technique first.
5. Spot 2 μL of 5×10^5 cell/mL fresh overnight culture into each well. Two to three replicates are advisable.
6. Incubate the plates for 2 days at 30 °C.
7. Inspect plates for individual colony growth. The lowest concentration that completely inhibits growth identifies the minimum inhibitory concentration (MIC).

Protocol 4: Liquid Phase Dose Response (IC_{50})

1. Using the serial dilution described in Protocol 3, perform a liquid-phase growth assay according to the method in Protocol 2.
2. The IC_{50} value can be quantified using dose response or graphing software such as SigmaPlot or PRISM.

3.3.1 Identifying Cellular Targets and Mechanism of Action

Mechanistic insights into small molecule activity can be deduced via a chemical genetic interaction profile (CGIP) assay. There are two primary approaches: pinning of the deletion mutant array (DMA) onto solid media containing a bioactive compound, or a competitive growth assay using a pooled sample of the DMA in liquid media containing the BASM.

3.4 Solid-Phase DMA Pinning Array

Solid-phase chemical genetic profiling utilizes the haploid, non-essential DMA pinned onto agar plates containing bioactive compound. This is effectively the reciprocal experiment to the solid-phase assay described previously in Protocol 1: Yeast halo assay. Instead of testing a single yeast strain against a library of small molecules, a single BASM is tested against a comprehensive yeast mutant strain library. Chemical genetic interaction profiling requires substantial amounts of compound as the deletion mutants are pinned on 14-80 plates, depending on the library format.

Protocol 5: Solid-Phase Chemical Genetic Interaction Profiling (CGIP)

1. Autoclave Agar (2×) and H-SC (2×) media buffered with HEPES (1 M, pH 7.2 to a final concentration 25 mM) sufficient to fill enough plates to cover the deletion library (≈ 40 mL/plate). Make the Agar (2×) fresh in a 2 L Erlenmeyer flask with a magnetic stir bar added, then autoclave. Combine both media components with stirring. Add the H-SC (2×) slowly to the melted agar by pouring gently down the side of the Erlenmeyer flask (H-SC can be made in advance and stored at RT away from light, but works best if made up and autoclaved with the agar). Stir on a heated magnetic stirring plate until the Schlieren lines disappear.
2. Cool the HEPES-buffered SC (H-SC) agar to 65 °C in a water bath and add compound dissolved in the carrier DMSO to a final concentration equal to the EC_{70} (*see* **Note 10**). Ensure the total volume does not exceed final concentration of 1 % DMSO. Pour and allow to set.
3. Repeat with buffered H-SC agar + DMSO (equivalent volume as for compound plates) in the absence of the BASM to normalize for mutant-specific growth defects.
4. Transfer the DMA onto H-SC agar + BASM by replica pinning using a pin transfer robot. Use a fresh pin pad for each library plate or include a pin sterilization step if using fixed pins.

Repeat with control H-SC agar + solvent control plates (*see* **Note 11**). Alternatively, a manual pinning tool can be used. When using a manual pinning tool, wash pins between library plates by dipping in 10 % bleach solution, followed by dH_2O, then dH_2O again and finally 70 % isopropanol. Flame pins to sterilize and remove isopropanol. Cool for 10 s before commencing pinning of next plate.

5. Seal the plates with parafilm, cling film or a Ziploc bag to minimize desiccation. Incubate the plates at 30 °C for 24 h.
6. Score the colony size for each deletion mutant in both BASM treated and untreated control conditions. Determine the synthetic lethal and synthetic sick interactions by comparing growth of each mutant in treated relative to the solvent control (Fig. 2).

3.5 Liquid-Phase Growth Competition (Microarray Analysis)

Since supply of compounds, especially natural products may be severely limited; economy of reaction volume is of critical importance. The principles of solid-phase chemical genetic interaction profiling analysis [7] (CGIP) can also be applied to liquid systems [8]. The following microarray assays are easily conducted in small volumes, making judicious use of a given BASM.

The presence of unique molecular barcodes within each deletion strain is integral to the assays. In each strain the open reading frame is replaced by a marker for Kanamycin resistance, flanked by two unique 20-nt barcodes (Up and Dn) adjoined by a universal primer site on either side. Thus, all deletion strains can be identified and their populations quantified within a mixed population using PCR amplification of the corresponding barcodes. The presence of the two tags makes for more robust data, while the universal primer sites allow tandem amplification of all barcodes.

In pooled CGIP assays, a mixed culture containing every deletion mutant is grown in the presence of the BASM at the IC_{30}, and the tags of each strain from the resulting growth competition are amplified from isolated genomic DNA in two tandem PCR reactions—one for Up tags and one for Dn tags. The amplicons are then hybridized to an oligonucleotide array carrying the complementary barcode sequences, and the relative abundance of each deletion strain can then be quantified via the signal intensity of their respective barcodes [3, 9, 10]. Mutants deleted for genes that are important for growth under a specific condition compete less effectively within the assay pool. This diminishes their population size and the signal intensity of their molecular barcode. Therefore, a single experiment can identify all the genes required for growth under a given condition, allowing ranking in order of their importance to fitness. The homozygous profiling (HOP) assay utilizes

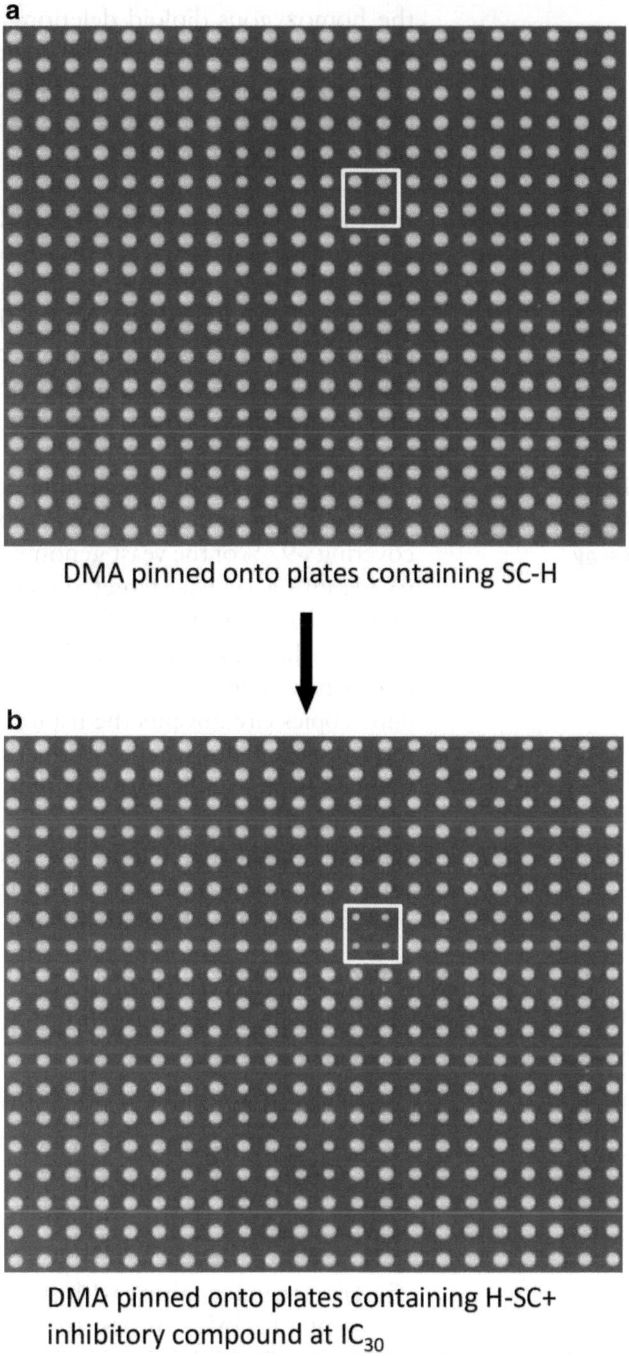

Fig. 2 Solid phase chemical genetic profiling. Each mutant is in quad replicates were pinned on to agar plates with either (**a**) SC-H or (**b**) SC-H + inhibitory compound at IC_{30}. The mutant marked by the *white box* grow normally in the absence of the inhibitory compound in (**a**) but failed to grow in the presence of the inhibitory compound in (**b**) therefore is sensitive to the inhibitory compound at a semi lethal concentration

the homozygous diploid deletion strains [8], allowing analysis of nonessential genes, while the analogous array for essential-gene analysis, haploinsufficiency profiling (HIP), is conducted using the genome-wide heterozygous diploid deletion encompassing ~97 % of the genome [9, 11, 12].

3.6 The Homozygous Profiling (HOP) Assay

Interactions uncovered by the HOP assay do not directly identify the primary cellular target of a bioactive molecule. Instead, synthetically sick or lethal interactions indicate genetic relationships between given DMA strains and the cellular target of the compound, such that loss of activity in both gene products generates a nonadditive phenotype. Thus, hypersensitive mutant strains identify "friends of the target" through guilt-by-association.

3.7 The Haploinsufficiency Profiling (HIP) Assay

Assessment of essential gene interactions is achieved by screening the heterozygous deletion mutant library, a near complete collection covering ~97 % of the yeast genome. This approach has the potential to identify a cellular target directly by the mechanism of drug-induced haploinsufficiency [13]. However, this is only applicable to BASMs where the primary cellular target encodes an essential gene that is not under dosage control. Transcriptional control of single gene copies circumvents the haploinsufficiency mechanism, rendering important drug–gene product interactions invisible. In drug-induced haploinsufficiency, the mutant heterozygous for the target is sensitized in comparison to wild-type yeast. Dosage-dependent synthetic lethal interactions are also revealed by HIP; therefore, using both homozygous and heterozygous mutant screens is a complementary approach for identifying a cellular target [14].

Protocol 6: Liquid-Phase Microarray Profiling

1. Inoculate a thawed 0.5 mL aliquot of the DMA pool in 10 mL SC.
2. Incubate overnight at 30 °C on a rotating drum.
3. Determine cell titre via hemocytometer or spectrophotometer
4. Seed a 20 mL culture of SC with 1×10^7 cells and divide into two separate 10 mL cultures. This ensures at least 1×10^3 representatives of each deletion mutant per 10 mL.
5. Treat the first culture with an IC_{30} concentration of the experimental BASM. It is best to determine the appropriate concentration through a liquid-phase dose–response assay, counting cells via haemocytometer (*see* **Note 10**). It is also possible to determine 70 % residual growth using a spectrophotometer, although this is less precise.

6. Treat the second of the parallel cultures (control) with an equivalent volume of carrier solvent.
7. Grow both cultures for 10 generations (~15 h) at 30 °C with shaking.
8. Dilute cultures to 5×10^5 cells/mL in 10 mL fresh SC containing an IC_{30} concentration of the experimental BASM. Repeat dilution for the DMSO control.
9. Grow both cultures for an additional 10 generations (~15 h) at 30 °C with shaking.
10. Extract and purify genomic DNA from 1.5 mL of each culture using a DNA purification kit, ensuring RNA is removed.
11. Quantify DNA using, for example, Hoechst 33258 dye and calf thymus DNA standard. 100 ng of DNA is needed per PCR reaction.

PCR Amplification

12. PCR reactions should be set up in a sterile environment, such as a laminar flow hood, with PCR product-free pipettes and consumables. No DNA solutions or PCR products should be brought into the hood. Take precautions to avoid cross-contamination of reactions with previously amplified PCR product.
13. Four PCR reactions are required for each experiment. Up and Dn tags are amplified in separate reactions for both the control and experimental DNA samples, which are labelled with Cy3 and Cy5 tagged primers respectively.

Control DNA (no compound)	Up tag PCR Dn tag PCR	Primers U1 and U2comp-Cy3 Primers D1 and D2comp-Cy3
Experimental (compound)	Up tag PCR Dn tag PCR	Primers U1 and U2comp-Cy5 Primers D1 and D2comp-Cy5

Primer sequences	
D1	5′-CGGTGTCGGTCTCGTAG
U1	5′-GATGTCCACGAGGTCTCT
D2comp-Cy3 or -Cy5	5′-Cy [3/5]-CGAGCTCGAATTCATCGAT
U2comp-Cy3 or -Cy5	5′-Cy [3/5]-GTCGACCTGCAGCGTACG

14. Perform PCR reactions in 60 μL final volume (1.5 mM $MgCl_2$, 0.2 mM dNTPs, and 1 μM primers).

15.

PCR master mix	1× (μL)	9× (1 array) (μL)
10× Platinum Taq buffer	6.0	54.0
50 mM $MgCl_2$	1.8	16.2
5 mM (each) dNTPs	2.4	21.6
Platinum taq (Invitrogen) (5 units/μL)	0.2	1.8
H_2O	40.8	367.2
Total volume	51.2	460.8

In 8×0.2 mL tubes	
Labelled primer (25 μM)	2.4 μL
Unlabelled primer (25 μM)	2.4 μL
Master mix	51.2 μL
DNA (25 ng/μL)	4.0 μL

16. Amplify negative control samples (dH_2O instead of DNA) for each reaction.

PCR conditions		
94 °C	3 min	
94 °C	30 s	38 cycles
50 °C	30 s	
72 °C	30 s	
72 °C	5 min	
10 °C	Hold	

17. Run 5 μL of each PCR on a high resolution 4 % MetaPhor Agarose gel with 10-bp marker. The size of the amplified product for both Up and Dn tags is 56-bp.
18. Combine 55 μL each PCR product and 20 μL blocking mix (12.5 μL 100 μM U1 primer; 12.5 μL D1 100 μM; 12.5 μL 100 μM U2 block; 12.5 μL 100 μM D2 block; 50 μL dH_2O). These complimentary oligonucleotide sequences bind with the priming regions of the PCR products preventing them from binding with each other and allowing them to hybridize to the microarray.
19. Add 24 μL 3 M sodium acetate pH 5.2 (1/10 volume), 600 μL absolute ethanol (2.5×volume), 1 μL 5 mg/mL linear acrylamide (acts as a carrier for precipitation). Vortex to mix and precipitate at −20 °C for 1–1.5 h.

Blocking primer sequences	
D2block	5′-ATCGATGAATTCGAGCTCG
U2block	5′-CGTACGCTGCAGGTCGAC

20. Spin 16,000 × *g* at 4 °C for 30 m.
21. Wash with 70 % ethanol—add 1 mL, spin 5 m at 4 °C.
22. Remove ethanol. Dry briefly in dark, at room temperature.
23. Redissolve in 50 μL dH_2O. Store at −20 °C if not hybridizing the same day.
24. Hybridize within 72–96 h. Fluorescent signal attenuates with longer storage.

Hybridizing PCR Products to Microarray

25. This protocol is optimized for the custom yeast tag Microarray from Agilent Technologies with Agilent hybridisation chambers and oven.
26. Combine 50 μL PCR products with 50 μL 2× hybridization buffer (2 M NaCl, 20 mM Tris HCl pH7.5, 1 % Triton X-100, and filter-sterilize. Add DTT (1 mM final concentration) immediately prior to use. Gently pipette mix, avoiding bubbles. Denature hybridization mix at 95 °C for 2 m.
27. Apply 100 μL of hybridization mix (avoid bubbles) to active side of the gasket slide sitting in the hybridization chamber base. Place microarray slide, active side down, on to gasket slide and assemble chamber, followed by gentle rotation about the horizontal axis to wet slide completely.
28. Place in preheated hybridization oven at 42 °C for 4 h at a moderate rotation speed.
29. Read slides in a microarray scanner at 532 nm (control) and 635 nm (experimental) (*see* **Note 12**).

Microarray Data Analysis

30. We use GenePix 6.0 to extract data from the microarray image file. Features are aligned and fluorescent signal data obtained within GenePix (refer to publisher's instructions).
31. Open the GenePix results file (.gpr file) in a spreadsheet program and remove any spots that should not be included in the analysis (e.g., control spots, spots flagged as poor quality). Remove data spots that correspond to essential gene deletions if using a homozygous deletion pool.
32. Large data sets obtained from microarray experiments need to be statistically manipulated to normalize the data. We recommend the Web-based SNOMAD software (http://pevsnerlab.

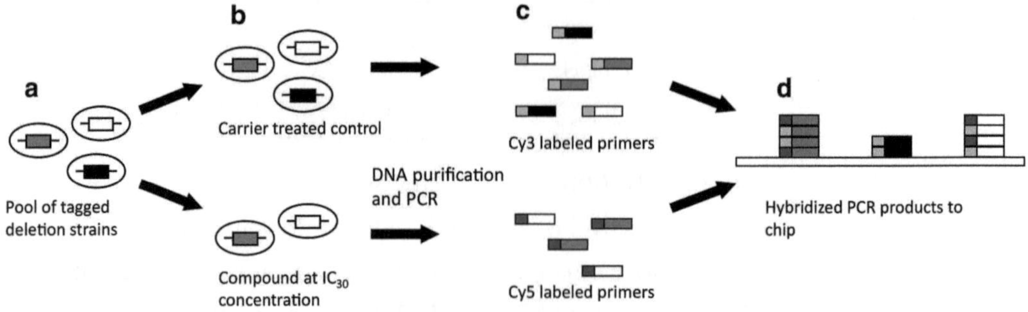

Fig. 3 Outline of microarray protocol. (**a**) Aliquots of the pooled YKO strains are split into control and experimental samples. (**b**) Cultures are competitively grown in parallel with carrier solvent and an IC_{30} concentration of the queried compound for 10 generations. The cultures are then diluted and grown again in the presence of solvent or compound for a further 10 generations. Strains without the genes required for growth under the selected drug condition will have lower relative abundance in the experimental sample. (**c**) Genomic DNA is isolated from each population and barcode regions of each strain amplified by PCR, utilizing Cy3 and Cy5 labelled primers for the control and experimental samples respectively. (**d**) After further processing, the combined samples are hybridized to the microarray chip. A scanner reads the ratio of Cy3 to Cy5 labelled product for each gene, indicating gene deletion strains that are underrepresented in the experimental sample and therefore likely important for growth under the selected condition

kennedykrieger.org/snomadinput.html) to normalize the data and correct for intensity-dependent biases in the Cy5/Cy3 ratios.

33. Separate the Up and Dn tags into separate worksheets and save as .txt files. Upload each .txt file into SNOMAD. Use the *median F635 (Cy5)* and *median F532 (Cy3)* columns of data for the data types "ONEintensities" and "TWOintensities" respectively. *Perform* and *Graph* the transformation in **step 3,** Log transformation (Log base 2); **step 4**, calculation of mean log (intensities) and log (ratios) and **step 5,** local mean normalization across element signal intensity (Span 0.3, Trim 0.1) [15].
34. Submit this data for processing and copy the results into a fresh worksheet. Remove data spots with low (<500) *median F532* scores to reduce the level of noise in the data.
35. Average the normalized log ratios for tag replicates and calculate *z*-scores ((x − mean)/SD). Sort by *z*-scores, scoring *z*-scores < −3 as a probable significant hit (*see* **Note 13**) (Fig. 3).

4 Notes

1. Careful thought should be given to yeast strain choice before embarking on a screening program. Yeast is naturally resistant to a wide variety of bioactive molecules through the pleiotropic drug resistance (PDR) network. The PDR network is a suite of membrane-bound protein pumps that mediate

ATP-dependent efflux of various compounds to overcome unfavorable environments [16–18]. Compounds that serve as substrates for the PDR network pumps are effluxed before they can accumulate to an intracellular a concentration to have an inhibitory effect. To overcome this PDR-mediated drug efflux, wild-type yeast cells must be flooded with high concentrations of small molecule inhibitors, requiring substantial amounts of compound which are not always readily available. Therefore, a strain deficient in PDR can be screened in place of the wild-type, requiring less compound to achieve the desired level of inhibition in wild-type. However, the pumps have promiscuous, overlapping specificity for substrates, complicating any attempt to sensitize cells with a single gene deletion. An alternative approach is to knock out the transcriptional master regulators of the PDR network: *PDR1* and *PDR3*. Gene deletions in *PDR1* and *PDR3* have demonstrated increased sensitivity to inhibitory compounds as well as sensitizing to compounds that do not demonstrate bioactivity in wild-type yeast [19]. We use an S288C-based *pdr1Δ, pdr3Δ* double mutant as a drug-sensitive wild type strain for preliminary screens.

2. The low pH of SC media attenuates the bioactivity of many pharmacologically active compounds. Buffering of the media to near neutrality reverses this effect with minimal expansion in doubling time. In our experience, MOPS buffer works best for solid phase assays. The commonly used HEPES buffer produces gas bubbles in the plates after incubation.
3. For maximum consistency and reproducibility we suggest inoculating saturated, overnight cultures. This allows uniform starting conditions in screens involving large numbers of plates or multiple hours of liquid handling.
4. Some thought should be given to screening concentration. A preliminary concentration that is too low risks missing active compounds through false negatives. Alternatively, too high a preliminary concentration risks unnecessarily wasting precious compounds, especially natural product isolates. Experimenters should first decide on the upper limit cutoff for bioactivity. We generally use a 1 mM working stock and dilute 100-fold into media for a final 10 μM screening concentration. Since yeasts are naturally resistant to many compounds, low micromolar sensitivity in yeast generally translates into mid-to-low nanomolar activity in mammalian cells. Finally, compound libraries are traditionally diluted in DMSO, so care should be taken regarding total concentration of solvent. DMSO concentrations >3 % inhibit yeast growth.
5. 96-well format provides an orderly layout of compounds, and for compound handling via a liquid-handling robot. Evaporation in the corner wells of liquid-phase experiments

can lead to false positive absorbance readings because of reduced path length. To avoid these corner plate effects and allow for a variety of controls, we leave columns 1 and 12 open. Hand-assembled libraries should be arrayed at 80 compounds per plate. Depending on library size, controls can be added to each plate in columns 1 and 12 or a separate control plate can be used iteratively with each library plate. In addition to the negative, solvent control, include a positive control such as 1 μM rapamycin or 10 μM cycloheximide for each plate to establish a baseline OD for complete inhibition. It is useful to have at least one well that receives no solvent so that solvent effects can be identified and quantified. If screening with mutant yeast strains, it is advisable to test each strain in advance for DMSO sensitivity.

6. To maximize the dynamic range of the assay, initial cell density should be just below the sensitivity threshold of the spectrophotometer used to measure turbidity. To identify compounds with mild growth inhibitory activity, the cell density should be measured within two doubling times of reaching saturation (from 15 to 18 h post inoculation).
7. It is important that all cultures are uniformly resuspended in bioactivity or dose–response assays. Mutant flocculant strains or compounds inducing aggregation can be difficult to resuspend homogeneously. Vortexers with a rotational diameter smaller than the well tend to swirl the plates, creating a "bull's-eye" spot in the middle that can be observed by holding the plate up to a light. This is especially problematic with plate readers that have a built-in vortexing capability.
8. Normalize the absorbance values by subtracting the absorbance for complete inhibition in the positive control from the entire plate (e.g., $Abs_{590}ALL - Abs_{590}CHX$). Alternatively, the plates can be read immediately upon inoculation to create a 0 time-point "plate blank."
9. In yeast, the maxima and minima of the dose–response curve lie within one logarithm. Thus, half-logarithm dilutions make it possible to capture at least a single point near the EC_{50}. In addition, half-logarithm series plot evenly on a semi-log graph. For dose–response assays serially dilute each BASM in DMSO to an appropriate concentration range. Always use a fresh, unopened bottle of DMSO to perform the dilutions. DMSO is VERY hygroscopic and will rapidly become diluted with atmospheric water after several openings.
10. Chemical genetic interaction profiling is most effective using small molecule concentrations at which the phenotype is first visible. This concentration represents the "sweet spot" of chemical genetic interaction, when cells are being "tickled" by the compound. This is the beginning of the steepest part of the

dose–response curve and is the concentration where small perturbations will have maximum observable effect, allowing the widest dynamic range for interacting mutations to be identified. For growth assays, this corresponds to the IC_{30} (the concentration at which growth is reduced by 30 % relative to the control). This value can be calculated using graphing software. Alternatively, the IC_{30} can be determined directly from the dose–response curve. This is straightforward if the *y*-axis is plotted as residual growth (%). For solid phase assays, the IC_{30} must be estimated from the MIC. If there is a concentration that produces intermediate sized colonies prior to the MIC, use this concentration (or a little less if the colonies are very small). Otherwise reduce the MIC by half. A special word of warning: BASMs may have dramatically different EC/MIC values on solid phase and liquid phase experiments. Do not assume that solid and liquid phase EC/MIC concentrations may be used interchangeably.

11. If compound quantities permit, the CGIP should be performed more than once to eliminate false positives. Pinning each mutant strain in quadruplicate also minimizes false positives derived from pinning errors, and noise associated with colony-size. The colonies at the outermost edges of a pinned array generate "edge effects" that pose a special problem. In solid-phase assays, lack of nutrient competition produces false negatives through enhanced colony growth. We use a deletion array that has been constructed with a "picture frame" border of wild type colonies (*URA3::Kan*^{R}) to provide uniform growth conditions for mutant strains.

12. A variety of scanners are available for the reading of slides. In our lab, slides were scanned on a GenePix 4000B microarray scanner at the Otago Genomics Facility, University of Otago, Dunedin, NZ.

13. Filtering of the profiles for genes conferring multidrug resistance [20] is routinely performed and the data analyzed in both filtered and unfiltered forms. These include genes whose deletion confers greater membrane permeability such as *ERG2, ERG3, ERG5*, and *ERG6*, and components of the pleiotropic drug resistance network: the transcription factor, *PDR1*, which regulates genes involved in multidrug resistance and *PDR5*, encoding a drug-efflux pump [8, 20]. There are several other PDR genes which display increased sensitivity to multiple compounds, such as those integral to vesicular transport pathways including *VPS4, VPS20, VPS24*, and *VPS36*. After filtering, mutant strains with a *z*-score < -3 are denoted as hits, and the profiles are analyzed using cluster-function programs such as Cytoscape, with the BiNGO plug-in, or the online resource FunSpec (http://funspec.med.utoronto.ca/). Hierarchical cluster analysis of HOP

data sets is widely used to compare the genetic profiles of bioactive molecules; this can aid mechanism definition in uncharacterized compounds. Compounds with similar cellular targets, display significant overlap in their respective sensitivity profiles [8].

Acknowledgments

We would like to thank Rosemary Heathcott for her invaluable assistance in the development and optimization of the microarray protocol.

References

1. O'Connor CJ, Laraia L, Spring DR (2011) Chemical genetics. Chem Soc Rev 40: 4332–4345
2. Guarente L (1993) Synthetic enhancement in gene interaction: a genetic tool come of age. Trends Genet 9:362–366
3. Boone C, Bussey H, Andrews BJ (2007) Exploring genetic interactions and networks with yeast. Nat Rev Genet 8:437–449
4. Stockwell BR (2000) Chemical genetics: ligand-based discovery of gene function. Nat Rev Genet 1:116–125
5. Bircham PW, Maass DR, Roberts CA et al (2011) Secretory pathway genes assessed by high-throughput microscopy and synthetic genetic array analysis. Mol Biosyst 7:2589–2598
6. Gassner NC, Tamble CM, Bock JE et al (2007) Accelerating the discovery of biologically active small molecules using a high-throughput yeast halo assay. J Nat Prod 70:383–390
7. Parsons AB, Brost RL, Ding H et al (2004) Integration of chemical-genetic and genetic interaction data links bioactive compounds to cellular target pathways. Nat Biotechnol 22: 62–69
8. Parsons AB, Lopez A, Givoni IE et al (2006) Exploring the mode-of-action of bioactive compounds by chemical-genetic profiling in yeast. Cell 126:611–625
9. Lum PY, Armour CD, Stepaniants SB et al (2004) Discovering modes of action for therapeutic compounds using a genome-wide screen of yeast heterozygotes. Cell 116:121–137
10. Brenner C (2004) Chemical genomics in yeast. Genome Biol 5:240
11. Baetz K, McHardy L, Gable K et al (2004) Yeast genome-wide drug-induced haploinsufficiency screen to determine drug mode of action. Proc Natl Acad Sci U S A 101:4525–4530
12. Giaever G, Chu AM, Ni L et al (2002) Functional profiling of the *Saccharomyces cerevisiae* genome. Nature 418:387–391
13. Giaever G, Shoemaker DD, Jones TW et al (1999) Genomic profiling of drug sensitivities via induced haploinsufficiency. Nat Genet 21: 278–283
14. Deutschbauer AM, Jaramillo DF, Proctor M et al (2005) Mechanisms of haploinsufficiency revealed by genome-wide profiling in yeast. Genetics 169:1915–1925
15. Peyser BD, Irizarry R, Spencer FA (2008) Statistical analysis of fitness data determined by TAG hybridization on microarrays. In: Walker JM (ed) Microbial gene essentiality: protocols and bioinformatics. Humana Press, New York, pp 369–381
16. Bauer BE, Wolfger H, Kuchler K (1999) Inventory and function of yeast ABC proteins: about sex, stress, pleiotropic drug and heavy metal resistance. Biochim Biophys Acta 1461: 217–236
17. Wolfger H, Mamnun YM, Kuchler K (2004) The yeast Pdr15p ATP-binding cassette (ABC) protein is a general stress response factor implicated in cellular detoxification. J Biol Chem 279:11593–11599
18. Jungwirth H, Kuchler K (2006) Yeast ABC transporters—a tale of sex, stress, drugs and aging. FEBS Lett 580:1131–1138
19. Emerson LR, Skillman BC, Wolfger H et al (2004) The sensitivities of yeast strains deficient in PDR ABC transporters, to quinoline-ring antimalarial drugs. Ann Trop Med Parasitol 98:643–649
20. Hillenmeyer ME, Fung E, Wildenhain J et al (2008) The chemical genomic portrait of yeast: uncovering a phenotype for all genes. Science 320:362–365

Chapter 12

Phenomic Assessment of Genetic Buffering by Kinetic Analysis of Cell Arrays

John Rodgers, Jingyu Guo, and John L. Hartman IV

Abstract

Quantitative high-throughput cell array phenotyping (Q-HTCP) is applied to the genomic collection of yeast gene deletion mutants for systematic, comprehensive assessment of the contribution of genes and gene combinations to any phenotype of interest (phenomic analysis). Interacting gene networks influence every phenotype. Genetic buffering refers to how gene interaction networks stabilize or destabilize a phenotype. Like genomics, phenomics varies in its resolution with there being a trade-off allocating a greater number of measurements per sample to enhance quantification of the phenotype vs. increasing the number of different samples by obtaining fewer measurements per sample. The Q-HTCP protocol we describe assesses 50,000–70,000 cultures per experiment by obtaining kinetic growth curves from time series imaging of agar cell arrays. This approach was developed for the yeast gene deletion strains, but it could be applied as well to other microbial mutant arrays grown on solid agar media. The methods we describe are for creation and maintenance of frozen stocks, liquid source array preparation, agar destination plate printing, image scanning, image analysis, curve fitting, and evaluation of gene interaction.

Key words Gene interaction, Genetic buffering, Quantitative high-throughput cell array phenotyping (Q-HTCP), Yeast mutant arrays, Happy arrays, Reference arrays, Source arrays, Dilution plates, Destination plates, Array map, Perturbation, EASY Phenomics image analysis and growth curve fitting software

1 Introduction

Phenotypes are polygenic and mediated by gene interaction, meaning that genetic and environmental variables that influence a phenotype further depend on the allele status of multiple loci across the genome. This combinatorial aspect of the phenome renders it complex, and understanding this complexity is fundamental to mapping functional genomic variation [1]. *Saccharomyces cerevisiae* is the best model for studying eukaryotic gene interaction due to the availability of yeast deletion mutant arrays, including deletions of nonessential genes and hypomorphic alleles of essential genes [2]. To investigate genetic buffering networks, mutant strain cell arrays can be challenged with drugs or environmental variables

Jeffrey S. Smith and Daniel J. Burke (eds.), *Yeast Genetics: Methods and Protocols*, Methods in Molecular Biology, vol. 1205, DOI 10.1007/978-1-4939-1363-3_12, © Springer Science+Business Media New York 2014

and/or systematically combined with mutations of interest [3–8]. Rigorous quantification of cell proliferation phenotypes enables accurate classification of gene interactions based on their strength of effect, thereby facilitating identification of genetic modules and buffering networks [3, 9, 10]. Development of quantitative cell array phenotyping for growth curve analysis is in a relatively early stage [11], as its importance came to light only after work with the yeast mutant arrays, constructed fairly recently [12], revealed a high frequency of gene interaction [3, 6, 13, 14].

In a general sense, gene interaction is defined as the difference between the expected and observed phenotypes resulting from a genetic mutation in combination with another biological perturbation. Yeast mutant arrays provide genomic coverage (*see* Subheading 3.6) study gene-drug, gene-environment, or gene-gene interaction. Q-HTCP is a protocol that involves cell array production, time-lapse imaging, image analysis, and fitting of the resulting data to a logistic growth function to obtain kinetic growth parameters that can be used to measure gene interaction across the entire genome [3, 15, 16]. Quantification of gene interaction is also variable depending on a phenotypic parameter chosen for analysis, and a neutrality function that establishes an expected phenotype [17].

The methods described here are envisioned for collection of kinetic cell array phenotype data in a systematic and scalable way so that it can be analyzed in a flexible and cumulative manner. Gene interaction varies as a function of cellular context, and thus the interpretation of gene interaction networks depends on the biological nature of the aggregated phenotypic data [9]. Q-HTCP can be used to obtain gene interaction profiles for any drug, environmental factor, or gene of interest. It is ideal if the perturbation of interest is rate limiting for cell proliferation so that both aggravating (enhancing) and alleviating (suppressing) interaction can be observed [16].

For gene-gene interaction studies, the synthetic genetic array (SGA) method, described elsewhere [6], should be carried out to construct double-mutant strains before Q-HTCP. For application of the Q-HTCP method, the following variation of the SGA method may be considered: by the standard SGA method, double mutants are assessed for gene interaction in the final step of strain construction. SGA can be done quantitatively, where phenotypes are measured as colony outgrowth area at a selected time point, and a multiplicative neutrality function is used [6, 14, 18–20]. However, if using Q-HTCP to quantify gene-gene interaction, a variation of the SGA procedure works better, where the mutation of interest is conditionally expressed, frozen stocks are made at the final step of SGA, and quantitative growth curve analysis is carried out as a subsequent step [16, 21]. By utilizing conditional (e.g., tetracycline-regulated) expression of the query gene, interactions can be assessed at multiple levels of query gene expression by shifting cells to media containing different concentrations of tetracycline [22]. Thus for Q-HTCP-based gene-gene interaction analysis, the SGA procedure is first used for strain construction (not phenotypic analysis) and subsequent experiments can incorporate a variety of media not compatible with SGA selection. Kinetic phenotyping and gradations of perturbation rigorously assess gene interaction, but in doing so, reduce the overall throughput in exchange for increased quantitative precision. Nevertheless, robust measurement of gene interaction can be performed on a phenomic scale [3, 15, 16, 21–23].

Future directions of Q-HTCP methodology include development of robotic cell array imaging equipment with supporting software [3, 15, 18, 24, 25], database development for storing and retrieving Q-HTCP-derived gene interaction data [26], and network analysis for developing yeast phenomic models informative of human disease [16, 27].

2 Materials

1. Yeast mutant arrays consist of deletion strains for nonessential yeast genes [7, 12], hypomorphic alleles for essential genes called DAmP (decreased abundance by mRNA perturbation) alleles where the 3′ untranslated region (UTR) of genes has been disrupted [5], and temperature-sensitive alleles [4]. In general, these are constructed in the same genetic background (e.g., BY4741, BY4742, or BY47473). These libraries provide nearly full coverage of the *S. cerevisiae* genome. The deletion strain project is described at http://yeastdeletion.stanford.edu/. Strains can be purchased from Open Biosystems at Thermo Scientific. Q-HTCP methodology is applicable to any yeast mutant array, as well as other organisms or cell types that grow to a lawn on agar media.
2. Yeast media: Use YPD liquid media to passage cells and for making glycerol stocks [28], adding selective antibiotics when useful (e.g., G418 and/or ClonNat/nourseothricin). It is ideal to grow cell arrays ("happy arrays") by inoculating from frozen glycerol stocks just prior to Q-HTCP. However, happy arrays can be stored at 4 °C and used for a few weeks. The composition of the agar media is typically integral to the study design and thus experiment specific. We use glass-distilled water, add amino acids after autoclaving, and use a Mediaclave (Integra Biosciences) and peristaltic pump for media consistency.
3. Pin tools: 96- and 384-pin tools can be obtained from V&P Scientific (http://www.vp-scientific.com/pin_tools.htm); ours

were customized for use with the Sciclone ALH3000 instrument (Caliper Life Sciences). There are many types of pin tools, including manual versions and different robotic adapters. The main technical parameter is the diameter of the pin. For Q-HTCP, we use the FP6 style pin (1/16 in. cut with blunt end); these provide an adequate area to create a lawn when transferring cells from liquid to agar yet are small enough to accommodate a 384 format. A floating pin style (not fixed/rigid) is best and is required for robotic printing.

4. Microtiter plates: Different microtiter plates are used throughout the protocol: (a) "Rectangular monowell" plates (greater interior plate area) are used for imaging purposes (Nunc Cat# 267060) (*see* **Note 1**). (b) "Notched monowell" plates (Nunc Cat# 242811) are preferred for SGA (because they are more affordable, provide orientation, and work well except for quantitative imaging), though the rectangular monowell plates also work. (c) "Standard profile 384-well" plates (Nunc Cat# 242757) are used for glycerol stocks for long-term storage. These provide greater well volume—up to ~65 μL working volume—and are stackable. (d) Low profile 384-well plates from Evergreen scientific (Evergreen Scientific Cat# 222-8210-01I and Cat# 290-8217-010) have cylindrical wells with a working volume of ~40 μL that yield more consistent volume of transfer, leading to better printing quality for the agar array. Therefore, Evergreen 384-well plates are used for "happy arrays" and "dilution arrays," where precision in the volume of transfer affects the uniformity of the results. (e) "96-Well plates" (Cat# 260860) are obtained from Nunc (*see* **Note 2**).
5. Tools for producing cell arrays: There are many robotics options, but manual array printing systems, with a mechanical guide to insure that arrays are printed precisely the same, can also be used successfully. We primarily use the Sciclone ALH3000 (Automated Liquid Handling with 20-position work deck) integrated with Twister II (transport arm and microplate storage) (http://www.caliperls.com/products/lab-automation/). Such systems are operated by a PC, with control software (e.g., Caliper "Maestro" software), and provide a drag-and-drop graphical interface to create automated liquid-handling methods.
6. Scanner for cell array imaging: The Epson Expression 10000 XL—Photo Scanner with Transparency Unit Lid.
7. Incubator: A floor model incubator is ideal given the high number (200 or more) of cell arrays one may wish to analyze simultaneously per experiment.
8. Centrifuge with swinging style rotor and microtiter plate adapters.

9. Plate shaker, e.g., Lab-Line Instruments, Inc.; titer plate shaker model No. 4625.
10. 3/32″ thick plexiglass cut to ½″ by 12″ and metal washers (for Scanner; *see* Subheading 3.4.3).
11. Adhesive aluminum foil (USA Scientific, Cat. # 2938-4100). We have found that the foil rated for −40 °C, rather than −80 °C, prevents adhesive sticking to the plates when foil is removed from glycerol stocks.
12. "Brayer" from USA Scientific (http://www.usascientific.com/filmaccessories.aspx) for sealing glycerol stocks with adhesive foils.
13. Pin cleaner: From V&P Scientific (http://www.vp-scientific.com/V&P_pincleaner.php).
14. For pin tool sterilization: 10 % Bleach, sterile water, 100 % ETOH, blotting paper, paint pad.
15. Roll of plastic wrap (e.g., Saran wrap).

Optional

16. Bench top MediaClave—from Integra Biosciences (http://www.integra-biosciences.com/sites/mediaclave_new_1_e.html)—facilitates media consistency through being programmable with continuous stirring and internal temperature monitoring.
17. Peristaltic pump—streamlines pouring media from Media Clave. For use with the MediaClave, a "homemade" system for media pouring is sometimes helpful (available upon request).
18. "Tough Tags" labels: These help at −80 °C for reading writing on frosty plates with ethanol; double labeling (on tag and directly on plate) is recommended for long-term storage.

3 Methods

3.1 Sterilizing the Pin Tool

The pin tool must be sterilized to avoid cross-contamination. Effective sterilization can be achieved by using 10 % bleach to kill contaminating cells, water to rinse, and a final dip in ethanol followed by air-drying (or ethanol with flaming if manually pinning). A "scrub pad" (made from a paint pad obtained from a hardware store, cut to size of the microtiter plate; this can also be bought from V&P Scientific) is useful if working with cell paste transfers. The scrub pad should be wet with 10 % bleach and is used at the start of the cleaning process to loosen the cell paste from the pins. Blotting paper is used to remove bleach before rinsing in water and to remove water before finally rinsing the pin tool in ethanol. Using two bleach baths provides additional sterilization, as cells will accumulate in the first bleach bath; 30–60 s at each step of the sterilization is adequate, and intervals can be empirically determined by experimental assessment of decontamination for a particular protocol.

3.2 Cleaning the Pin Tool

Whereas pin tool "sterilization" is performed multiple times during every protocol, pin tool "cleaning" is needed at the beginning of an entire experiment (e.g., every few days or so) to remove proteins, lipids, and other debris that can build up on the pins. Prolonged use without cleaning can lead to loss of liquid transfer efficiency and uniformity. We use V&P Pintool cleaner (proprietary solution) and follow their directions. It is a good habit to clean the pin tool before each Q-HTCP printing (we reuse the solution), and consider pin tool cleaning anytime a decline or variability in spot quality is experienced. Sanding or polishing the pins may also improve the quality of cell array printing.

3.3 Creating and Working with Frozen Glycerol Cell Array Stocks

There are many scenarios for creating and handling cell array frozen stocks; examples are described below. When creating new cell arrays (e.g., from hand-picking colonies off agar media, or to consolidate existing arrays through a "hit-picking" process), it is generally best to first inoculate cultures in a 96-well liquid array format. By contrast, it is easier to replicate glycerol stocks already in 384 format by first growing the array on agar (e.g., similar to the final step of SGA). Slow-growing cultures will reach a lower density on agar than in a liquid array; however very-slow-growing strains are difficult to work with regardless of how they are maintained. In the case of consolidating from 4×96-array format to 1×384-array format, propagating arrays in liquid or agar may work equally well. Whether on agar or in liquid, arrays are grown to near stationary phase (2–3 days) before glycerol stocks are made.

3.3.1 Create New 96-Well-Format Cell Array Frozen Stocks

Strains are provided commercially as 96-well glycerol stocks. To improve Q-HTCP throughput, strains should be consolidated into a 384-well format. Before consolidating however, making multiple copies of the library in 96-well format will help preserve it by providing a means of recovery in case of future cross-contamination or other accidental loss. Multiple copies can be made at once; using a multichannel pipette (robotic 96 channel if possible), draw an amount of volume of stock culture desired to inoculate all copies (e.g., 1–2 μL/copy) and distribute evenly. Alternatively, new 96-well arrays can be made by inoculating wells individually, such as consolidating hits for retesting and/or streaking the original library for single colonies to assess single-colony clones for reproducibility. After 2–3-day incubation at 30°, autoclaved 80 % glycerol is added to a final concentration of 20 %; arrays are sealed with adhesive foil and shaken to mix glycerol and cells. To make copies:

1. Label and pre-fill plates with 150 μL (or less) YPD (G418 if appropriate for selection/sterility).
2. Resuspend cultures of the existing 96 array (if copying an existing glycerol stock) and inoculate, or individually inoculate each well of the 96-well plate. Consider placing reference strains (*ydl227c-Δ0*::G418^{R}) on each array when possible (*see* **Notes 5** and **9**).
3. Incubate at 30 °C for 48–72 h.
4. Resuspend cells by orbital shaking (this will facilitate final cell suspension after adding glycerol). Add sterile 80 % glycerol to final concentration of 20 % (50 μL of autoclaved 80 % glycerol; use 80 % glycerol to reduce viscosity).
5. Apply adhesive foil and affix with roller/brayer.
6. Mix by orbital shaking until the glycerol/water interface is gone and cells are evenly suspended. Pipetting is usually inadequate to suspend cells or mix glycerol. Store at −70 °C immediately.

3.3.2 Consolidate 96-Well Frozen Glycerol Stock to 384-Well Arrays

To transfer yeast strain cultures from four 96-well plates into the corresponding quadrants of a 384-well plate, positions A1 from the first, second, third, and fourth 96-well plates correspond (clockwise) to positions A1, A2, B2, and B1 on the 384-well plate, respectively. The working volume for 96-well plates is 100–200 μL, meaning cells can be suspended by orbital shaking without well spillover. Max volume for 96-well plates is 300 μL; thus after orbital shaking of cell arrays, additional volume can be added and suspension mixed by pipetting before transfer. The optimal volume for Evergreen plates is 30–40 μL or 40–65 μL for standard 384-well plate. Orbital shaking will not suspend cells or glycerol in 384-well plates; however this can be achieved by stirring with a pin tool. In general, glycerol should be mixed before adding to 384-well plate. 384-Array glycerol stocks can be created simultaneously with making new 96 glycerol stocks (Subheading 3.3.1), or by freezing/thawing 96 glycerol stocks. To consolidate from 96 to 384 arrays:

1. Prepare a sterile 20 % solution of glycerol in water, YPD, or YPD supplemented with G418 if diluting/expanding existing stock prior to consolidation. Use 80 % sterile glycerol if making new 96-well glycerol stocks and consolidating to 384-well at the same time (described in Subheading 3.3.1).
2. Thaw (if necessary) four 96-well frozen glycerol stock plates, shake to suspend cells, and then remove foil (*see* **Note 3**).

3. Add to a 96-well plate a volume of 20 % glycerol that will provide adequate volume to transfer to 384-well plates (e.g., 100 μL so that 50 μL can be transferred to each of the two 384-well plates); more than one 384-well array can be made at a time. Mix by pipetting if well volume exceeds 200 μL (orbital shaking will cause spillover). Use a volume adequate to preserve the original 96-well stocks (slightly diluted) as the new 384-well stocks are created (*see* **Note 4**).
4. Seal plates with adhesive foil, and move promptly to −70 °C to minimize time thawed. Retrieve another set of four plates to thaw while next set of thawed plates is processed.
5. For downstream application, as a way of controlling for plate effects, add the reference strain (same genetic background with no gene deletion; we use the hoΔ0::G418^{R} strain for this) to three of the four empty wells in Rows O and P of a 384 array. These are arranged in a square (e.g., O3, O4, P4, P3 for the MAT**a** collection), resulting from an empty well on Row H of each 96-well array of deletion mutants. Leaving one of the four wells empty provides a way of distinguishing the MAT**a** from a MAT**alpha** library array (*see* **Note 5**).

3.3.3 Create/Replicate Frozen Stocks from Agar Plates

This protocol is used to preserve double mutants after SGA, to make additional copies of 384-array glycerol stocks, or as an alternative method of consolidating 96-well liquid arrays to a 384-well format (*see* **Note 6**).

1. Start by making a 384-agar cell array (e.g., endpoint of the SGA method or after inoculating liquid array to agar to replicate 384 stocks).
2. Label and pre-fill standard profile 384-well plate(s) with 60 μL YPD/20 % glycerol.
3. Transfer cells with sterile pin tool directly from the agar media to the glycerol stock media (*see* **Note 7**).
4. Seal with adhesive aluminum foils, and store at −80 °C.
5. Sterilize pin tool, and repeat the above steps for each cell array stock.

3.4 Quantitative High-Throughput Cell Array Phenotyping

Overview: An "Experiment," by definition, should have a completely matrixed design (Fig. 1); one axis of the matrix will be gene mutations (i.e., the identities of the strains on the yeast mutant arrays), and the other axis will be the perturbations systematically tested for every yeast mutant array. Perturbations can be drugs, genes (e.g., double mutants), or a combination (e.g., tetracycline-regulated gene in a query strain). Thus defined, multiple Q-HTCP experiments can be performed simultaneously, but the arrays should be grouped as separate experiments (do not place plates from different experiments on the same scan).

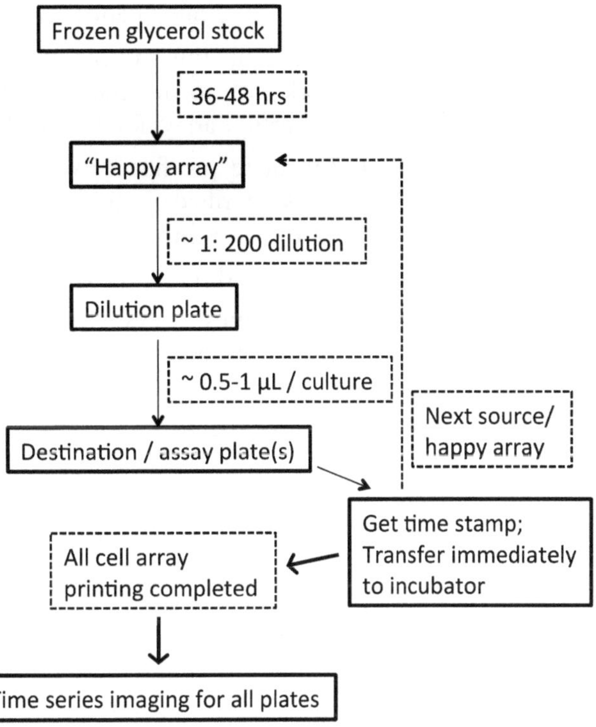

Fig. 1 Q-HTCP workflow

Performing experiments in this way streamlines downstream analysis after the imaging step. Within an experiment, every *source yeast mutant array* will be tested with the same number and type of *perturbations*, which are defined by the makeup of the media in the *destination* plates to which the yeast mutant arrays are transferred prior to imaging. Thus, the number of *source arrays* multiplied by the number of *destination plate types* is equal to the total number of arrays to be imaged as part of one *experiment* (*see* **Notes 8** and **9**).

3.4.1 Making Happy Arrays

"Happy arrays" are freshly grown liquid culture arrays inoculated from glycerol stocks. Happy arrays increase reproducibility by reducing expression of confounding phenotypes, for example, gene-specific strain variability in freeze-thaw viability or survival in stationary phase. Happy arrays can be used after storage at 4 °C, and uniform handling of the entire happy array collection can only help (e.g., time held at each temperature). It is not recommended to inoculate destination plates directly from glycerol stocks. To make happy arrays:

1. Label and fill Evergreen 384-well plates with 40 μL YPD (G418 or another selective antibiotic can be added to the media if desired to maintain mutant selection or prevent contamination). Consider filling a couple of extra plates for backup in case of error/need for extra plate during protocol.

2. Centrifuge filled plates for 30 s at ~400 × *g* (*see* **Note 10**).
3. Thaw cell array frozen stocks (groups of four are convenient to work with to minimize the time the arrays are out of freezer). Centrifuge arrays for 30 s at ~400 × *g* and remove foil (*see* **Note 9**).
4. Inoculate pre-filled Evergreen 384-well plates with glycerol stock (sterilize pin tool, stir pin tool on bottom of stock, transfer to happy array, repeat) (*see* **Note 3**).
5. Reseal and return glycerol stocks to −80 °C.
6. Wrap happy arrays with cellophane in packs of 5–10 plates to reduce evaporation and provide easier handling (Evergreen plates are not interlocking and do not stack well).
7. Incubate at 30 °C for 48–72 h.
8. Optional: Observe happy arrays and record cultures that fail to grow. Alternatively, or in addition, vegetative growth can be assessed by printing to YPD agar as a control in the Q-HTCP experiment. Happy arrays can be stored at 4 °C and used as short-term stocks (a few weeks).

3.4.2 Print Happy Arrays to Test Media

After they are fully grown (48–72 h), happy arrays serve as "source" arrays and are printed (i.e., transferred/inoculated) to "destination" plates consisting of rectangular monowell plates filled with agar media. To optimally discriminate growth differences, a dilution step is recommended. The purpose of the "dilution" plate is to reduce the cell density before transfer, similar to serial dilution in a traditional spot test. As diluted cultures go through more generations to reach the endpoint of growth, smaller differences in cell proliferation can be detected [3]. From the dilution plate, the cell arrays are printed to agar media containing drugs or other variables of interest that constitute the assessment of gene interaction (*see* **Note 11**).

1. Prepare agar plates for an experiment and number them consecutively. The plates are printed in mini-series with each happy array being printed to all destination plate types (e.g., destination plates 1 thru 5 = *happy array source plate 1* printed to no-drug-control destination agar media, and media with high and low concentrations of Drug 1 and Drug 2; plates 6 thru 10 = *happy array 2*, plated to the next series of 5 destination plates) (*see* **Notes 8–12**).
2. Prepare dilution plates by filling Evergreen 384-well plates with 40 μL sterile distilled water or media (it is a good practice to make a couple of extra dilution plates). Media will usually yield larger spots than water (a given drop volume will spread out over a larger area); approximate volume to spot is 0.5–1 μL; approximate number of cells to spot is 200–1,000 (*see* **Notes 11** and **12**).

3. Centrifuge dilution plates for 30 s at ~400 × *g* (to remove air bubbles; if forgotten, cultures may not be diluted correctly).
4. Transfer cells in series from the happy array to the dilution plate and then to the destination agar plates. As a rule of thumb, transfer ~0.5 μL from happy array to dilution plate (~1:100 dilution), and about ~0.5 μL from destination to agar plate (~500 cells). Use pin tool to stir wells on source and dilution plates. Agar printing should be precise and reproducible with array squarely aligned with plate (use a robot or a mechanical guide, *see* **Note 12**).
5. Record the time that each happy array is processed in the workflow (this will be needed for growth curve fitting; the same time stamp can be used for all of the destination plates of a given source array as they are all moved to the incubator at the same time). Automated liquid handlers should have capability to collect print times, but this data must be recorded manually if not.
6. As the printing is completed for each set of agar destination plates corresponding to a happy array, move these to the incubator before processing the subsequent happy array (an important detail to equalize times of incubation for all arrays in the experiment as small temperature differences can have large effects on growth rate) (*see* **Note 14**).

3.4.3 Acquire Images

1. Use the Epson Expression 10000 XL—Photo Scanner with Transparency Unit Lid: Make sure a 3/32″ sheet of clear Plexiglas cut to ½″ × 12″ is used as a guide along the left side of the scanner, and that it is squarely flush against the glass border at top left. This is required to properly position the plates for imaging. The scanner transparency unit lid must be raised in the z direction so that it can rest evenly across the plates; metal washers should be used for the purpose of holding the transparency unit lid in this optimal position. Remove the lids from the cell array agar plates prior to scanning. The scanner should accommodate ten plates, oriented as shown in Fig. 2. Careful positioning of plates prior to scanning is essential, because the same array map will be used to analyze images from an entire experiment of scans (*see* **Note 13**).

   Scanner Settings

   Mode: Professional mode

   Document type: Film

   Film type: Positive film

   Image type: 8-Bit gray scale

   Resolution: 280 dpi

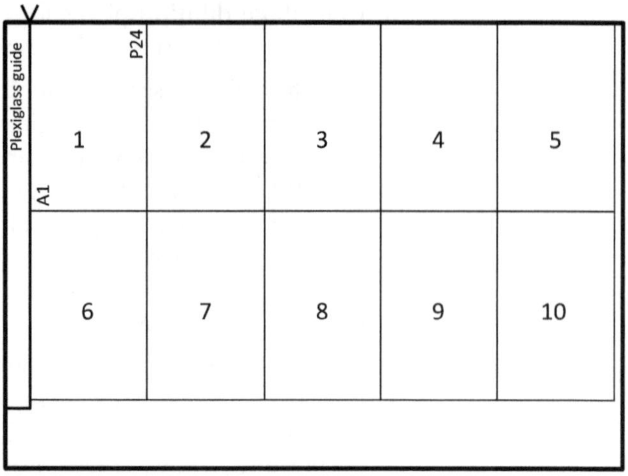

Fig. 2 Q-HTCP scan layout

Document size: 12.20 in. × 16.50 in.

Target size: 12.20 in. × 16.50 in.

Scale: 100 %

(Uncheck all options):

No unsharp mask filter

No grain reduction

No dust removal

Scan file type = Tiff

2. Create a new "experiment folder," so that it can be selected from the scan driver as the location to save the images. The "film" document-type mode of scanning is used, in which the images if opened and viewed will appear to be flipped horizontally, representing the view looking upward from beneath the glass surface of the scanner (this is normal, and the image analysis software assumes this orientation). Using the "positive film" mode, the cultures will appear gray and the background white. Begin scanning as soon as printing of all destination plates in an experiment is completed (this may be several hours after printing the first source to destination plate series). Scanning will be in groups of ten, and typically in the order the plates were printed. The file-naming scheme utilizes a format provided by the Epson scan driver and the resulting file names are important for proper calling by the image analysis software described below. In the Epson Scan driver, choose the "experiment folder" to store images. Name each scan *set* according to the approximate hours after the first scan set was taken. Use three digits followed by "hr" (e.g., the first scan set will be "000hr"). The driver will automatically increment each scan

file name and save to the selected experiment folder (e.g., 000hr001, 000hr002). The order for scanning and the order of plates on each scan must be maintained the same for every time point, and the placement of plates on the scanner must be precise for all scan sets throughout the experiment (Fig. 2). Scan sets should be acquired at 2–6-h intervals for 3–5 days, until all cultures have reached their plateau phase of growth, or carrying capacity.

3. The first scan set ("000hr") should be obtained immediately after all plates are printed. Typically, the second scan set should be obtained 12–16 h after the first plates are printed if grown on glucose media (*see* **Note 14**). If spotting at higher density (e.g., without dilution), earlier scans will be needed; or alternatively, if the growth rate on the media (e.g., non-fermentable media) is slow, it may take longer before proliferation of the cell culture arrays is detectable by scanning. In any case, once the image threshold is reached (it generally correlates with when spots can be seen by careful visual inspection), scanning should be more frequent during early/exponential growth (e.g., once per doubling time), and can be less frequent as cultures begin reaching carrying capacity and growth slows down. It is ideal to have six or more readings during the dynamic growth of each culture. This will occur in different time windows on different scans, but the reference arrays serve as a guide (*see* **Note 9**).

Scanning Protocol

1. Group plates in the incubator in sets of ten, in the order to be scanned. If performing more than one experiment at a time (*see* **Note 8**), start each new experiment as a separate scan set (i.e., do not combine plates from two different experiments on a single scan). Leave empty positions on the final scan of an experiment; do not fill them with plates from the subsequent experiment. Otherwise, always include ten cell arrays per scan.
2. Remove from the incubator one scan/stack of plates at a time (to avoid temperature effects).
3. Name scan images following the convention "000hr001" where the first five digits are entered into the Scan driver file name box and represent an approximate elapsed time from the first scan and the last three digits represent the scan number. The Epson scanner software will auto-increment and append the last digits as an auto-indexing feature (deselect the option to automatically overwrite file names).
4. Scans from all time points can go in the same experiment folder (e.g., "000hr001.... 000hr008; 010hr001... 010hr008"). Each scan time series will be later separated prior to image analysis.
5. Continue scanning until all cultures appear to have reached a growth plateau at which time additional imaging is non-informative (*see* **Note 15**).

3.5 Kinetic Analysis of Growth from Time Series of Cell Array Images

After imaging has been completed, image analysis and growth curve fitting are carried out using custom software. An earlier version of image analysis software for agar cell arrays was described [15]. Time series images of agar cell arrays are analogous to OD readings of a liquid culture array. Compared to OD monitoring of liquid cultures, agar cell array image analysis reports over a greater dynamic range and does not saturate at high growth density; these data fit very well to a logistic growth function [15]. For Q-HTCP the density of each spot culture is assessed as a pixel average over its area on the cell array. A newer version of the software (EASY Phenomics) takes multiple input files for image analysis (*see* **Note 16**). These include (1) the raw images reorganized by placing each scan time series in a separate folder; (2) an array map file defining the approximate location of each spot culture on a scan (*see* **Note 17**); (3) a print times file, which defines "time zero" when each proliferating agar cell array was created (by printing/inoculating from the happy array and dilution plate to the destination plate); and (4) the Master Plate and Drug/Media files, which contain information associated with each cell array in the experiment and their organization in the plate/scan layout. The latter information can be incorporated after image analysis and growth curve fitting have been satisfactorily completed, essentially providing the experimental labels for each growth curve. To perform Q-HTCP analysis with Easy Phenomics software (*see* **Note 16**):

1. Create the necessary files.
   (a) Create a directory of Q-HTCP experiments and an "Experiment Folder" for each experiment. Sort the time series for each scan in the experiment into its own folder (000h001, 010hr001… etc. in a folder called "Scan1"; 000hr002, 010hr002… etc. in a folder called "Scan2"… , etc.), and place all the Scan folders into the experiment folder.
   (b) Create a folder named "PT Map Scans" within the Experiment Folder. Select a few good-quality images, and copy them into the PT Map Scans folder. Use scan images containing cell arrays at all of the ten positions and where the plates are perfectly placed. Choose well-aligned scans without shifted or angulated plates, and with most spot cultures having growth (*see* **Note 17**).
   (c) Make a Print Times file, called "PrintTimes.txt" and place it in the Experiment Folder. The print times file contains the time at which each Source Array was printed, marking the beginning of the growth curve for each set of destination plates; a single print time applies to all destination plates associated with a single source array. All plates associated with the same print time should be moved as a

group to the incubator prior to printing the subsequent source array. The number of print times will be equal to the number of source arrays in the master plate file. The "PrintTimes.txt" has the format:

3/1/2012 14.45

3/1/2013 14.52

(d) Create a folder in the Experiment Folder called "MasterPlateFiles." This folder contains the information about the Source Arrays and Destination Plates comprising the experiment, and will be incorporated to correctly label the cell array growth curve data (*see* **Note 16**).

(i) Make a "Master Plate File," consisting of the identities of strains on each source array file and store it in the "MasterPlateFiles" folder (templates will be provided with the software).

(ii) Make a "DrugMedia" file and store it in the MasterPlateFiles folder (templates will be provided with the software). The DrugMedia file contains information about the media and perturbations that are being tested in the experiment (i.e., the series of destination plates that each source array is printed too).

2. After all files have been placed in their correct location in the Experiment Folder, run the EASY Phenomics software (*see* **Note 16**):

(a) From the "File" drop-down menu, select "New Analysis Folder" and browse to the correct Experiment Folder to create an Analysis Folder that will contain several folders used for image analysis, curve fitting, and report generation. The Analysis Folder will automatically be named with the current date with any additional name information input appended (the same experiment can be analyzed multiple times and ways). Results will be output to this Analysis Folder.

(b) From the "Run" drop-down menu, select "Create PinTool Map," and then select images from the "PT Map Scans" folder to create an array map locating all 3,840 spots on a scan. Inspect the array map, choose other scan images if needed to improve the array map, and continue when array map is satisfactory.

(c) From the "Run" drop-down menu, select "Run Image Analysis," and then choose one or more scan series to be analyzed. A pop-up box will indicate when the image analysis is complete.

(d) From the "Run" drop-down menu, select "Run Curve Fit." The image density data for a single spot culture can be chosen and displayed as a plot with the fitted growth curve, or entire scans (3,840 spot cultures per scan) can be analyzed in which case the growth parameters are output to a spreadsheet and saved in the Analysis Folder. For curve fitting, the data are fit to a logistic growth function:

$$G(t) = K / \left(1 + e^{-r(t-l)}\right)$$

$G(t)$ is the size of the population at time, t. K is the carrying capacity, or final yield of the culture; r is the maximum specific rate (biologically, this is a population size-normalized rate, when there is nutrient excess); l is the time at which half the carrying capacity is reached, which also corresponds to the maximum growth rate (absolute population size increase, not normalized to population size). The parameters K, r, and l are cell proliferation phenotypes, which can be used to quantify interaction from multiple growth curves [15]. A method for quantifying gene interaction is described in the next section, though the growth curve parameters could be employed in other models of gene interaction [16, 17].

3. Compile the final results.
   (a) From the "Generate Reports" drop-down menu, select "Import MP and DrugMedia Data," browse to the MasterPlate and DrugMedia files in the MasterPlateFiles folder within the Experiment Folder, and choose them. The data in these files will be associated with curve fits obtained previously.
   (b) From the "Generate Reports" drop-down menus, select "Gen Results File." This will incorporate all together, the labels, curve fit parameters, time series data used for curve fitting, statistical quality metrics, and diagnostic criteria. The final result files are found in the PrintResults folder within the Experiment Analysis folder. This ResultsFile.txt should be saved as ResultsFile.xls file to further utilize the EASY Phenomics plotting and figure generation utilities (*see* **Note 16**).

3.6 Quantify Interactions

Conceptually, gene interactions provide experimental insight to predict how a phenotype resulting from biological perturbations can unexpectedly change when the perturbations occur in combination [1, 23]. The biological perturbations must include at least one genetic mutation, where gene interaction can be generally represented as

$$\text{Phenotype}_{\text{observed}} = \text{Phentoype}_{\text{expected}} + \text{Gene interaction}$$
$$\text{Gene interaction} = \text{Phenotype}_{\text{observed}} - \text{Phenotype}_{\text{expected}}$$

$Phenotype_{observed}$ is the directly measured phenotype in the context of a perturbation combination (e.g., double mutant or single mutant additionally treated with drug). $Phenotype_{expected}$ incorporates the directly measured phenotypes in the context of no perturbation and each single perturbation, and depends in part on a neutrality function [17], which reflects the assumptions inherent to determining the $Phenotype_{expected}$ (*see* **Note 18**). An approach we have recently found to be useful employs "ΔL" (*l* is a parameter of the logistic growth function known as time to maximum growth (TMR), or the time at which half the carrying capacity (K/2) is achieved); it is used as an indication of a time shift in the growth curve [15, 16]. Using the *l* values to compare growth of cultures, we (a) determine the distribution of control strain phenotypes over a range of perturbation intensity; (b) normalize the effect of the yeast mutant array strain mutation in the absence of a second perturbation (i.e., the single perturbations not treated with drug or dial down of a query gene); (c) subtract the *l* value for each perturbation combination from the median single-perturbation state of wild-type strain; d) fit the differences between the single- and double-perturbation states across a range of perturbation intensities to a quadratic equation; and e) use the fitted difference at the highest perturbation intensity observed [16]. Other neutrality models of interaction and/or incorporating other cell proliferation phenotypes (e.g., other parameters of the logistic function) can be investigated with Q-HTCP data. We expect that the utility of particular cell proliferation phenotypes (i.e., *K*, *l*, or *r*) will be different for different perturbations. For example, the phenotype induced by some drugs may be best reflected in the rate, while others are better reflected by the carrying capacity or L parameter.

With the goal of Q-HTCP providing growth curve-derived phenotypes that support high-resolution modeling of genetic buffering networks, it is important to establish protocols that can be applied across different laboratories that will lead to an accumulation of gene interaction data that can be assessed in an integrative manner to advance our understanding of complex phenotypes in yeast, which may also be evolutionarily relevant for phenomic analysis of human disease [16].

3.7 Plate Sterilization

To help research funds go further and preserve the environment, plasticware can be reused by scooping out agar and rinsing plates, soaking plates in 7 % bleach in deionized water (with a dash of dishwashing liquid) for several hours, rinsing with deionized water, drying, rinsing with 100 % ETOH, and storing in sealed plastic bins (more details available upon request).

4 Notes

1. Monowell plates should be labeled on the left side (A1 being in the top left corner), looking down on the agar array (*see* Fig. 2), and the same volume of agar (38–40 mL) should be poured in every array. Use a striping pattern to indicate media type at the time of pouring, but number plates consecutively just prior to cell array printing, when each source array will be printed to an identical series of different media types. Use 2 % agar and allow it to cool for ~45 min unlidded after pouring; this will improve printing by facilitating absorption of liquid after printing (*see* **Note 8**).
2. Except for the Evergreen plates, we use all Nunc brand to maintain lid compatibility (lids are non-interchangeable between brands).
3. Work with 8–12 plates out of freezer at a time (in sets of four), one or two sets thawing while other is being transferred.
4. Rinse tips with water and sterilize by autoclaving to recycle/reuse and reduce cost.
5. In addition to placing the reference strain on each 384-well yeast mutant array to assess plate effects (e.g., mislabeling or other problem with media production), create full "Reference Arrays" (all wells have the reference strain) to assess position effects and print them at the beginning and end of each experiment. The Ref Arrays, if printed first and last in an experiment, provide assessment of technical variation from beginning to end of experiment, and also provide data to assess array position effects (e.g., variation in pin tool transfers or edge effects on cell proliferation). We recommend using the *ho*-Δ*0* deletion strain (*ydl227c*-Δ*0*::G418^R), which can also be manipulated identically to the yeast mutant arrays for SGA-construction of double-mutant array controls [16].
6. Agar arrays are easier to work with, but slow-growing strains may be outcompeted on the agar array (since the media is shared), thus yielding low-density stocks. Such strains can be difficult to analyze subsequently; however low-density stocks can be partially restored to higher density at the "happy array" step of Q-HTCP (where media is not shared and each culture is grown to saturation). Since it is not possible to vortex in 384-well format, glycerol should be premixed before adding to plates. Cells will not grow appreciably in the high glycerol media, so move glycerol stocks to −70 °C storage soon after transferring cell paste.
7. Before freezing the new glycerol stock, consider inoculating fresh media for happy arrays that can be used to proceed directly to Q-HTCP without the need for a later freeze-thaw step.

8. Source-destination combinations should be unique within an "experiment" (no duplicates). For example if an experiment is being performed in replicate (e.g., every source to destination is printed in duplicate), replicate arrays should be grouped and scanned as two separate series, rather than a single "experiment." In general, plates from each experiment are scanned together in successive groups of ten, and plates from different experiments should not be grouped on the same scan (if the last scan of one experiment contains less than ten plates, do not include the first plate of the subsequent experiment on that scan). For example, if in a single experiment there are seven source arrays and four different destination/perturbation plates, with two replicates of each plate, the first replicate set would be imaged on scans 1–3, and the second replicate set on scans 4–6; scans 3 and 6 would consist of only 8 arrays.

9. Use a full "Reference Array" at the beginning and end of each experiment. The Ref Arrays provide assessment of technical variation from beginning to end of experiment and also provide data to assess "array position effects." For the reference arrays, instead of the true reference strain (e.g., BY4741 or BY4742), which is functionally homothallic, we use the G418-sensitive *ho-Δ0* deletion strain (*ydl227c-Δ0*::G418^R), which can be manipulated identically to the yeast mutant arrays for growth on G418 and SGA construction of double-mutant array controls [16]. In addition to the reference arrays, control strains can be added to three of the eight empty positions on the yeast mutant arrays to aid identification of plate-to-plate variation among destination plates of the same type (*see* **Note 5**).

10. It is important to centrifuge the media-filled plates to remove trapped air bubbles that may occur during filling; if forgotten, bubbles can expand by trapping CO_2 and forcing cultures out of well.

11. The exact amount of dilution matters less than consistency in the dilution factor for cultures across the array. Higher dilution enables discrimination between smaller growth differences, but it is important to have a lawn density high enough (100–1,000 cells per spot) for confluent population growth, rather than distinct colonies, which introduce artifactual differences. Have the plates at room temp before printing and make sure that the agar surface is dry. Printed cultures will run together where the agar surface is wet. To avoid this problem, allow agar plates to solidify with lid off (*see* **Note 1**), and store plates agar side up. It is ok to invert agar arrays after liquid drops have been absorbed into agar, but not before. Move sets of competed destination plates to incubator immediately and record print time before proceeding with cell array printing. If spots on destination arrays are not uniform, pin tool cleaning may help (Subheading 3.2).

12. Using V&P Scientific FP6 (1/16″ diameter) style pins, three transfers will yield about the desired 0.5 μL. One can empirically determine the transfer volumes needed to obtain a light lawn of cells at each spot of the array. If there are spot quality/uniformity issues on the cell arrays, troubleshooting includes cleaning the pin tool with V&P cleaner or another suitable chemical wash. In addition, different agar media have different surface tension properties and water has greater surface tension than media. How well plates are dried also affects spot characteristics on array. If spots are not of desirable size consider using media instead of water for dilution plates.
13. Create a directory to contain all Q-HTCP experiments. Within the Q-HTCP image folder, create a separate folder for each experiment and name accordingly (date, initials of experimenter, brief description of experiment, etc.). Scan ten plates at a time unless it is the final scan of an experiment. Remove only ten plates at a time from incubator. Pay careful attention to the placement of cell arrays on the scanner. Correctly position the plexiglass guide, and make sure that all ten cell arrays are flush against the glass surface border of the scanner, bases of plates are not overlapping, and all sides of arrays are squarely in contact (Fig. 2). Do not have plates from different experiments on the same scan. If working quickly, it will take ~2 min per scan; for example, for experiment with 150 plates, expect it will take about 30 min to collect each time point of data.
14. Return plates quickly to incubator (only ten at a time should be out for scanning) to avoid plate effects due to temperature variation.
15. If plate contamination occurs during the course of scanning, in general it is easier to leave it alone, and remove the images prior to image analysis rather than trying to remove the contamination from the plate. If a plate must be removed during scanning, use an empty plate as a placeholder to maintain the scan positions for all other plates in the series. Keep a tray of water inside the incubator to increase humidity, which will slow cracking in the agar. Additionally, it may help to reduce airflow, depending on the incubator model.
16. The software is still under development, but we have compiled an executable prototype for use with this protocol. It is available upon request, along with a more detailed user's manual.
17. Exemplary scan images from the experiment, where nearly all cultures on the scan are growing, should be used to create the array map. This is typically the reference arrays and arrays that are unperturbed (control media). Scans can be taken with ten exemplary cell arrays solely for the purpose of creating the array map file if there are not exemplary scans within the experiment.

Multiple scans can be overlaid so that a composite image maps all cell array positions of a scan (*see* Subheading 3.5, **step 2**). The array map file may be similar from experiment to experiment, but it is good practice to obtain exemplary scan images specific for each experiment in case something about the cell array printing or placement of cell arrays changes from one experiment to another.

18. For quantifying interactions, it is recommended to incorporate summary statistics from the relevant cell proliferation parameter for the Reference Arrays to assess the full range of assay variation and to more confidently assess whether a particular gene mutant influences the phenotype.

References

1. Hartman JL IV, Garvik B, Hartwell L (2001) Principles for the buffering of genetic variation. Science 291:1001–1004
2. Dixon SJ, Costanzo M, Baryshnikova A et al (2009) Systematic mapping of genetic interaction networks. Annu Rev Genet 43:601–625
3. Hartman JL IV, Tippery NP (2004) Systematic quantification of gene interactions by phenotypic array analysis. Genome Biol 5:R49
4. Li Z, Vizeacoumar FJ, Bahr S et al (2011) Systematic exploration of essential yeast gene function with temperature-sensitive mutants. Nat Biotechnol 29:361–367
5. Schuldiner M, Collins SR, Thompson NJ et al (2005) Exploration of the function and organization of the yeast early secretory pathway through an epistatic miniarray profile. Cell 123:507–519
6. Tong AH, Evangelista M, Parsons AB et al (2001) Systematic genetic analysis with ordered arrays of yeast deletion mutants. Science 294:2364–2368
7. Winzeler EA, Shoemaker DD, Astromoff A et al (1999) Functional characterization of the S. cerevisiae genome by gene deletion and parallel analysis. Science 285:901–906
8. St Onge RP, Mani R, Oh J et al (2007) Systematic pathway analysis using high-resolution fitness profiling of combinatorial gene deletions. Nat Genet 39:199–206
9. Guo J, Tian D, McKinney BA et al (2010) Recursive expectation-maximization clustering: a method for identifying buffering mechanisms composed of phenomic modules. Chaos 20:026103
10. Hartwell LH, Hopfield JJ, Leibler S et al (1999) From molecular to modular cell biology. Nature 402:C47–C52
11. Blomberg A (2011) Measuring growth rate in high-throughput growth phenotyping. Curr Opin Biotechnol 22:94–102
12. Giaever G, Chu AM, Ni L et al (2002) Functional profiling of the *Saccharomyces cerevisiae* genome. Nature 418:387–391
13. Parsons AB, Brost RL, Ding H et al (2004) Integration of chemical-genetic and genetic interaction data links bioactive compounds to cellular target pathways. Nat Biotechnol 22:62–69
14. Tong AH, Lesage G, Bader GD et al (2004) Global mapping of the yeast genetic interaction network. Science 303:808–813
15. Shah NA, Laws RJ, Wardman B et al (2007) Accurate, precise modeling of cell proliferation kinetics from time-lapse imaging and automated image analysis of agar yeast culture arrays. BMC Syst Biol 1:3
16. Louie RJ, Guo J, Rodgers JW et al (2012) A yeast phenomic model for the gene interaction network modulating CFTR-ΔF508 protein biogenesis. Genome Med 4:103
17. Mani R, St Onge RP, Hartman JL IV et al (2008) Defining genetic interaction. Proc Natl Acad Sci U S A 105:3461–3466
18. Baryshnikova A, Costanzo M, Kim Y et al (2010) Quantitative analysis of fitness and genetic interactions in yeast on a genome scale. Nat Methods 7:1017–1024
19. Costanzo M, Baryshnikova A, Bellay J et al (2010) The genetic landscape of a cell. Science 327:425–431
20. Tong AH, Boone C (2006) Synthetic genetic array analysis in *Saccharomyces cerevisiae.* Methods Mol Biol 313:171–192
21. Singh I, Pass R, Togay SO et al (2009) Stringent mating-type-regulated auxotrophy increases the accuracy of systematic genetic

interaction screens with *Saccharomyces cerevisiae* mutant arrays. Genetics 181:289–300
22. Hartman JL IV (2007) Buffering of deoxyribonucleotide pool homeostasis by threonine metabolism. Proc Natl Acad Sci U S A 104: 11700–11705
23. Hartman JL IV (2006) Genetic and molecular buffering of phenotypes. In: Rodriguez R, Kaput J (eds) Nutritional genomics: discovering the path to personalized nutrition. Wiley, Hoboken, NJ, p 496
24. Collins SR, Schuldiner M, Krogan NJ et al (2006) A strategy for extracting and analyzing large-scale quantitative epistatic interaction data. Genome Biol 7:R63
25. Lawless C, Wilkinson DJ, Young A et al (2010) Colonyzer: automated quantification of micro-organism growth characteristics on solid agar. BMC Bioinformatics 11:287
26. Koh JL, Ding H, Costanzo M et al (2009) DRYGIN: a database of quantitative genetic interaction networks in yeast. Nucleic Acids Res 38:D502–D507
27. Califano A, Butte AJ, Friend S et al (2012) Leveraging models of cell regulation and GWAS data in integrative network-based association studies. Nat Genet 44:841–847
28. Burke D, Dawson D, Stearns T (2000) Methods in yeast genetics. CSHL Press, Cold Spring Harbor

Chapter 13

Detection of Short-Range Chromatin Interactions by Chromosome Conformation Capture (3C) in Yeast

Badri Nath Singh and Michael Hampsey

Abstract

We describe a modified 3C ("chromosome conformation capture") protocol for detection of transient, short-range chromatin interactions in the yeast *Saccharomyces cerevisiae.* 3C was initially described by Job Dekker and involves formaldehyde cross-linking to stabilize transient chromatin interactions, followed by restriction digestion, ligation, and locus-specific PCR. As such, 3C reveals complex three-dimensional interactions between distal genetic elements within intact cells at high resolution. Using a modified version of Dekker's protocol, we are able to detect gene loops that juxtapose promoter and terminator regions of yeast genes with ORFs as short as 1 kb. We are using this technique to define the *cis-* and *trans-*acting requirements for the formation and maintenance of gene loops, and to elucidate their physiological consequences. We anticipate that this method will be generally applicable to detect dynamic, short-range chromatin interactions, not limited to gene loops.

Key words Yeast, *Saccharomyces cerevisiae*, Chromatin, Transcription, 3C, Gene loop, Short-range interaction

1 Introduction

Chromosome conformation capture (3C) is a powerful technique to detect and quantify the frequency of physical interactions between any two regions of a genome [1]. 3C involves stabilization of transient chromatin interactions by formaldehyde cross-linking, followed by chromatin extraction and subsequent digestion with a specific restriction enzyme. DNA fragments are then ligated in dilute solution under conditions that favor intramolecular ligation. Following reversal of cross-links, specific ligation products are quantified by PCR (Fig. 1). Accordingly, 3C converts physical chromatin interactions in vivo into specific ligation products in vitro with the abundance of PCR products representing the frequency of interaction. Variations on 3C, including "circular 3C" and "3C on ChIP" (both denoted 4C), "3C carbon copy" (5C), and Hi-C, have been developed to expand the scope of detection of chromatin interactions [2–10].

Jeffrey S. Smith and Daniel J. Burke (eds.), *Yeast Genetics: Methods and Protocols*, Methods in Molecular Biology, vol. 1205, DOI 10.1007/978-1-4939-1363-3_13,

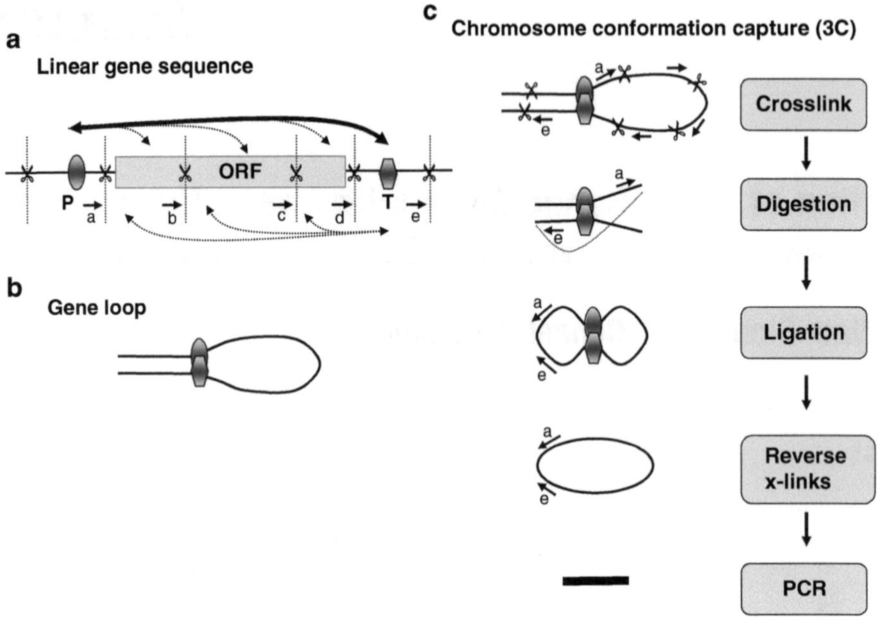

Fig. 1 Schematic depiction of the 3C analysis. (**a**) Linear depiction of a gene with promoter on the *left* (P), terminator on the *right* (T). Primers are denoted by lower case letters a–e. Restriction sites are denoted by *dotted vertical lines* with scissors. (**b**) A transient gene loop formed by juxtaposition of the promoter and terminator. (**c**) Schematic depiction of 3C method to detect gene loops. PCR products from tandem primer pairs a and e are specific for juxtaposition of fragments P and T

3C was developed to detect long-range interactions, on a scale of a few kilobase pairs to a few hundred kilobase pairs. We have adapted 3C to explore dynamic interactions at single-gene loci in the yeast *S. cerevisiae.* A key feature of our protocol is formaldehyde cross-linking in whole cells, rather than isolated nuclei, a condition that we find to be essential for stabilization of transcription-dependent chromatin interactions. Using this protocol, we have detected physical interactions between promoter and terminator regions of genes with ORFs as short as 1 kb in length, and have begun to define the structural requirements for the formation and persistence of gene loops. An earlier version of our protocol was published previously [11].

We have successfully used 3C to detect gene loops using derivatives of the commonly used laboratory strains S288C and W303. Detection of gene loops by 3C in wild-type strains is unaffected by elevated growth temperature; thus, loops can be detected with equal efficiency in cells incubated at 30 °C or 37 °C. This observation is important because it allows for the use of temperature-sensitive mutants to identify looping requirements. For example, we were able to demonstrate that looping is transcription dependent by observing diminished looping in a temperature-sensitive *rpb1-1* mutant at 37 °C; the critical control showed that looping was unaffected at 37 °C in the wild-type strain [12–14].

2 Materials

2.1 Buffers

1. FA lysis buffer: 50 mM HEPES·KOH (pH 7.9), 140 mM NaCl, 1 mM EDTA (pH 8.0), 1 % Triton X-100, 0.1 % sodium deoxycholate, 1 mM PMSF. PMSF must be freshly added to the FA lysis buffer just before use from a 100 mM stock solution.
2. TBS: 10 mM Tris–HCl (pH 7.5), 200 mM NaCl.
3. TBE: 90 mM Tris–HCl (pH 7.5), 90 mM boric acid, 2 mM EDTA.
4. TE: 10 mM Tris–HCl (pH 8.0 or pH 8.9), 1 mM EDTA.
5. 6× Sample buffer: 0.25 % xylene cyanol FF, 30 % glycerol in water. Do not include bromophenol blue as the dye can obscure detection of short PCR products.

2.2 Selection of Genes and Restriction Enzymes for 3C Analysis

Resolution of chromatin interactions by 3C is defined by restriction digestion (Fig. 1). Accordingly, target genes must be chosen based on the distribution of appropriate restriction sites. Restriction sites must flank the putative interacting regions, designated fragments P and T in Fig. 1. The closer the restriction sites are to the region of interest, the greater the degree of resolution by 3C.

More than a single restriction enzyme can also be used as long as these enzymes generate compatible cohesive ends for subsequent ligation. For example, *Bam*HI, *Bcl*I, and *Bgl*II can be used simultaneously as all three enzymes generate identical GATC cohesive ends. Restriction enzymes with four-base pair recognition sites are likely to yield higher 3C resolution as these enzymes cut the genome on average every 256 base pairs. We have obtained especially good results (high resolution and efficient ligation) with the 4-bp recognition enzymes *Msp*I and *Nla*III. Restriction enzymes that cut at 65 °C cannot be used in 3C analysis of yeast because prolonged incubation at 65 °C disrupts DNA-protein cross-linking.

2.3 Design of 3C Primer Pairs

1. Primer length should be typically 20–30 bp with 40–50 % GC content and not more than 5 °C difference in melting temperature among the primers.
2. Each primer should be analyzed to confirm that its sequence is unique in the genome.
3. New primers should be designed if multiple PCR products are observed.
4. The size of PCR products should be relatively small (150–350 bp), allowing for detection of quantitative differences among PCR products.
5. Primers should be positioned 50–200 bp from the adjacent restriction sites (Fig. 1).

6. Primer pairs can be designed in either tandem or divergent orientation, although tandemly oriented primers are preferable (Fig. 1). Divergent primer pairs (for example, if primer "a" were oriented in the opposite direction) have the potential to yield PCR products without the DNA template necessarily having been cut between the primer pairs. Accordingly, meaningful 3C data using divergent primer pairs necessitates control PCR reactions using convergent primer pairs to monitor digestion at internal restriction sites [6, 11].

2.4 Controls

Proper consideration and execution of controls are essential to obtaining meaningful data from 3C analysis [15].

2.4.1 Primer Pair Efficiency

When comparing 3C data using multiple PCR primers, it is critical to control for primer pair efficiency. Template DNA to be used as a control for primer pair efficiency can readily be generated by digesting and ligating PCR-amplified DNA encompassing the region of interest. Quantitative data can then be obtained by normalizing 3C PCR products with PCR products from randomly ligated DNA. Template DNA can also be generated by digestion and random ligation of purified genomic DNA, although template DNA generated in this manner is more likely to include random insertions of genomic DNA between the restriction fragments to which the primers anneal. In our hands, control PCR products generated from template DNA derived from total genomic DNA often include the anticipated product as the smallest PCR product, as well as multiple, larger products.

2.4.2 Random Interactions

To conclude from 3C data that DNA loops juxtapose distal regions of the gene, it is essential to demonstrate that these two regions do not randomly associate with DNA along the entire length of the gene. This issue can be addressed by "walking" along the gene using one common primer paired with primers that anneal to each of the internal restriction sites. For example, if chromatin fragments encompassing the promoter (P) and terminator (T) reflect a gene loop (Fig. 1), then primers a and e would yield PCR products at levels significantly above primer a or e paired with primers b, c, and d [11]. Note, however, that primers from adjacent DNA fragments (e.g., primers a and b or d and e) can yield relatively high levels of PCR products, presumably due to cross-linking of proximal DNA (e.g., proximal nucleosomes) that is independent of looping.

2.4.3 Strain Comparisons

Comparison of 3C data between two different strains requires data normalization. This can be done by PCR amplification of input control template DNA. In our experiments, we usually amplify a non-transcribed region of chromosome V employing the same primer pair used in our chromatin immunoprecipitation experiments [12–14]. 3C PCR data is then normalized to this input control PCR product. Accordingly, 3C data can be directly compared among different strains or different growth conditions.

2.4.4 Linearity

1. 3C is designed such that PCR levels reflect the amount of ligated template DNA as a measure of the frequency of physical interaction between specific regions of the genome. This outcome requires that template DNA be titrated to determine the linear range of PCR amplification. The most reliable method to quantify PCR data is to use real-time, quantitative PCR (Q-PCR), which assures that data is obtained in the linear range where PCR products reflect DNA template concentrations [16]. If Q-PCR is used to quantify 3C PCR products, care must be taken to confirm amplification of a single product, either by gel electrophoresis or by melting curve analysis (*see* Subheading 3.6, **step 4**).
2. Reliable quantitative data can also be generated by standard PCR where products are visualized in an agarose gel and quantified using a PhotoImager.
3. Alternatively, PCR amplification can be performed in the presence of a radioactive substrate (e.g., α-[^{32}P]dATP); radioactive PCR products are then resolved in 8 % polyacrylamide-TBE gels and quantified using a PhosphorImager.

3 Methods

3.1 Cell Growth

1. Inoculate a 5 ml seed culture in appropriate growth medium from a freshly growing colony (Petri dish). Incubate overnight, with shaking, at 30 °C.
2. From the overnight culture, inoculate 50 ml of growth medium into a 250 ml flask to OD_{600} ~0.15. Incubate cells with shaking at 30 °C to OD_{600} = 0.65–0.8. For a wild-type strain growing in rich medium at 30 °C this will take approximately 5 h. When using temperature-sensitive mutants, harvest cells at mid-log phase and resuspend in pre-warmed (37 °C) medium, followed by incubation at the restrictive temperature (37 °C), typically for 1 h, depending upon the mutant (*see* **Note 1**).

3.2 Chromatin Cross-Linking

1. Add 1.4 ml of 37 % formaldehyde (final concentration = 1 %) to 50 ml culture. Mix thoroughly and incubate for 15–25 min at room temperature, either on a shaker or by swirling briefly every 5 min (*see* **Note 2**).
2. Add 5.4 ml of 1.25 M glycine to quench excess formaldehyde. Mix and incubate for 5 min at room temperature either on shaker or by swirling gently every 1 min (*see* **Note 3**).
3. Transfer cells to a 50 ml centrifuge tube. Pellet cells by centrifugation at 3,000 × *g* for 10 min at 4 °C using a Sorvall SH3000 centrifuge or equivalent.
4. Wash cells once with 10 ml of ice-cold TBS containing 1 % Triton X-100. Pellet cells and resuspend in 1 ml of ice-cold

TBS containing 1 % Triton X-100 and transfer to a 1.5 ml Eppendorf tube.

5. Pellet cells by centrifugation at 10,000 × *g* for 10 min at 4 °C in a microcentrifuge; aspirate buffer. Resuspend the pellet thoroughly in 500 μl of FA lysis buffer by pipetting up and down. Freeze the cells in liquid nitrogen and store at −80 °C.
6. Cells can be stored at this point at −80 °C for up to 1 year.

3.3 Chromatin Extraction

1. Thaw cells on ice and add an equal volume (~500 μl) of acid-washed glass beads. Vortex the cells vigorously (in the cold room) for two cycles of 20 min with a 10 min break on ice. Be sure that the beads do not settle during vortexing since a large amount of foaming occurs. Use ice-cold glass beads to prevent heat formation during vortexing, which can cause physical distortion of chromatin structure.
2. Puncture the Eppendorf tube at the bottom using a heated 22 gauge needle and insert into a 15 ml centrifuge tube. Centrifuge in a Sorvall SH3000 at 3,000 × *g* for 10 min at 4 °C. Collect the lysate in a 15 ml tube. Discard the Eppendorf tube containing the glass beads.
3. Resuspend the loose pellet and transfer the entire lysate to an Eppendorf tube. Centrifuge at 13,000 × *g* in the microfuge for 10 min at 4 °C. In a good chromatin preparation, expect to see a thin transparent layer of chromatin on the top of a turbid pellet of cell debris.
4. Discard the supernatant and resuspend the pellet in 1 ml FA lysis buffer. Centrifuge at 13,000 × *g* in the microfuge for 10 min at 4 °C. Discard the supernatant, and resuspend the pellet thoroughly in 500 μl 10 mM Tris–HCl (pH 7.5).
5. Chromatin samples at this stage can be stored in 50 μl aliquots at −80 °C for several months.

3.4 Chromatin Digestion, Ligation, and Reversal of Cross-Links

1. To 50 μl of the cross-linked chromatin sample, add 10 μl of 10× restriction enzyme buffer and 30 μl of ddH_2O. Mix well by pipetting up and down.
2. Add restriction enzyme, mix well, and incubate for 5 h at 37 °C with gentle shaking. We often use 10 μl of *Eco*RI (20 U/μl) or *Hin*dIII (20 U/μl). We have also obtained excellent results using the 4-bp recognition enzymes *Nla*III and *Msp*I. Monitor the extent of restriction digestion by convergent PCR (*see* **Note 4**).
3. Add 10 μl of 10 % SDS and heat the sample at 65 °C for 20 min to inactivate the restriction enzyme. Add 565 μl of ddH_2O and 75 μl of 10 % Triton X-100 to sequester the SDS.

4. Centrifuge the sample at 13,000 × *g* for 20 min at room temperature. Discard the supernatant and resuspend the pellet thoroughly in 100 μl 10 mM Tris–HCl (pH 7.5) by pipetting up and down (*see* **Note 5**).
5. To 100 μl of digested chromatin add 350 μl of Quick Ligase Buffer, 245 μl of ddH_2O, and 5 μl of Quick T4 DNA ligase (total volume = 700 μl). Mix thoroughly and incubate the sample for 1 h at 25 °C (*see* **Note 6**).
6. To ensure complete removal of RNA, add 2 μl of 10 mg/ml DNase-free RNase to the reaction mixture. Incubate at 37 °C for 10 min.
7. To reverse cross-links, add 7 μl of 10 % SDS and 5 μl of 10 mg/ml proteinase K to the ligation mixture, mix well, and incubate overnight at 65 °C in a water bath.

3.5 Precipitation of Nucleosome-Free DNA

1. Add an equal volume (700 μl) of phenol:chloroform:isoamyl alcohol (25:24:1) to the reaction mixture. Vortex the sample for 20 s, and centrifuge at 10,000 × *g* for 10 min at room temperature.
2. Collect the upper aqueous phase. Be careful to avoid the interface layer as the aqueous phase will appear turbid.
3. Repeat the phenol:chloroform:isoamyl alcohol extraction one time, followed by chloroform/isoamyl alcohol extraction one time to remove the residual phenol.
4. Transfer the aqueous phase to a fresh Eppendorf tube, and add 1/10 volume of 3 M sodium acetate (pH 5.2) and 2 μl of glycogen (20 mg/ml). Vortex briefly. Add 2.5 volumes of ethanol and mix gently by inverting the tube. Incubate for 20 min at room temperature and centrifuge for 20 min at 13,000 × *g* at room temperature (*see* **Note 7**).
5. Decant the supernatant by gently inverting the tube to avoid dislocating the tiny pellet. Recentrifuge at 13,000 × *g* for 10 min at room temperature. Carefully withdraw the remaining supernatant using a 100 μl pipette. Air-dry the pellet completely and dissolve in 50 μl TE (pH 8.0). This is now the 3C DNA template.
6. Determine the DNA concentration by absorption spectroscopy at 260 nm. Typically 250–300 ng of the DNA template is required for 3C PCR reactions (*see* **Note 8**).
7. The 3C template can be stored at this stage at −20 °C for several months.

3.6 PCR

1. Thaw the reaction components on ice and vortex well before use. For each PCR reaction mix 5 μl 10× PCR buffer, 1 μl 10 mM dNTP mix, 1 μl of each primer pair (25 pmol), 0.5 μl of Taq DNA polymerase (5 U/μl), 300 ng of 3C DNA

template, and ddH_2O to bring the total volume to 50 μl. Tap to mix, and then centrifuge briefly in a microcentrifuge to collect reagents at the bottom of the tube.

2. Program the thermal cycler for hot-start, typically 94 °C for 2 min, followed by 30 cycles at 94 °C for 30 s, 52 °C for 30 s, and 72 °C for 30 s, and then final extension at 72 °C for 5 min.
3. When the PCR amplification is complete, add 10 μl of 6× gel loading buffer without bromophenol blue and briefly centrifuge to mix. Separate the PCR products in a 1.5 % agarose gel containing 1× TBE and 0.5 μg/ml ethidium bromide, typically running the gel at 100 V for 2 h (*see* **Notes 9** and **10**).
4. Quantify the PCR products using a gel documentation system containing appropriate software. The 3C interaction frequency for each primer pair is determined as the ratio of the 3C PCR product to the control PCR product. Set up the PCR reactions in triplicate to determine average interaction frequency. Accordingly, the relative abundance of the PCR products is proportional to the frequency with which the chromatin regions interact (*see* **Note 11**).

4 Notes

1. Cell density (A_{600}) can significantly affect chromatin yield and efficiency of restriction digestion. Accordingly, care must be taken to harvest different cell cultures at comparable A_{600} values. If the cell density is higher (OD_{600} ~0.9–1.0), use less (e.g., 20 μl) chromatin for restriction digestion (*see* Subheading 3.1, **step 2**).
2. For detection of short-range gene loops it is important that the time of cross-linking does not exceed 20–25 min. We are able to detect gene loops that specifically juxtapose 5′ and 3′ ends at *HEM3* (ORF = 984 bp) following 20 min of cross-linking [13, 14]. Longer periods of exposure to formaldehyde result in enhanced cross-linking along the length of the gene. Use formaldehyde that is less than 1 year old for efficient cross-linking (*see* Subheading 3.2, **step 1**).
3. Glycine at 1.25 M often precipitates. In this case, redissolve crystals by heating at >50 °C with stirring before use. Alternatively, make the solution fresh before each experiment (*see* Subheading 3.2, **step 2**).
4. It is essential that the cross-linked chromatin be digested thoroughly. Chromatin digestion can be monitored by convergent PCR across a restriction site prior to and following digestion. If digestion is incomplete, increase the time of incubation and/or amount of restriction enzyme. As noted in Subheading 2.3, extreme care must be taken regarding the

choice, storage, and use of restriction enzymes for 3C analysis. Be aware that different restriction enzymes cut chromatin with different efficiencies. To detect looping at genes with short ORFs, it is critical to perform restriction digestions for 4–5 h. Digestion of cross-linked chromatin for less than 4 h is not adequate for detection of short-range interactions. Robustness of restriction enzyme activity is also extremely important. In our experience, enzymes that are shipped in dry ice give better and more reproducible results than the enzymes shipped at 4 °C. Enzymes for 3C analysis should be stored with utmost care; the use of older enzymes can adversely affect results (*see* Subheading 3.4, **step 2**).

5. The supernatant could be also used to prepare 3C samples, but the presence of restriction buffer can adversely affect ligation efficiency. The pellet at this stage contains a sufficient amount of digested chromatin for subsequent analyses (*see* Subheading 3.4, **step 4**).
6. The ligation reaction must be done in dilute solution to minimize intermolecular ligation of non-cross-linked DNA. Quick DNA ligase gives better and more reproducible results than regular T4 ligase and improves the sensitivity of the 3C signal for detection of short-range interactions (*see* Subheading 3.4, **step 5**).
7. Precipitation of DNA below room temperature increases salt precipitation. It is therefore preferable to precipitate the DNA at room temperature. Ethanol precipitation of DNA at the final step should be performed in the presence of a carrier. Nuclease-free glycogen gives consistent results (*see* Subheading 3.5, **step 4**).
8. It is critical to accurately determine the concentration of 3C template DNA. As DNA concentration is determined from A_{260} absorbance, care must be taken to digest all RNA. Be sure to use DNase-free RNase. Also be aware that A_{260} readings can be affected by the presence of salt during DNA precipitation. Titration of the DNA template must be done to assure a linear relationship between the amount of DNA and the resulting PCR products. The same amount of template DNA should be used in all subsequent PCR reactions (*see* Subheading 3.5, **step 7**).
9. 3C PCR and control PCR should be performed in the same PCR run and analyzed in the same gel to avoid variation in DNA staining between gels. Use gel loading buffer which does not contain bromophenol blue if the PCR product is ~150–500 bp as bromophenol blue interferes with visualization of DNA (*see* Subheading 3.6, **step 3**).
10. If not getting PCR product, DNA might be lost during precipitation or A_{260} readings are inaccurate due to the presence of salt or RNA (*see* Subheading 3.6, **step 3**).

11. We do not assay 3C PCR products by Q-PCR because we often see multiple PCR bands by gel electrophoresis. If Q-PCR is used to quantify 3C PCR products, care must be taken to confirm amplification of a single product, either by gel electrophoresis or by melting curve analysis (*see* Subheading 3.6, **step 4**).

Acknowledgments

We are grateful to Athar Ansari (Wayne State University) for initially developing the 3C protocol in our laboratory and for his insight into the application of 3C to study gene loops. The work in our laboratory is supported by NIH RO1 grants GM39484 (to M.H.) and GM068887 (to Claire Moore (Tufts Medical School) and M. H.).

References

1. Dekker J, Rippe K, Dekker M et al (2002) Capturing chromosome conformation. Science 295:1306–1311
2. Dostie J, Richmond TA, Arnaout RA et al (2006) Chromosome conformation capture carbon copy (5C): a massively parallel solution for mapping interactions between genomic elements. Genome Res 16(10):1299–1309
3. Dostie J, Dekker J (2007) Mapping networks of physical interactions between genomic elements using 5C technology. Nat Protoc 2: 988–1002
4. Dostie J, Zhan Y, Dekker J (2007) Chromosome conformation capture carbon copy technology. Curr Protoc Mol Biol Chapter 21:Unit 21 14
5. Simonis M, Klous P, Splinter E et al (2006) Nuclear organization of active and inactive chromatin domains uncovered by chromosome conformation capture-on-chip (4C). Nat Genet 38:1348–1354
6. Simonis M, Kooren J, de Laat W (2007) An evaluation of 3C-based methods to capture DNA interactions. Nat Methods 4: 895–901
7. Zhao Z, Tavoosidana G, Sjölinder M et al (2006) Circular chromosome conformation capture (4C) uncovers extensive networks of epigenetically regulated intra- and interchromosomal interactions. Nat Genet 38: 1341–1347
8. Gondor A, Rougier C, Ohlsson R (2008) High-resolution circular chromosome conformation capture assay. Nat Protoc 3:303–313
9. Fullwood MJ, Ruan Y (2009) ChIP-based methods for the identification of long-range chromatin interactions. J Cell Biochem 107: 30–39
10. Lieberman-Aiden E, van Berkum NL, Williams L et al (2009) Comprehensive mapping of long-range interactions reveals folding principles of the human genome. Science 326:289–293
11. Singh BN, Ansari A, Hampsey M (2009) Detection of gene loops by 3C in yeast. Methods 48:361–367
12. Ansari A, Hampsey M (2005) A role for the CPF 3′-end processing machinery in RNAP II-dependent gene looping. Genes Dev 19: 2969–2978
13. Singh BN, Hampsey M (2007) A transcription-independent role for TFIIB in gene looping. Mol Cell 27:806–816
14. Laine JP, Singh BN, Krishnamurthy S et al (2009) A physiological role for gene loops in yeast. Genes Dev 23:2604–2609
15. Dekker J (2006) The three "C" s of chromosome conformation capture: controls, controls, controls. Nat Methods 3:17–21
16. Miele A, Gheldof N, Tabuchi TM et al (2006) Mapping chromatin interactions by chromosome conformation capture. Curr Protoc Mol Biol Chapter 21:Unit 21 11

Chapter 14

Chromosome Conformation Capture (3C) of Tandem Arrays in Yeast

Maria D. Mayán and Luis Aragón

Abstract

Studying interphase chromosome arrangements at the molecular level can provide important details on the function and coordination of many metabolic processes that take place on DNA, such as transcription or DNA repair. The chromosome conformation capture (3C) methodology was originally developed in yeast to study the interphase organization of a single-yeast chromosome. 3C assays allow the identification of physical interactions between distant DNA segments and thus provide detailed information on the folding of chromatin in the native cellular state. Since its initial development, the technique has been used to advance our understanding of the folding of gene loci and has yielded important discoveries, like the demonstration that distant transcriptional regulatory elements interact with their target promoters through chromatin loops. The 3C method uses formaldehyde cross-linking to covalently link interacting chromatin segments in intact cells. After chromatin digestion and ligation of cross-linked fragments, the frequency of interaction between distant DNA elements can be quantified. However, the design, analysis, and controls used are critical when using 3C. In this chapter, we describe the general protocol for 3C analysis and demonstrate it through analysis of interactions in repetitive sequences of the ribosomal gene array of *Saccharomyces cerevisiae*.

Key words Chromatin interactions, Repetitive DNA, Transcription, DNA loops

1 Introduction

The chromosome conformation capture (3C) methodology was initially developed to analyze how yeast chromosome III is arranged during interphase [1]. Later the methodology was adapted to investigate the interactions between distant elements which regulate the B-globin locus activity in mammalian cells [2]. 3C and 3C-based technologies are powerful tools for the analysis of the organization of interphase chromosomes in the cell's natural environment. Presently, 3C has become a standard research protocol in certain research areas, such as nuclear organization, transcriptional architecture, and other DNA-related processes like DNA replication and repair [3]. Importantly, many of the proteins required for these processes have been identified and their function can be

Jeffrey S. Smith and Daniel J. Burke (eds.), *Yeast Genetics: Methods and Protocols*, Methods in Molecular Biology, vol. 1205, DOI 10.1007/978-1-4939-1363-3_14, © Springer Science+Business Media New York 2014

investigated using genetics and biochemistry. The use of 3C techniques in this context can provide a wealth of information about the sequences and the proteins that provide the chromatin architecture during different steps of these metabolic DNA processes.

It is important to note that due to the flexibility of the chromatin fiber, DNA segments are engaged in random collisions. Therefore, the detection of a ligation product does not necessarily reveal a specific interaction. Determining that an interaction is specifically occurring in the native state requires the demonstration that two DNA sites interact more frequently with each other than expected randomly. Furthermore, chromatin interactions are highly dynamic, varying among different cell types as well as stages of the cell cycle. The design of the 3C protocol and the election of restriction enzymes are two important aspects to take into consideration for the success of the technique. In brief, the 3C technique involves cross-linking exponentially growing cells with low concentrations of formaldehyde to cross-link proteins to other proteins and to DNA. Intact nuclei are then isolated, solubilized, and digested. Digested chromatin fragments are then subjected to ligation using very low DNA concentrations, to favor intramolecular ligation. Reversal of cross-links is then followed by PCR analysis to reveal whether or not ligation products were formed (Fig. 1). Therefore, using 3C, physical interactions are translated into positive PCR amplification of ligation products whose abundance can measure the strength of such interactions and provide information about how chromatin is dynamically arranged during interphase processes and what are the protein factors that are crucial to establish and maintain such interactions.

A large number of protocols have been described before on 3C techniques. These include those designed for the detection of chromatin loops in metazoan organisms [4], and the detection of long-range [1, 5] or short-range interactions in yeast [6]. They are all designed to investigate interactions between DNA elements that differ in DNA sequence. Here we describe an adaptation of 3C methodology which can be used to study short- and long-range interactions in tandem repeats, such as those of the ribosomal gene cluster of *S. cerevisiae* (Fig. 2).

Fig. 1 Schematic representation of the 3C methodology using divergent primers. Divergent primers are denoted as P1 and P2. Restriction sites are denoted as RE. DNA fragments "Site 1" and "Site 2" (*red* and *green squares*) become juxtaposed by binding factor (*black circle*). Formaldehyde treatment cross-links DNA segments to proteins and proteins with each other, which leads to cross-linking of interacting DNA segments ("Site 1" to "Site 2"). Restriction digestion then separates the non-cross-linked DNA from the cross-linked chromatin. This is followed by ligation at low concentration to favor intramolecular digestion. The ligation then results in DNA fragment that can be amplified using divergent and convergent primers when the cross-links are reversed

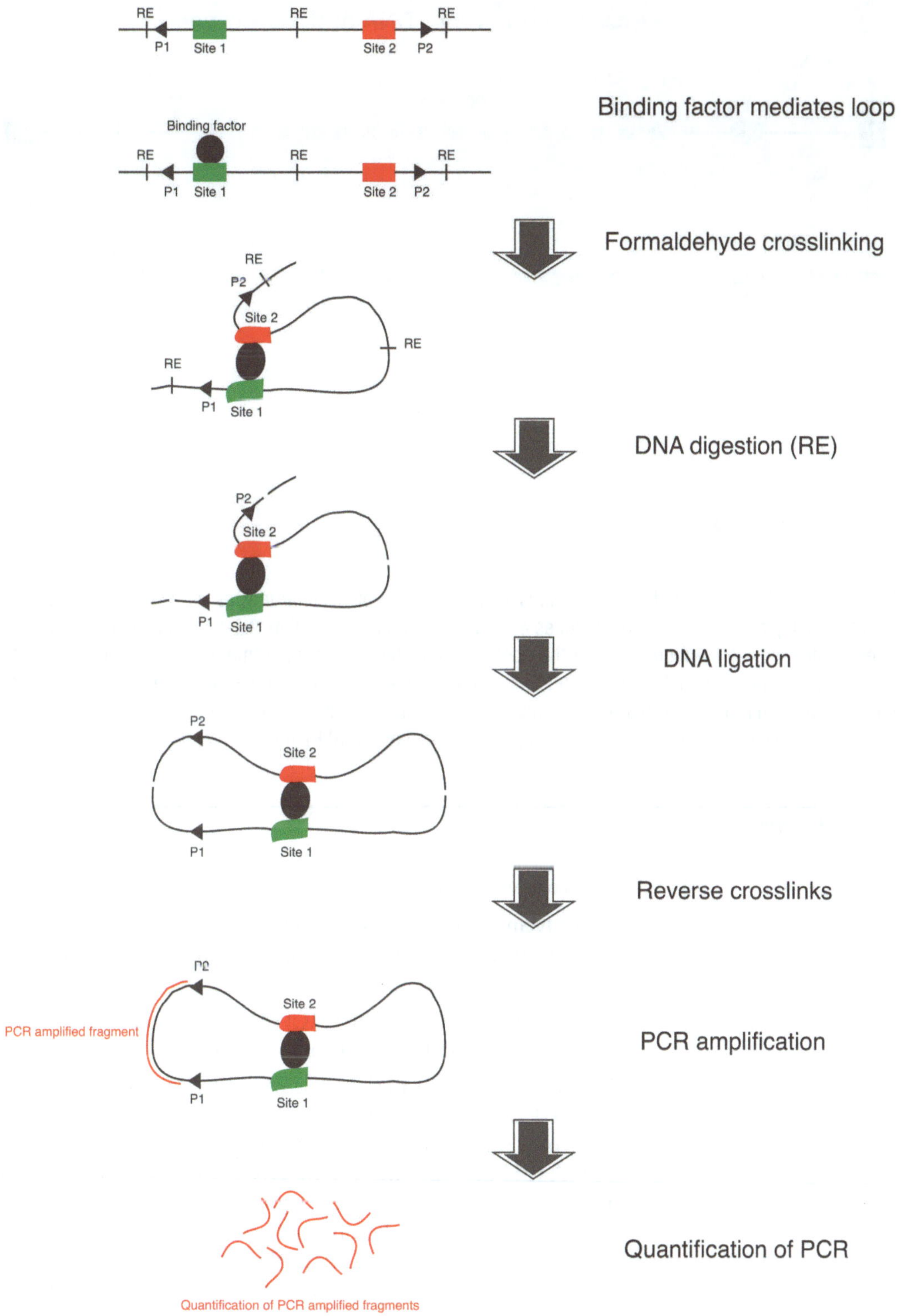
RE
RE
RE
P1
Site 1
Site 2
P2
Binding factor mediates loop
Binding factor
Formaldehyde crosslinking
DNA digestion (RE)
DNA ligation
Reverse crosslinks
PCR amplified fragment
PCR amplification
Quantification of PCR
Quantification of PCR amplified fragments

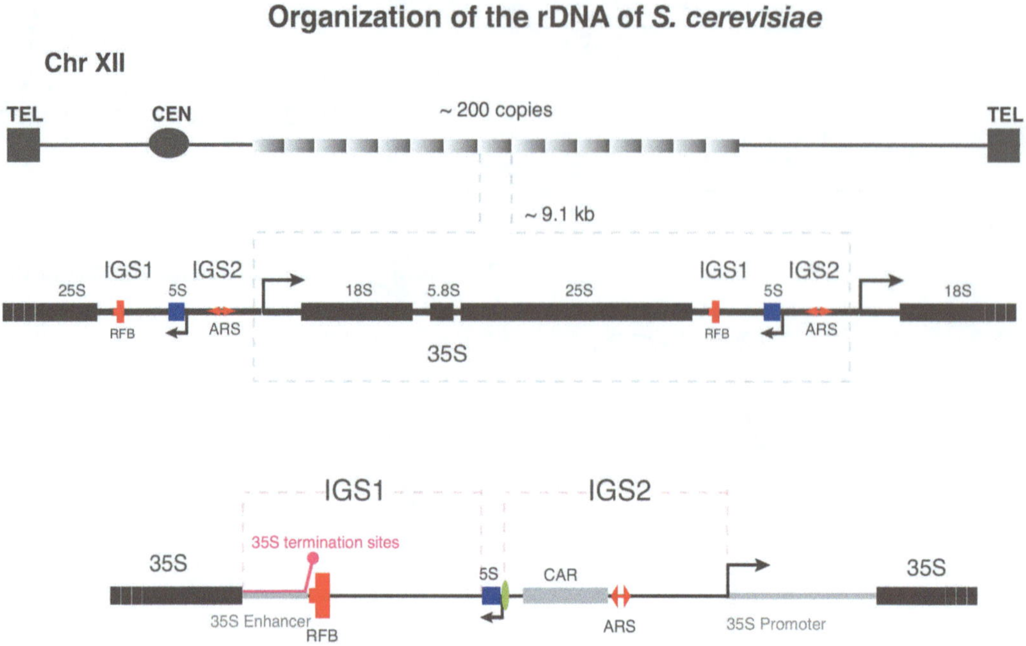

Fig. 2 Schematic representation of the ribosomal gene array of *Saccharomyces cerevisiae*. Budding yeast contains multiple copies of ribosomal genes arranged on chromosome XII as 9.1-kb units tandemly repeated. These genes provide the foundation for the ribosome-manufacturing compartment, the nucleolus. Each ribosomal copy contains the 35S (containing 25, 18, and 5.8S), and 5S rRNA genes separated by two intergenic spacers (IGS1 and IGS2). Specialized features in IGS include a cohesin-binding sequence (CAR) (*grey box*), a replication origin (ARS) in IGS2 (*red*), and a replication fork barrier (RFB) in IGS1 (*red*) [8]

2 Materials

1. 37% (wt/vol) formaldehyde.
2. 1 M glycine: Dissolve 18.8 g glycine in ddH$_2$O and bring up to 250 mL (may require gentle heating). Store at room temperature.
3. 1× TBS buffer: 10 mM Tris–HCl [pH 7.5] and 200 mM NaCl containing 1 % Triton X-100 and 1 mM PMSF.
4. IP lysis buffer: 150 mM NaCl, 50 mM Tris–HCl (pH 7.5), 5 mM EDTA, NP-40 (0.5 % vol/vol), Triton X-100 (1.0 % vol/vol). For 500 mL, add 4.383 g NaCl, 25 mL of 100 mM EDTA (pH 8.0), 25 mL of 1 M Tris–HCl (pH 7.5), 25 mL of 10 % (vol/vol) NP-40, and 50 mL of 10 % (vol/vol) Triton X-100. Store at 4 °C. Before using prepare IP buffer containing 1 mM PMSF. Per 1 mL IP buffer, add the following immediately before use and keep on ice: 5 μL of 0.1 M PMSF in isopropanol (–20 °C).
5. 10 % (wt/vol) Chelex 100: Add 1 g Chelex 100 resin to water (MilliQ or NANOpure) and bring up to a final volume of 10 mL. Store at room temperature.

Table 1
Primer sequences used

Name	Sequence	Product size
*Nsp*I 1F1 R1	CCGGGGCCTAGTTTAGAGAG CCCTTCTCTTTCAACCCATC	241 bp
*Nsp*I 2F2 R2	AGGGCTTTCACAAAGCTTCC TCCCCACTGTTCACTGTTCA	245 bp
*Nsp*I 3F3 R3	GCGTATGGTCACCCACTACA TTCTCCACATCACAATGCAC	269 bp
*Nsp*I 4F4 R4	GAGGTGTTATGGGTGGAGGA GCCACCATCCATTTGTCTTT	232 bp
*Nsp*I 5F5 R5	CAGAGAGACCCGAAAAAGCA GAGCCATTCGCAGTTTCACT	251 bp
*Nsp*I 6F6 R6	TGCGAATGGCTCATTAAATC TGAAACCATGGTAGGCCACT	265 bp
*Nsp*I 7F7 R7	GGCCCAGAGGTAGCAAACAC GGAAATGACGCTCAAACAGG	253 bp
EF2 ER2	AGCCAGCGAGTCTAACCTTG TTGTCCAAATTCTCCGCTCT	233 bp
35S P3f 35S P3r	TGCGACGTAAGTCAAGGATG CTGGCTTCACCCTATTCAGG	229 bp
ChrVIf ChrVIr	TGCCCTCAAAGCATGAGTAG ATGGAACTGCCTTACGGTTG	252 bp

6. 20 μg/μL proteinase K.
7. Specific primers (*see* Table 1 for primer sequences) (*see* Fig. 3 for primer locations within the ribosomal DNA of *S. cerevisiae*).

3 Methods

3.1 Cell Culture

1. Inoculate a 25 mL culture in the desired culture medium from a single freshly growing colony of a Petri dish.
2. Incubate overnight at 30 °C with vigorous shaking.
3. From the overnight culture, inoculate 50 mL of growth medium in a 200 mL flask to OD_{600} ~0.2–0.3 (*see* **Note 1**).
4. Incubate cells with shaking at 30 °C until they reach an OD_{600} ~0.6–0.8. In the case of wild-type strains this is around 4–6 h.

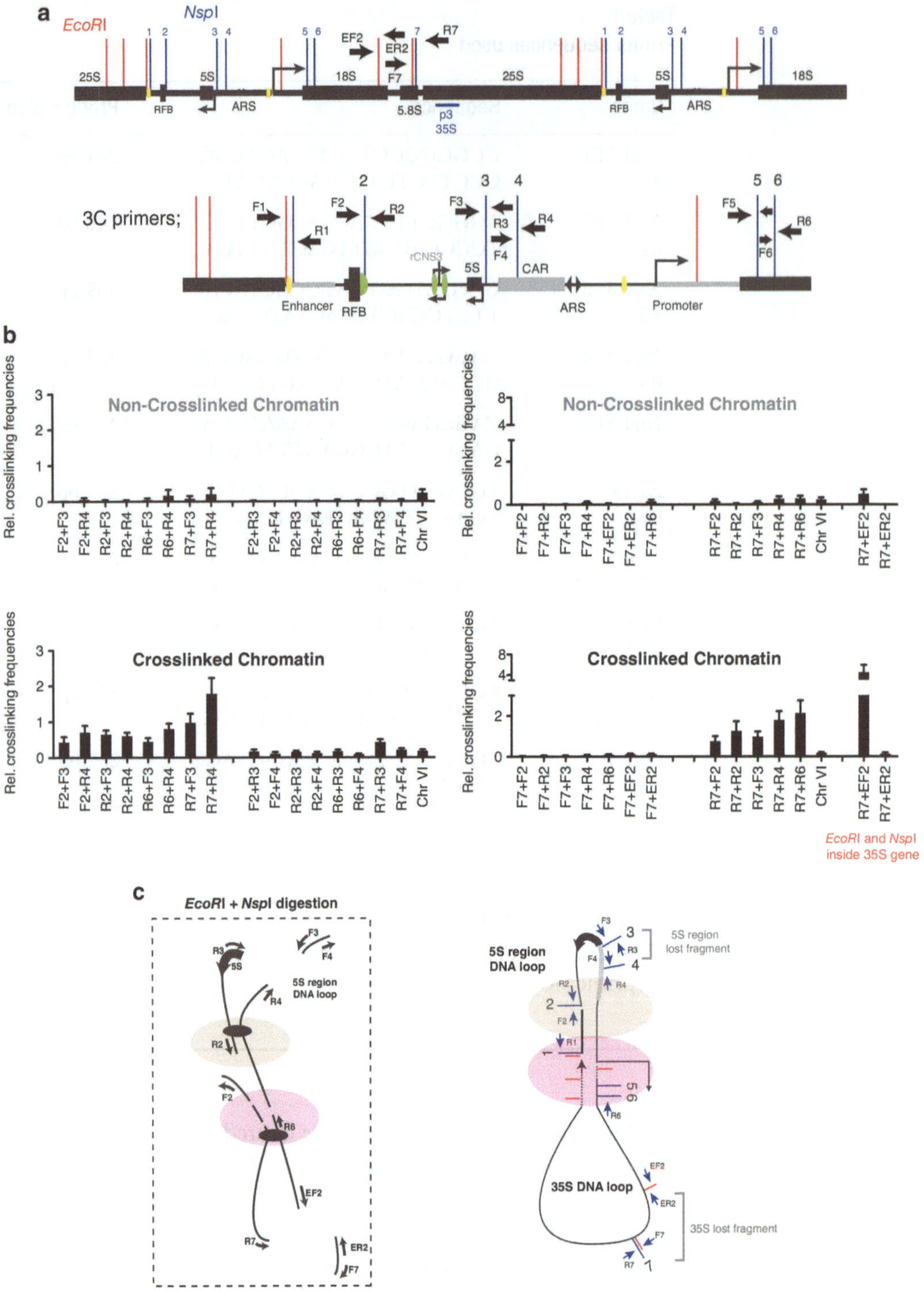

Fig. 3 Mapping long-range interactions between yeast rDNA intergenic sequences (IGS) using chromosome conformation capture (3C) with *EcoRI* and *NspI* digestion. (**a**) Schematic representation of the location of *EcoRI* [8] and *NspI* (*blue*) sites along the 9.1-kb rDNA repeat (including zoomed-in IGS regions). The position of the primers used in the analysis is also shown (2). (**b**) Quantification of interactions by SYBR *green* real-time

3.2 Cross-Linking

1. Add formaldehyde to a final concentration of 1.42 % (40 μL 37 % formaldehyde per mL of culture) (*see* **Note 2**).
2. Incubate for 10–15 min at room temperature. Mix thoroughly either on a shaker or by swirling every 5 min. Do not allow the time of cross-linking to exceed 20 min.
3. Add 125 mM glycine to quench excess formaldehyde (141 μL 1 M glycine per mL of culture).
4. Mix thoroughly and incubate for an additional 5 min at room temperature.
5. Transfer cells to a 50 mL centrifuge tube.
6. Pellet cells by centrifugation at 1,000×*g* for 5 min at 4 °C using an Eppendorf 5810R centrifuge (A-4-81 rotor) or equivalent.
7. Wash cells with 10 mL of ice-cold TBS containing 1 % Triton X-100.
8. Pellet cells and resuspend in 1 mL of ice-cold IP buffer containing PMSF.
9. Transfer to a 1.5 mL Eppendorf tube.
10. Pellet cells by centrifugation at 10,600×*g* for 10 min at 4 °C in an Eppendorf microcentrifuge, and aspirate buffer.
11. Resuspend the pellet thoroughly in 500 μL of IP buffer containing PMSF by pipeting up and down.
12. Freeze cells in liquid nitrogen and store at −80 °C. Cells can be stored at this point for up to 1 year.

3.3 Isolation of Cross-Linked Chromatin

1. Thaw cells gently by placing tubes on ice.
2. Add ice-cold acid washed glass beads (500 μL).
3. Break the cells by incubating them in a FastPrep Machine or equivalent twice for a cycle of 1 min at 4 °C. Alternatively, you can vortex the cells vigorously (in the cold room) for two cycles of 10 min with a 5–10-min break (*see* **Note 3**).

Fig. 3 (continued) PCR analysis using different primer combinations as indicated. Cross-linking efficiencies are expressed relative to an undigested fragment from the 35S gene (p3). The relative cross-linking efficiencies from cells that were not cross-linked with formaldehyde are shown as control for random ligations. (**c**) Schematic representation of the long-range interactions between yeast rDNA intergenic sequences. The left diagram shows the products generated after digestion. Note that while amplification is obtained when primer R7 is used in combination with a series of other primers, no amplification is obtained with F7 primer indicating the formation of a loop that encompasses the 35S gene region. Note that the middle fragment (on the 35S between *NspI* and *EcoRI* sites), which forms part of the loop, is not ligated (and thus is lost) during the protocol (for more details *see* ref. 7). The PCR reaction conditions are designed for amplification of short fragments (around 200 bp) (Table 1). All primers were designed close to the restriction enzyme sites to amplify fragments of 200 ± 50 bp. R7 + EF2 fragment corresponds to the 3C-ligation fragment

4. Puncture Eppendorf tube using a needle and insert into a 15 mL centrifuge tube.
5. Centrifuge in the Eppendorf 5810R centrifuge or equivalent at 1,000 ×*g* for 10 min at 4 °C.
6. Collect the lysate in a 15 mL tube (*see* **Note 4**).
7. Discard the Eppendorf tube containing the glass beads.
8. Resuspend the loose pellet and transfer the entire lysate to new Eppendorf tube (*see* **Note 2**).
9. Centrifuge at 18,000 ×*g* in the Eppendorf microcentrifuge for 5 min at 4 °C.
10. Discard the supernatant.
11. Resuspend the pellet in 1 mL IP buffer.
12. Centrifuge at 18,000 ×*g* in the Eppendorf microcentrifuge for 10 min at 4 °C.
13. Discard the supernatant.
14. Resuspend the pellet in 500 μL 10 mM Tris–HCl (pH 7.5).
15. Chromatin samples at this stage can be stored in 50–100 μL aliquots at −80 °C for several months.

3.4 Restriction Digestion of Cross-Linked Chromatin

1. Thaw aliquots of chromatin samples gently by placing tubes on ice.
2. Add 500 μL 10 mM Tris–HCl (pH 7.5), mix gently, and centrifuge at 15,000 ×*g* for 5 min at 4 °C.
3. Discard the supernatant.
4. Resuspend the pellet in 100 μL 1× digestion buffer.
5. Centrifuge and discard supernatant.
6. Resuspend the pellet in 250 μL 1× digestion buffer.
7. Add 500–1,000 units of restriction enzymes (*see* **Note 5**) and incubate at 37 °C for at least 5 h. The restriction digestion should be carried out with occasional mixing O/N at 37 °C.
8. Stop the reaction by adding 10 % SDS (1 % final concentration).
9. Incubate for 20 min at 65 °C.
10. To sequester SDS and allow the subsequent ligation reaction, add 10 % Triton X-100 (1 % final concentration).
11. Mix and incubate tubes for 2 min at room temperature.
12. Centrifuge at 15,000 ×*g* for 5 min at 4 °C.
13. Discard the supernatant.
14. Resuspend the pellet thoroughly in 100 μL 10 mM Tris–HCl (pH 7.5).

3.5 Ligation of Cross-Linked Chromatin

1. Add Quick Ligase Buffer, ddH_2O, and Quick T4 DNA ligase (total volume = 700 μL to 5 mL) (*see* **Note 6**).
2. Mix and incubate the sample for 1 h at 25 °C.

3.6 Reversal of Cross-Links on Chromatin

1. To ensure complete removal of RNA add 5–15 mL of DNase-free RNase (100 mg/mL) to the reaction mixture.
2. Incubate at 37 °C for 15 min.
3. Centrifuge at 15,000 × *g* for 5 min at 4 °C.
4. Discard the supernatant.
5. Resuspend the pellet thoroughly in 100 μL 10 mM Tris–HCl (pH 7.5).
6. Add 10–30 μL of 20 mg/mL proteinase K. Incubate the mix for 4–6 h at 65 °C.
7. Add 100 mL 10 % (wt/vol) Chelex 100. Briefly (10 s) vortex samples to mix (*see* **Note 7**).
8. Incubate at 100 °C for 30 min.
9. Centrifuge at 1,700 × *g* for 1 min.
10. Transfer supernatant (190 mL) to a new tube. Be careful to avoid transferring any chelex resin as it can lead to a loss of PCR signal.
11. Add 50 mL of water to beads.
12. Vortex for 10 s.
13. Centrifuge and collect 50 mL of supernatant and pool with the previous supernatant.
14. The DNA template can be stored at this stage at −20 °C.

3.7 PCR Reactions on rDNA

1. For each PCR reaction mix 5 μL of DNA template, 5 μL of the primer mix (forward and reverse for each case), and 10 μL of SYBr-green PCR Master Mix to bring the total volume to 20 μL.
2. Set the conditions of the PCR for 95 °C 10 min, 1 cycle; 95 °C 15 s, 58 °C 30 s, and 72 °C 30 s, 30 cycles; and 72 °C 10 min, 1 cycle.
3. Quantify PCR products using Sybr Green real time PCR (CFX96TM Real-Time PCR Detection System) relative to an undigested fragment and normalized to a ligation fragment which are used as a loading and ligation control [7] (as shown in Fig. 3).
4. PCR reactions should be done in duplicate or triplicate, to determine average interaction frequency. The abundance of the PCR products (relative to control) provides the frequency with which those two regions interact (*see* **Notes 8–12** and Fig. 3).

4 Notes

1. When working with temperature-sensitive mutants, harvest cells at mid-log phase and resuspend in warm (37 °C) medium for desired time. If working with several samples make sure to harvest different strains, or different cultures of the same strain, at the same OD_{600}.
2. The described protocol can be adapted to analyze chromatin interactions from mammalian cells by starting in **step 1** (3.2 Cross-Linking) and omitting **steps 2–9** (3.2 Isolation of Cross-Linked Chromatin).
3. Ensure that sample does not overheat during vortexing, which can cause distortion of chromatin structure.
4. Cell lysates can be directly collected using a pipet. The beads should be washed twice with IP buffer. The filtrate is transferred into a 15-mL Falcon tube and spun for 15 min at 4 °C.
5. The choice of restriction enzyme will depend on the sequence to be analyzed. For small or repetitive sequences, frequently cutting restriction enzymes are used. Because of the repetitiveness of the rDNA locus, with hundreds of identical sequences, multiple restriction enzymes (*Nsp*I and *Eco*RI) need to be used during the protocol to ensure full digestion of target sequences. For this case we have used 1,000 units of *Eco*RI during 5 h at 37 °C and 1,000 units of *Nsp*I overnight. Not all enzymes digest chromatin with the same efficiency. *Eco*RI, *Bgl*II, or *Hind*III digest equally well cross-linked chromatin. To compare the frequency of 3C products between different samples or different conditions, the chromatin should be digested at least 50–80 % before continuing with the ligation.
6. The ligation reaction is done in diluted solution to minimize ligation of non-cross-linked DNA. The conditions of ligation should be standardized depending on the quantity of chromatin and restriction enzyme used.
7. Using Chelex 100 to isolate DNA instead of phenol-chloroform reduces the variability between samples, improving the quantification of the ligation frequency of the 3C-ligated fragments.
8. An important consideration in 3C is the choice of restriction enzymes. It is important that they are distributed along the target regions so that they flank the putative interacting regions. We have used *Eco*RI to investigate interactions on the IGS regions of rDNA. When working with divergent primers, the occurrence of sites between the fragments of interest is necessary to ensure that ligation products are cross-link dependent. In the rDNA, the repetitive of the locus requires multiple restriction enzymes (we used *NspI* and *EcoRI*, *see* Fig. 3) to ensure full digestion of target sequences.

9. The efficiencies of the different primer pairs must be assessed (as in ref. 7).
10. An important control in the 3C methodology is to ensure that the PCR products are dependent on formaldehyde cross-linking. The method to control for this is to carry out a "no-cross-link sample" where the material is not treated with formaldehyde (Fig. 3b). The absence or significant reduction of PCR amplifications in this control demonstrates that the amplification is not due to random interaction/ligations but rather stabilization of transient interaction by formaldehyde cross-linking.
11. Another important control is to carry out a no-ligation control.
12. To conclude from 3C data about the formation of DNA loops it is important to "walk" along the DNA segment that forms the loop using primers anchored at one end in combination with primers adjacent to internal restriction sites to demonstrate that segments within the loop are lost due to the presence of RE sites in the loop (*see* Fig. 3b, c).

References

1. Dekker J, Rippe K, Dekker M et al (2002) Capturing chromosome conformation. Science 295:1306–1311
2. Tolhuis B, Palstra RJ, Splinter E et al (2002) Looping and interaction between hypersensitive sites in the active beta-globin locus. Mol Cell 10:1453–1465
3. Oza P, Jaspersen SL, Miele A et al (2009) Mechanisms that regulate localization of a DNA double-strand break to the nuclear periphery. Genes Dev 23:912–927
4. Simonis M, Klous P, Splinter E et al (2006) Nuclear organization of active and inactive chromatin domains uncovered by chromosome conformation capture-on-chip (4C). Nat Genet 38:1348–1354
5. Dekker J (2006) The three "C" s of chromosome conformation capture: controls, controls, controls. Nat Methods 3:17–21
6. Singh BN, Ansari A, Hampsey M (2009) Detection of gene loops by 3C in yeast. Methods 48:361–367
7. Mayan M, Aragon L (2010) Cis-interactions between non-coding ribosomal spacers dependent on RNAP-II separate RNAP-I and RNAP-III transcription domains. Cell Cycle 9: 4328
8. Paredes S, Maggert KA (2009) Ribosomal DNA contributes to global chromatin regulation. Proc Natl Acad Sci U S A 106:17829–17834

Chapter 15

Global Analysis of Transcription Factor-Binding Sites in Yeast Using ChIP-Seq

Philippe Lefrançois, Jennifer E.G. Gallagher, and Michael Snyder

Abstract

Transcription factors influence gene expression through their ability to bind DNA at specific regulatory elements. Specific DNA-protein interactions can be isolated through the chromatin immunoprecipitation (ChIP) procedure, in which DNA fragments bound by the protein of interest are recovered. ChIP is followed by high-throughput DNA sequencing (Seq) to determine the genomic provenance of ChIP DNA fragments and their relative abundance in the sample. This chapter describes a ChIP-Seq strategy adapted for budding yeast to enable the genome-wide characterization of binding sites of transcription factors (TFs) and other DNA-binding proteins in an efficient and cost-effective way.

Yeast strains with epitope-tagged TFs are most commonly used for ChIP-Seq, along with their matching untagged control strains. The initial step of ChIP involves the cross-linking of DNA and proteins. Next, yeast cells are lysed and sonicated to shear chromatin into smaller fragments. An antibody against an epitope-tagged TF is used to pull down chromatin complexes containing DNA and the TF of interest. DNA is then purified and proteins degraded. Specific barcoded adapters for multiplex DNA sequencing are ligated to ChIP DNA. Short DNA sequence reads (28–36 base pairs) are parsed according to the barcode and aligned against the yeast reference genome, thus generating a nucleotide-resolution map of transcription factor-binding sites and their occupancy.

Key words ChIP-Seq, ChIP, Yeast, Chromatin, Transcription factor, Binding site, Genomics, Multiplex

1 Introduction

Gene expression is often regulated by the binding of transcription factors (TFs) to specific DNA sequences within intergenic regions termed transcription factor-binding sites. The genome of the budding yeast *Saccharomyces cerevisiae* contains about 6,000 predicted ORFs, of which 200–300 encode TFs [1]. Transcription factors bind preferentially to regions containing a consensus motif, enabling computational prediction of putative binding sites. However, these predictions must be validated experimentally [2], as many regions with perfect consensus motifs can remain unbound

Jeffrey S. Smith and Daniel J. Burke (eds.), *Yeast Genetics: Methods and Protocols*, Methods in Molecular Biology, vol. 1205, DOI 10.1007/978-1-4939-1363-3_15,

while those displaying imperfect motifs can show high level of protein binding [3]. The method of choice for validation of TF binding to DNA, chromatin immunoprecipitation (ChIP), was first developed to characterize RNA polymerase II binding in bacteria [4]. Briefly, DNA-protein complexes are covalently cross-linked by formaldehyde. Cross-linked yeast cells are lysed, and the lysates are then sonicated to shear chromatin fragments into smaller pieces, amenable to subsequent immunoprecipitation (IP) [5]. Antibodies raised against the TF of interest, or against a specific epitope (if the TF is epitope tagged) are used to recover DNA-protein complexes containing the TF of interest. DNA is purified from proteins by reversing the cross-links using heat, followed by proteinase K protein degradation of the proteins. Enrichment for regions bound by a particular TF can be determined by PCR quantification, comparing a yeast strain with a TF-epitope fusion to its isogenic control strain, either to a ChIP in the untagged parental strain or to a mock IP if using a TF-specific antibody. PCR detection is not suitable to discovery of novel binding regions given the low throughput and need for specific primers for amplification. With the development of DNA microarrays, it became possible to query the entire genome for sites bound by a particular TF, using a ChIP approach coupled to hybridization of the recovered DNA to microarrays [6]. This technology, called ChIP-chip, has been successful to identify globally transcription factor-binding profiles [7]. Recently, massively parallel, high-throughput sequencing technologies such as Roche's 454, Illumina's Genome Analyzer and HiSeq, LifeTechnologies' SOLiD and IonTorrent, Helicos' HeliScope, Pacific Biosciences' PacBio RS, and Complete Genomics' DNA nanoball sequencing have revolutionized large-scale genomics projects by generating millions of short DNA sequence reads in a few days, at single-nucleotide resolution. ChIP followed by high-throughput sequencing (ChIP-Seq) has emerged as a powerful method to discover and characterize functional elements of any genome, and was first developed for mammalian applications [8, 9]. The reduced background, decreasing cost of sequencing, lack of cross-hybridization, increased sensitivity, single-nucleotide resolution, and high dynamic range are among the advantages that helped to establish ChIP-Seq over ChIP-chip as the current gold standard in gene regulation studies [10]. The multinational consortiums, ENCODE in humans [11], and modENCODE in worms and flies [12], have taken advantage of novel sequencing technologies to characterize the entire repertoire of functional genomic elements. While the initial transition from ChIP-chip to ChIP-Seq quickly gained momentum in higher eukaryotes, ChIP-Seq studies in organisms with smaller genome remained rare, given high cost per sample and excessive generation of sequence reads compared to the number required to map binding sites at high confidence [13]. Our group developed a multiplex ChIP-Seq strategy to process multiple samples simultaneously and in a cost-effective way, which was used to characterize the distribution of several DNA-binding proteins, including RNA polymerase II, the centromeric histone H3-variant Cse4, and the TF Ste12 [13]. ChIP-Seq experiments in yeast have been valuable in determining the effect of chromatin structures during ChIP that affect several organisms [14, 15]. Novel binding sites were found for key transcription factors involved in various important cellular functions, including ribosome biogenesis [16, 17], transcriptional silencing [18], noncoding RNA regulation [19], ubiquitination [20], stress response [21], metabolism [22], and DNA replication [23] among others. In particular, the sensitivity of yeast ChIP-Seq can be exploited to characterize on a wide spectrum the variation among individuals, at the level of transcription factor binding [24]. ChIP-Seq studies in other fungi, such as *Saccharomyces bayanus* [18], *Cryptococcus neoformans* [25], and *Neurospora crassa* [26], are also feasible given that high-quality genome assemblies exist. In the future, given the rapid development in sequencing technologies, more ChIP-Seq samples could be analyzed in parallel using increased multiplexing [27], all at a lower cost. Challenges still remain for data storage and, very importantly, for the development of simple and fast, yet efficient and precise, computational tools and algorithms that can be used daily by bench scientists [28].

This chapter details a strategy to perform ChIP-Seq in yeast (Fig. 1). It first describes chromatin immunoprecipitation in yeast to prepare ChIP DNA from yeast strains with epitope-tagged TFs. Next, we present the protocol to generate high-quality Illumina DNA sequencing libraries for subsequent high-throughput sequencing on Illumina's Genome Analyzer IIx, with a special focus on a multiplexing strategy. Briefly, purified ChIP DNA is ligated to barcoded DNA adapters and then PCR-amplified for a few cycles. Multiplex sequencing libraries are mixed in equimolar ratio. This mixture is added to a single Illumina flowcell lane using a cluster station. During this step, single-DNA molecules are amplified to form a cluster containing a 1,000 identical clones. Clusters are then submitted to sequencing-by-synthesis, during which fluorescently labeled nucleotides with a reversible terminator are incorporated at each sequencing cycle. Finally, many aspects of ChIP-Seq data analysis are covered, including alignment of short sequence reads, peak calling for identification of TF-binding sites, data visualization, and some downstream analyses.

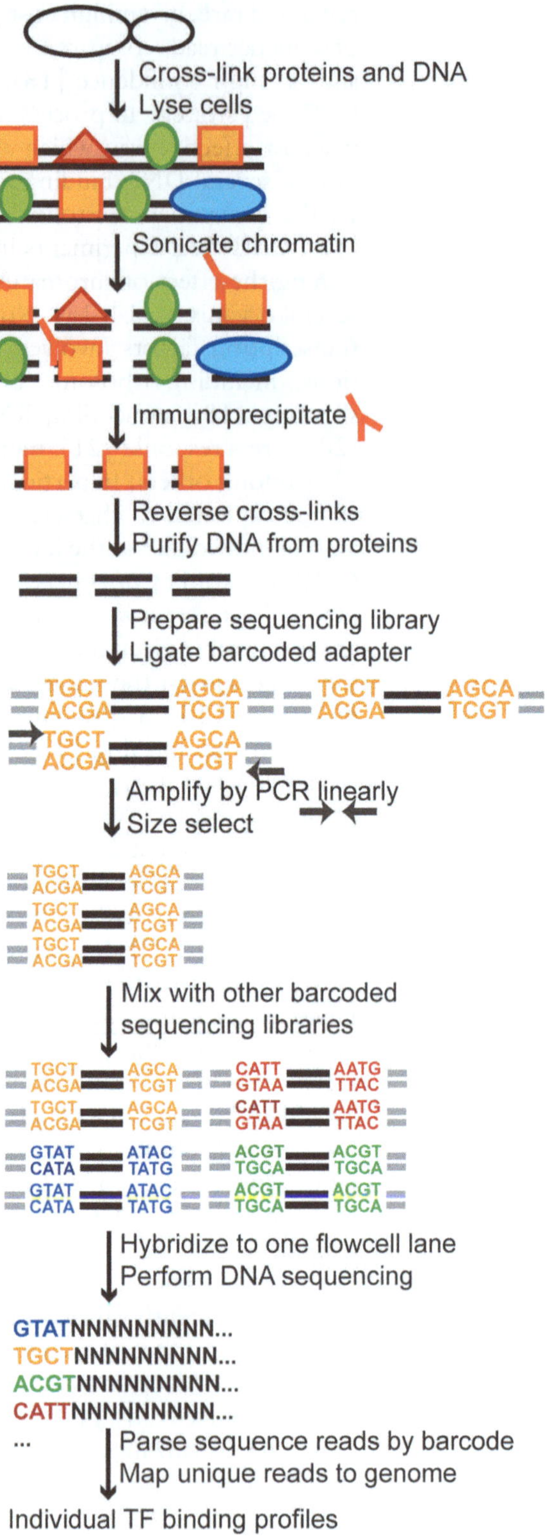

Fig. 1 Overview of the ChIP-Seq procedure in budding yeast, focusing on a multiplex high-throughput DNA sequencing approach on the Illumina platform

2 Materials

1. Yeast cells grown in the appropriate liquid medium, in a 2 L Erlenmeyer flask.
2. 37 % formaldehyde.
3. 2.5 M glycine (sterile).
4. Nuclease-free water.
5. 1 L PES 0.2 μm filter unit.
6. 0.5 mm zirconium/silica beads.
7. Syringe needle.
8. Lysis/IP buffer: 50 mM Hepes/KOH pH 7.5, 140 mM NaCl, 1 mM EDTA, 1 % Triton X-100, 9. 0.1 % sodium deoxycholate.
9. 100 mM PMSF (phenylmethanesulfonylfluoride).
10. Roche Complete protease inhibitor cocktail tablet (Roche).
11. FastPrep-24 machine (MP Biomedicals).
12. Branson Digital 450 Sonifier (Branson).
13. Sigma EZview anti-Myc or anti-HA affinity gel (Sigma) or Pan mouse IgG dynabeads (Invitrogen).
14. Lysis/500 mL NaCl buffer: 18 mL 5 M NaCl, 232 mL of lysis/IP buffer.
15. IP wash buffer: 10 mM Tris–HCl, 0.25 M LiCl, 0.5 % NP-40, 0.5 % sodium deoxycholate, 1 mM EDTA.
16. 1× TE: 50 mM Tris–HCl, 10 mM EDTA pH 8.0.
17. 1× TE/1 % SDS solution.
18. 1× TE/0.67 % SDS solution.
19. Rocking mixer.
20. 20 mg/mL proteinase K.
21. 45 and 65 °C water baths or heat blocks.
22. 100 and 70 % ethanol.
23. 20 mg/mL glycogen.
24. Pellet paint Co-Precipitant (Novagen).
25. 5 M LiCl.
26. MinElute PCR purification kit (Qiagen), including EB buffer and MinElute column.
27. Nanodrop spectrophotometer (Thermo Scientific) or Qubit fluorometer (Invitrogen).
28. 15 and 50 mL Falcon conical tubes.
29. 5 mL Falcon snap-cap conical tubes.
30. 2 mL screw-cap microcentrifuge tubes.
31. 1.5 mL microcentrifuge tubes.

32. Reagents for generation of barcoded sequencing libraries.
33. High-throughput DNA sequencer and its data analysis suite.

For ChIP-Seq analysis on the Illumina Genome Analyzer IIx:

34. Annealing buffer: 10 mM Tris–HCl pH 7.5, 50 mM NaCl, 1 mM EDTA.
35. Agarose gel electrophoresis apparatus.
36. Clean scalpel or razor blades or 18 cm × 18 cm glass cover slips.
37. Gibco RNase-free DNase-free water.
38. End-It DNA end repair kit (Epicentre), including End-repair enzyme mix, 10× End-repair buffer, 10 mM ATP, 2.5 mM dNTP mix.
39. QIAquick PCR purification kit (Qiagen), including EB buffer and QIAquick column.
40. Klenow fragment of DNA polymerase I (3′→5′ exonuclease activity minus) and 10× NEBuffer2 (New England Biolabs).
41. 100 mM dATP.
42. 96-Well PCR plates (ABgene).
43. Microseal A Film sealing microfilms (MJ Research).
44. Illumina genomic DNA adapter oligonucleotides augmented by a short index (or barcode).
45. LigaFast T4 DNA ligase and 2× DNA ligase buffer (Promega).
46. Track-It 50 bp DNA ladder and Track-It Cyan/Orange loading buffer (Invitrogen).
47. 2 % agarose single-comb 12-well E-Gel, stained with ethidium bromide or SYBR safe (Invitrogen).
48. E-Gel electrophoresis apparatus and visualization system (Invitrogen).
49. Illumina genomic DNA primers 1.1 and 2.1.
50. 2× Phusion master mix with HF buffer (New England Biolabs).
51. MinElute and QIAquick gel extraction kits (Qiagen).
52. Illumina Genome Analyzer sequencer, with reagents, flowcell, and cluster station.
53. Genome Analyzer Pipeline (Illumina) and other DNA sequencing analysis software and tools.

3 Methods

3.1 Chromatin Immunoprecipitation

1. Tag the transcription factor of interest by fusing a Myc or HA epitope at the C-terminus (*see* **Note 1**).
2. Grow 500 mL cultures of yeast cells in the appropriate media to exponential phase (OD600 between 0.6 and 1.0). This represents

about 4.5×10^9 cells per sample. Perform experiments in biological triplicates and process in parallel samples from epitope-tagged TF strains and untagged strains as an experimental control. **Note 2** lists several modifications or alternative manipulations to the ChIP protocol described within Subheading 3.1.

3. Cross-link DNA and proteins by adding 14 mL of 37 % formaldehyde for 15 min, with vigorous swirling every 5 min (*see* **Note 3**).
4. Add 27 mL of 2.5 M glycine for 10 min to quench the reaction. Swirl at least twice vigorously.
5. Collect cells by filtration in 1 L filter unit. Wash cells twice with 100 mL nuclease-free water.
6. Transfer cells from the filter to a 50 mL Falcon conical tube with 10 mL sterile water and repeat this step. Centrifuge cells at room temperature for 10 min at 3,500 RPM in tabletop centrifuge and discard the supernatant.
7. Resuspend the pelleted cells in 1 mL sterile water and transfer them to 2 mL screw-cap tubes. Repeat this step. Both tubes should contain similar volumes.
8. Spin down cells at full speed (~$16,000 \times g$) in a microfuge for 3 min and aspire the supernatant out. Weigh cells. Each tube should contain about 0.2–0.3 g of cells. Add approximately 1 mL of zirconium/silica beads. At this point, cell pellets can be frozen at −70°C for subsequent use.
9. Combine 50 mL lysis/IP buffer solution with 1 Roche Complete protease inhibitor tablet. Mix well until tablet is resuspended. Prior to use, add 0.5 mL 100 mM PMSF and keep on ice. 50 mL of this mixture can be used for six ChIP samples.
10. Add 1 mL of the lysis/IP solution from **step 9** to each cell pellet.
11. Lyse cells with the FastPrep-24 machine (*see* **Note 4**). Perform lysis using five 1-min bursts at a speed of 6.0 m/s. Keep cells in an ice-water bath during the 5-min rest period.
12. Pierce the bottom of the 2 mL screw-cap tube with a needle and place in a cap-less 5 mL snap-cap Falcon tube. Recover lysates by centrifuging for 3 min at $400 \times g$ in tabletop centrifuge with swinging bucket rotor. Add 0.5 mL of lysis/IP solution from **step 9** and spin again. Collect the lysate from the other 2 mL tube originating from the same replicate using the same procedure.
13. Transfer lysates to a 15 mL Falcon conical tube. Rinse the 5 mL tube with 1 mL lysis/IP solution and pool to the same 15 mL tube. The volume prior to sonication should be slightly above 4 mL.

14. Shear chromatin by sonicating lysates five times using a Branson Digital 450 sonifier (*see* **Note 5**). Set the amplitude at 50 % and time ON and OFF at 30 s. Keep the sonicator tip 0.5–1.0 cm from the bottom of the tube to reduce foaming. Perform sonication for groups of biological replicates together (e.g., alternate between the three untagged replicates for all cycles and then alternate between the three tagged replicates). Put samples on ice for at least 2 min between sonication rounds.
15. Clarify sonicated lysates by spinning at 4 °C for 5 min at 1,620 × *g* in tabletop centrifuge and transfer each replicate to three 1.5 mL microcentrifuge tubes.
16. Clarify lysates at 16,000 × *g* in microfuge for 10 min at 4 °C and pool all supernatants to a 15 mL conical tube. Make sure to avoid cell debris at the bottom of the tube. Add 2 mL of lysis/IP buffer. The total volume of each sample should be around 6 mL.
17. Set aside 250 μL of clarified, sonicated lysate before immunoprecipitation to process as input DNA, a reference sample for ChIP-Seq enriched for open chromatin (*see* **Note 6**) [14, 15].
18. Wash the entire bottle of Sigma EZview anti-Myc or anti-HA affinity gel with 1 mL lysis/IP buffer. Transfer antibody-coupled beads to a 15 mL Falcon tube using a broadened 1 mL pipette or large orifice tips. Wash the bottle with 1 mL lysis/IP solution three times and transfer to the 15 mL tube. Vortex and centrifuge for 2 min at 720 × *g* at 4 °C in tabletop centrifuge. Discard supernatant. Add 5 mL lysis/IP buffer and repeat vortexing, centrifugation, and removal of supernatant three times. Resuspend beads in 1 mL fresh lysis/IP solution.
19. Pipet 150–300 μL of pre-washed antibody bead solution to each clarified sonicated lysate from the end of **step 16**. Perform immunoprecipitation at 4 °C on a rocking mixer overnight (between 12 and 16 h) for Myc- or HA-tagged TF strains and their untagged control strains (*see* **Note 7**).
20. After immunoprecipitation, fill Falcon tubes with lysis/IP buffer, pellet antibody-coupled beads by centrifugation at 720 × *g* for 5 min at 4 °C, and remove supernatant.
21. Resuspend pelleted beads in 600 μL of lysis/IP buffer and transfer to a 1.5 mL microfuge tube using a broadened 1 mL pipette. Repeat procedure and pool in the same tube. Mix on a rocker in the cold room for 5 min and discard carefully the supernatant, with a small pipette attached to an aspirator.
22. Perform the following washes sequentially with 1 mL of the appropriate buffer for 5–10 min on a rocking mixer at 4 °C, followed by a 1-min, 845 × *g* centrifugation at 4 °C and gentle removal of the supernatant by aspiration. Wash twice in lysis/

IP buffer, once in lysis/500 mM NaCl buffer, twice in IP/wash buffer, and once in 1× TE buffer. Remove completely the small amount of TE from the beads.

23. Add 100–150 μL of 1× TE/1 % SDS solution to the immunoprecipitate to allow elution of chromatin complexes from the beads and mix. Incubate in a 65 °C water bath for 15 min, mixing briefly after 10 min (*see* **Note 8**). Pellet beads at 16,000×*g* for 1 min and transfer supernatant (the eluate) to a 1.5 mL tube.
24. Repeat elution of the immunoprecipitate by adding 150–200 μL of 1× TE/0.67 % SDS solution and mix. Incubate in a 65 °C water bath for 10 min. Perform centrifugation and eluate transfer as described in **step 23**.
25. Spin down the eluate at 16,000×*g* in microfuge for 2 min to send any residual antibody-coupled bead to the bottom of the tube and transfer the supernatant to a 2 mL screw-cap tube. Great care should be taken to avoid the last 5–20 μL fraction that may contain antibody beads (*see* **Note 8**).
26. Incubate eluates at 65 °C at least for 6–8 h or overnight. Cross-link reversal by heat treatment is crucial for isolating DNA from covalently bound DNA-protein complexes.
27. Briefly chill samples on ice for a few seconds. Dilute proteinase K solution in 1× TE to a final concentration of 0.4 mg/mL. Add 250 μL of the proteinase K/TE solution and incubate samples in a 42 °C water bath (37–50 °C works as well) for 2–4 h to ensure sufficient digestion of proteins.
28. After proteinase K treatment, add 3 μL pellet paint, 3 μL of 20 mg/mL glycogen, and 45 μL 5 M LiCl. Mix briefly. Add 1 mL of 100 % ethanol (more if it fits in 2 mL tube) and mix very well. Precipitate DNA overnight or several hours at −20 °C.
29. Finish DNA precipitation by placing samples at −70 °C for 1 h and centrifuge for 20 min at 16,000×*g* at 4 °C in microfuge. A pinkish-white pellet should be visible. Discard supernatant.
30. Wash pellet with 1 mL 70 % ethanol for 5 min. Spin for 10 min at 16,000×*g* at 4 °C and remove ethanol thoroughly.
31. Air-dry for 10 min or vacuum dry for 3–4 min. Resuspend completely in 100 μL TE.
32. Purify ChIP samples through MinElute columns. Each ChIP DNA sample is split between two MinElute columns. Elute with 21 μL EB and pool eluates from the same sample to a single tube.
33. Measure DNA concentration using a Nanodrop spectrophotometer (*see* **Note 9**).

34. Perform qPCR analysis of ChIP samples to ensure that the ChIP procedure worked properly, prior to the generation of a sequencing library and further computational analysis of ChIP-Seq samples (*see* **Note 10**).

3.2 Generation of Illumina DNA Sequencing Libraries from ChIP Samples

1. Use at least 100 ng of ChIP DNA from **step 32** of previous protocol (or input DNA isolated from **step 17** and processed as described in **Note 6**) to generate a sequencing library for a given sample. 150–250 ng of ChIP DNA (equivalent to about a quarter to half of the total volume of ChIP sample), or input DNA, is sufficient for preparation of high-quality Illumina sequencing libraries. The amount of starting DNA can be increased if previous generation of a sequencing library failed.
2. Optional: Isolate by agarose gel electrophoresis a ChIP DNA smear between 100 and 700 bp (*see* **Note 11**). We run samples on a 2 % gel at 100–110 V for about 20 min to minimize gel volume. Purify DNA using a QIAquick gel extraction procedure. Elute DNA on the QIAquick column with 34 μL EB.
3. End-repair ChIP DNA using the End-It DNA end repair kit, by combining 34 μL ChIP DNA, 5 μL 10× End-It repair buffer, 5 μL 10 mM ATP, 5 μL 2.5 mM dNTP mix, and 1 μL End-It repair enzyme (*see* **Note 12**). Incubate at room temperature for 45 min.
4. After the incubation period, purify end-repaired DNA through a QIAquick column using the QIAquick PCR purification kit protocol. Elute in 34 μL EB.
5. Incorporate a single "A" nucleotide to the 3′ ends of end-repaired ChIP fragments using the Klenow fragment of DNA polymerase I (3′–5′ exo minus), by mixing in a PCR plate 34 μL DNA from previous step, 5 μL 10× NEBuffer2, 10 μL 1 mM dATP, and 1 μL Klenow (3′→5′ exo minus) (*see* **Note 13**). Seal with a microfilm and incubate at 37 °C for 30 min in a PCR machine, without a heated lid (*see* **Note 13**).
6. After the reaction is completed, purify DNA through a MinElute column using the MinElute PCR purification kit procedure. Elute in 12.5 μL EB.
7. Dilute standard Illumina genomic DNA adapter oligonucleotide mix 1:40 (for ChIP samples, this is particularly important, use 1 μL of ~0.025–0.050 μM annealed barcoded adapters). For custom-made barcoded adapters for multiplex high-throughput sequencing, dilute annealed barcoded adapters to the working concentration used for standard Illumina genomic DNA adapters [13] (*see* **Note 14** for barcoded adapter design, annealing, and determination of oligo concentration).
8. Ligate diluted adapters to each sample, with the following components: 12.5 μL DNA from **step 6**, 15 μL 2× DNA ligase

buffer, 1 μL diluted barcoded adapter oligo mix, and 1.5 μL LigaFast T4 DNA ligase enzyme. Incubate for 15 min at room temperature.

9. Remove enzyme and buffers using the MinElute PCR purification kit and elute in 12 μL EB.
10. Size-select DNA between 150 and 500 bp, getting rid of adapter-adapter dimers. Pre-run a 2 % agarose E-Gel, leaving the combs in, if needed. Make a 1:10 dilution of the Track-It 50 bp ladder and Track-It loading buffer. Add 3 μL of diluted Track-It loading buffer to each sample from the previous step. Remove E-Gel comb and load 20 μL diluted 50 bp ladder. Pipet samples containing the loading dye in individual wells, separating them by at least 1 or 2 empty rows and avoiding the last row. Load 15 μL of Gibco nuclease-free water in all empty wells. Perform gel electrophoresis for 18–20 min. Visualize gel and mark the location of 150–500 bp fragments (*see* **Note 15**). Open E-Gel with a large spatula and excise each gel fragment with a clean disposable scalpel blade. Extract DNA using the QIAquick gel extraction kit (*see* **Note 15**) and elute in 28 μL EB.
11. PCR-amplify ChIP DNA with ligated adapters using Illumina genomic DNA primers 1.1 and 2.1. Combine and mix well 28 μL DNA from previous step, 1 μL of 1:1 diluted Illumina genomic DNA primer 1.1, 1 μL of 1:1 diluted Illumina genomic DNA primer 2.1, and 30 μL 2× Phusion HF polymerase master mix, and then transfer to a PCR plate. Amplify for 15–17 cycles using the following PCR parameters (*see* **Note 16**): 1 cycle of initial denaturation for 30 s at 98 °C, 15–17 cycles of amplification (denaturation for 10 s at 98 °C, annealing for 30 s at 65 °C, and extension for 30 s at 72 °C), a final extension for 5 min at 72 °C, and a cool down held indefinitely at 4 °C.
12. Purify PCR products through a MinElute column and elute in 12–15 μL EB.
13. Perform size selection of the final Illumina DNA sequencing library between 150 and 350 bp, using a 2 % agarose E-gel and the gel loading and band excision procedures described in **step 10**. Run the gel for 18–20 min. A bright DNA smear should appear from 150 bp to slightly below 500 bp (*see* **Note 17** for additional details). Extract DNA from agarose gel band with a MinElute gel extraction kit and elute in 20–25 μL EB, respecting gel extraction tips given in **Note 15**.
14. Measure DNA concentration using a Nanodrop spectrophotometer or a Qubit fluorometer (*see* **Note 18**). After this step, store sequencing libraries at −70 °C if sample submission to a sequencing facility will occur later.
15. Mix barcoded sequencing libraries in equimolar ratios, using DNA concentrations from previous step, to ensure appropriate

representation of samples sequenced simultaneously in the same Illumina flowcell lane (*see* **Note 18**).

16. Run library on Bioanalyzer, and submit library, comprising a mixture of individual barcoded libraries, for multiplex high-throughput DNA sequencing on the Illumina Genome Analyzer IIx platform (*see* **Note 19**). Sequence the amplified, adapter-ligated ChIP DNA, which will be preceded by a unique short identifier (barcode).

3.3 ChIP-Seq Data Analysis

1. For Illumina Genome Analyzer IIx, run the Genome Analyzer Pipeline software, which comprised three modules. First, Firecrest performs image analysis and fine-tune cluster locations. Next, Bustard is the base caller: it calculates the occurrence probability of a nucleotide at a given cluster, considering the pixel intensities of four images taken during each sequencing cycle (A, C, G, and T). A final sequence read of 32–36 bp is generated, consisting of the sequence of nucleotides that were called with maximal likelihood. Of particular importance are quality scores generated by Bustard to assess the quality of individual reads and exclude reads for downstream analysis. Finally, Gerald uses the ELAND algorithm to align reads against the reference genome, allowing up to two mismatches. Prior to barcode parsing and their subsequent removal, one should not pay attention to alignment metrics.
2. For barcoded samples, parse sequence reads from the ELAND query file according to the index, into distinct bins. Discard sequence reads without an intact barcode into a separate bin, but include their number in the calculation of mapping values. After reads have been separated by barcode, remove this short tag. Align the residual bases (~26–30) against the latest version of the yeast genome, using ELAND's stand-alone mode and allowing up to two mismatches (*see* **Note 20** for typical mapping values). Open-source programs for short-read alignments often contain additional capabilities, such as SNP calling (MAQ [29]) or short gapped alignments (SOAP [30]). The popular short-read aligners Bowtie [31] and BWA [32] are part of open-source high-throughput sequencing program suites that can perform various tasks, such as analysis of gene expression, SNP calling, isoform assembly, or TF-binding site peak calling (*see* **Note 21**).
3. Visualize ChIP-Seq profiles in the Integrated Genome Browser [33]. First, convert the eland_results file to a suitable file type (.sgr or .wig is common, .bam can be loaded too). This operation determines the number of sequence reads mapping to every nucleotide position, thus generating a signal file. Select the correct *S. cerevisiae* genome version from the pull-down menu or import a genome sequence file. Annotations and

other genome features from the *Saccharomyces* Genome Database (SGD) can be loaded as well. Explore and compare ChIP-Seq profiles (experimental TF-tagged strains vs. control untagged strains, or other relevant control) to RNA-Seq data, and open chromatin regions and other omics data that can be loaded in IGB (*see* **Note 22**).

4. Determine transcription factor-binding sites using a peak scoring algorithm. These programs were designed for determination of ChIP-Seq-binding regions across mammalian genomes, but simple modifications of key parameters can usually enable yeast-specific ChIP-Seq analysis (*see* **Note 23**). Distinctions between different algorithms usually concern how background is modeled or extracted from ChIP-Seq data, as well as the consideration of nonrandom artifacts and systematic biases. Given their different statistical treatment of random background and distinct criteria and thresholds for peak detection, the performance of peak-calling algorithm varies when dealing with identical samples [34]. In yeast, experimental ChIP-Seq data from epitope-tagged TF strains are scored either against a set of matched control ChIP-Seq experiments from untagged strains [16, 27], against a mock IP control if using an antibody directly against the TF of interest, or against Sono-Seq/Input-Seq experiments generated from input DNA, representing non-immunoprecipitated, cross-linked, sheared chromatin [14, 15, 24]. Here is a standard procedure to call transcription factor-binding sites using the PeakSeq algorithm [35]. Generate a mappability map of the yeast reference genome using PeakSeq and modify the following parameters to account for the smaller genome size: window size of 10 kb during the normalization step and bin size of 1 kb during the linear regression step. Considering only uniquely mapping reads, pool all replicates from a control ChIP-Seq experiment, such as ChIP in an untagged strain, into a scoring set (*see* **Note 24**). For a particular TF grown in the same medium, run the PeakSeq algorithm to score individual biological replicates against the scoring set, and to score pooled biological replicates against the same control. Score Sono-Seq/Input-Seq sample against the appropriate scoring set, generating a reference sample marking open chromatin. This will generate PeakSeq output files that contain significant regions.

5. Filter output files with a P-value or Q-value threshold to obtain statistically significant transcription factor-binding sites. For visualization of binding sites in IGB, convert the peak-calling output file into a .bed file of the following format: chromosome, chromosomal start position of peak, and chromosomal end position of peak. Inspect binding regions in IGB (*see* **Note 25**) and increase thresholds if needed. A minimal Q-value or

P-value threshold of 0.05 should be used, although it is very common and recommended to use more stringent thresholds, such as 0.01, 0.001, or 10^{-5} [24]. To filter out binding sites that might be significant at the *Q*-value level only due to an extremely elevated number of sequence reads, albeit presenting low enrichment experimental vs. control reads and therefore likely not biologically relevant, set up other filtering criteria of the PeakSeq output file, such as the difference between the number of experimental and control (background) PeakSeq sequence reads, the ratio between experimental and control PeakSeq reads, or the number of background (or experimental) reads [13, 35, 36]. This procedure usually removes sequencing artifacts as well.

6. Compare ranked lists of binding sites from biological duplicates. If the response between replicates is expected to be similar and non-stochastic, >90 % overlap is expected, depending on the number of binding sites in each list [35]. Performing ChIP-Seq experiments in biological duplicates is usually enough when reproducibility between replicates is high; otherwise, ChIP-Seq experiments should be done in biological triplicates. Target agreement plots are great diagnostic tools to determine if duplicates have sufficient reproducibility. A target agreement plot determines the fraction of overlapping targets when comparing similar fractions of ranked target lists from individual replicates. When comparing multiple transcription factors, it is often simpler to overlap binding sites from two individual replicates and combine them into a single peak list for comparative analysis.
7. If possible and when available, compare binding sites obtained by ChIP-Seq to previously published ChIP-chip regions. In our experience, ChIP-Seq analysis recovers 62–74 % of ChIP-chip peaks and identifies numerous novel binding regions, even doubling the number of binding sites or more for a particular TF.
8. If needed, validate a selection of novel binding sites by qPCR, as discussed in **Note 10**.
9. Perform high-level bioinformatics analysis on the ChIP-Seq data and final target lists. Such analyses include annotation of binding sites to neighboring genes, discovery of TF binding motifs enriched among significant ChIP-Seq-binding regions, generation of ChIP-Seq signal aggregation plots around specific genomic features, determination of enriched biological processes by gene ontology (GO) analysis, and various multivariate statistical analyses (clustering, discriminant analysis, principal component analysis, linear regressions, etc.). The R package ChIPpeakAnno [37] enables several analyses, including target

list overlap, annotation of binding region to nearest or overlapping genes, importing sequence data from genomic coordinates, and GO analysis. CisGenome is a biologist-friendly open-source analysis suite with a nice graphical user interface (GUI) and working under Windows [38]. CisGenome can identify peaks from standard alignment output files (ELAND or bowtie) with its built-in peak-calling algorithm seqpeak, load genome annotations, get DNA sequences from a target list, determine the genomic location of binding regions in comparison to transcription start sites, intergenic regions, and others, discover new TF consensus site motifs, and search for known TF-binding site motifs. A directory of available bioinformatics tools for high-throughput sequencing analysis can be found at the following web page: http://seqanswers.com/wiki/Software/list.

10. Set up and perform functional analyses, which may be additional biochemical assays, microscopy work, genetic manipulations, or other experimental validation.

4 Notes

1. We recommend using the 3-HA or 9-Myc epitopes from the PCR toolkit [39]. Correct TF-epitope fusions should be tested by sequencing the TF-epitope junctions and/or by western blot analysis to confirm the tag and the expression and/or by functional assays (cell viability, growth defects, etc.).
2. This note describes variations and alternatives to certain steps, or different manipulations, potentially useful if a lesser amount of starting material is used. At **step 2**, grow 50 mL of cells to collect 35–40 OD units of cells. In **steps 5–8**, spin in a 50 mL Falcon tube for 1–2 min at 2,880 × *g*, discard supernatant, resuspend in 1 mL of water, transfer to a 2 mL screw-capped tube, spin for 15 s at 16,000 × *g*, pipet out water, and flash freeze in liquid nitrogen. After **step 8**, pellets can be frozen at −70 °C for later use. At **step 11**, given the lesser amounts, three FastPrep bursts of at least 20 s are enough. At **step 12**, to recover lysates from the 2 mL screw-cap tubes, transfer into a 2 mL Eppendorf tube by spinning for 1 min with a ramp from 100 to 600 RPM; the additional wash is optional. There should be about 1–1.2 mL of lysate after transfer. A 3-min total sonication, using amplitude 20 % and time ON and OFF at 1 s, can be performed at **step 14**. At **step 17**, save only 10 μL input DNA. For the immunoprecipitation at **step 18**, use about 35 μL bead slurry per IP. If using a blocking procedure, first block the anti-Myc or anti-HA beads for 1 h at 4 °C with 0.5 μL 10 mg/mL BSA and 30 μL 10 mg/mL sonicated salmon sperm DNA. Using similarly blocked protein A beads,

pre-clear lysates for 30 min using those blocked protein A beads, which is followed by the immunoprecipitation for 2 h at 4 °C, adding the blocked anti-Myc or anti-HA beads and 25 μL of 10 mg/mL BSA. At **step 23**, a gel loading tip that the beads do not fit in might be used for transferring the eluates. In addition, if using smaller elution volumes (for example, 100 and 90 μL TE/SDS solution), eluates can be transferred to a 0.2 mL PCR tube at **step 23**. In the latter case, for reversal of cross-links and proteinase K treatment (**steps 26** and **27**), add 10 μL of 20 mg/mL proteinase K and incubate in the PCR machine for at least 2 h at 37–50 °C and then overnight at 65 °C; skip to **step 32** instead of performing the precipitation described in **steps 28–31**.

3. Cross-linking times could be increased if the protein of interest is indirectly binding DNA, through other proteins. Too much cross-linking might reduce the epitope availability, thus hindering the ChIP procedure. Too little cross-linking might not allow the recovery of enough material for ChIP analysis. In this case, it is possible to add DMA (dimethyladipimidate dihydrochloride) at 10 μM in 0.25 % DMSO for 10–45 min to increase cross-linking. Optimal cross-linking times could be about 10–15 min for histones and histone marks, 15–20 min for TFs, and 15 min up to 45 min for chromatin remodelers. Cross-linking can also be performed on a platform shaker.
4. This procedure using the FastPrep machine lyses over 95 % of yeast cells. Lysis can be monitored using regular light microscopy. Alternatively, a paint shaker can be used for cell lysis, but only 40 % of cells are typically lysed after a 30-min treatment.
5. Five cycles of sonication per replicate should shear chromatin to a median size of 450–500 bp. To optimize this procedure, samples with various numbers of sonication cycles should be clarified as described in **steps 15–16**. Take 250 μL of this sonicated lysate to degrade proteins, reverse cross-links, and purify DNA, as described in **steps 26–31**. Load on a 2 % agarose gel. A smear should appear between 100 and 1,000 bp, with a stronger intensity zone around the expected median size (450–500 bp). Using a Branson Analog 250 sonicator, 5–7 cycles of 30 s at constant (100 % duty cycle) and power setting about 6 should be equivalent.
6. Input DNA can be isolated for peak scoring purposes, but can also be as a reference sample as it marks open chromatin. Input DNA represents non-IP, cross-linked, sonicated chromatin. It is prepared similarly to ChIP DNA, skipping IP and IP washes. Briefly, add 250 μL of 1× TE/1 % SDS to the sample from **step 17**, perform **steps 26–31**, perform an RNase A treatment at 37 °C for 30 min by adding 2 μL 10 mg/mL RNase A, purify DNA through a MinElute column, and elute in 21 μL EB.

7. Antibody quantities could be optimized through preliminary IP experiments. It is possible to perform a two-step IP on untagged strains if ChIP-grade antibodies are available. In this case, incubate overnight samples with 5–200 μL of primary antibody, pre-wash in lysis/IP buffer Protein G or Protein A or Protein A/G agarose bead slurry, add 250 μL of 50 % agarose slurry for 1–2 h at 4 °C, and perform washes described in **step 22** in 15 mL Falcon tube at 4 °C with 10 mL of the appropriate buffer.
8. The temperature of the water bath must be above 65 °C, as determined with a thermometer. We sometimes have to set up the temperature to 67.5 or 68 °C to ensure that water reaches the appropriate temperature. Do not carry over residual beads prior to the reversal of cross-links. Residual beads might reduce the efficiency of this step and decrease the final amount of DNA recovered after the ChIP procedure.
9. This procedure yields 100–500 ng of ChIP DNA, but greater (up to 1 μg) or lower amounts (50 ng) are possible. Our range most often lies between 200 and 350 ng. DNA quantification could also be determined using the PicoGreen dsDNA assay (Invitrogen). To use Nanodrop, wash the probe with water and EB buffer, blank with 2 μL EB centered on the probe, and measure the DNA concentration and $A_{260/280}$ ratio with 2 μL ChIP sample.
10. ChIP-qPCR can be useful to verify the ChIP efficiency prior to investing time and money in further experiments, if previously characterized binding sites exist. Setting up reactions in triplicates, biological replicates of experimental (tagged) and control (untagged) strains are analyzed by qPCR to determine the enrichment at 2–4 known binding regions vs. enrichment around at least one negative control region where the TF of interest is not expected to bind. A dilution series of genomic DNA should be used as a normalization factor for PCR efficiency [40].
11. This optional gel extraction is recommended for the exclusion of very short and longer ChIP fragments that are outside the size range acceptable on Illumina's Genome Analyzer sequencer. At this step, input DNA should appear as a bright smear while the smear from ChIP DNA is usually visible but much fainter. During gel extraction, after agarose has been dissolved in QG buffer, samples should be chilled on ice for a few seconds (less than 30) to reduce GC bias in Illumina sequencing. Alternatively, one might perform a cold dissolution of agarose at room temperature to 37 °C instead of 50 °C, to eliminate such biases [41].
12. End repair blunts DNA fragments. It also ensures that all 5′ DNA ends are phosphorylated.

13. To prevent degradation of dATPs, cycles of freeze-thaw should be avoided. 20–25 μL aliquots of 1 mM dATP are prepared from the stock solution and frozen until single use. This step, given the low concentration of dATP, incorporates a single "A" nucleotide to the 3′ends of blunted ChIP fragments for subsequent ligation of Illumina-specific adapters.
14. Level of multiplexing should be determined computationally by simulations to ensure that sufficient reads for each barcoded sample will be obtained. This can be achieved by taking into account the number of expected binding sites, the nature of the DNA-binding factor (point TF vs. broader binding observed for RNA polymerase II or histone modifications), and the relative enrichment of the binding site in the ChIP sample [13]. Ideally, at least a million reads per biological replicate should be obtained, for about a 2.5-fold whole-genome coverage. Multiplex Illumina sequencing relies on the introduction of short, 3–6 bp index (or barcode) on Illumina genomic DNA adapters, directly following the Illumina adapter sequence, which is required for later steps of PCR amplification [13, 42, 43]. Barcoded adapters contain a final "T" overhang for ligation to end-repaired ChIP DNA, to which had been incorporated a 3′ "A" overhang. Examples of oligonucleotides, with 4 bp barcodes for 4-plex sequencing [13] or with 7 bp for 12-plex sequencing [44], are shown in Table 1. The forward primer contains the index with the final "T," at the 3′ end. It is highly advisable to add a phosphorothioate bond for this 3′ "T" overhang, although we have successfully generated barcoded libraries without this modification. The reverse primer has a 5′ phosphate group and the reverse-complement of the barcode at the 5′ end. In the design of barcodes, two criteria should be respected. First, according to the manufacturer's guidelines, barcoded adapters should have a balanced nucleotide design. Second, barcodes should be unique enough so that sequencing errors (especially one- or two-base) would not misassign a sequencing read to the wrong sample. When performing analysis, only reads with intact barcodes should be considering. After sequencing, all sequence reads will start with the short barcode, for identification to the original sample, followed by ChIP DNA. After sorting reads by barcode, barcode removal, and read alignment, individual ChIP-Seq profiles will be generated for each barcoded sample. Recent studies have taken advantage of the increase in the number of sequencing reads by sequencing 20 ChIP samples simultaneously in a single flowcell lane [27]. More multiplexing would be possible if comparing ChIP-Seq profiles of point TFs with few binding sites when sequencing on the Illumina HiSeq machine. Here are a few steps to anneal and generate

Table 1
Oligonucleotide sequences for barcoded ChIP-Seq, with 4-plex [13] and 12-plex [44] sequencing designs

Barcode	Forward/ reverse	Sequence (5′ → 3′)
ACGT	Forward[a]	ACACTCTTTCCCTACACGACGCTCTTCCGATCTACGT
	Reverse[b]	CGTAGATCGGAAGAGCTCGTATGCCGTCTTCTGCTTG
CATT	Forward[a]	ACACTCTTTCCCTACACGACGCTCTTCCGATCTCATT
	Reverse[b]	ATGAGATCGGAAGAGCTCGTATGCCGTCTTCTGCTTG
GTAT	Forward[a]	ACACTCTTTCCCTACACGACGCTCTTCCGATCTGTAT
	Reverse[b]	TACAGATCGGAAGAGCTCGTATGCCGTCTTCTGCTTG
TGCT	Forward[a]	ACACTCTTTCCCTACACGACGCTCTTCCGATCTTGCT
	Reverse[b]	GCAAGATCGGAAGAGCTCGTATGCCGTCTTCTGCTTG
ATCACGT	Forward[a]	ACACTCTTTCCCTACACGACGCTCTTCCGATCTATCACGT
	Reverse[b]	CGTGATAGATCGGAAGAGCGGTTCAGCAGGAATGCCGAG
CGATGTT	Forward[a]	ACACTCTTTCCCTACACGACGCTCTTCCGATCTCGATGTT
	Reverse[b]	ACATCGAGATCGGAAGAGCGGTTCAGCAGGAATGCCGAG
TTAGGCT	Forward[a]	ACACTCTTTCCCTACACGACGCTCTTCCGATCTTTAGGCT
	Reverse[b]	GCCTAAAGATCGGAAGAGCGGTTCAGCAGGAATGCCGAG
TGACCAT	Forward[a]	ACACTCTTTCCCTACACGACGCTCTTCCGATCTTGACCAT
	Reverse[b]	TGGTCAAGATCGGAAGAGCGGTTCAGCAGGAATGCCGAG
ACAGTGT	Forward[a]	ACACTCTTTCCCTACACGACGCTCTTCCGATCTACAGTGT
	Reverse[b]	CACTGTAGATCGGAAGAGCGGTTCAGCAGGAATGCCGAG
GCCAATT	Forward[a]	ACACTCTTTCCCTACACGACGCTCTTCCGATCTGCCAATT
	Reverse[b]	ATTGGCAGATCGGAAGAGCGGTTCAGCAGGAATGCCGAG
CAGATCT	Forward[a]	ACACTCTTTCCCTACACGACGCTCTTCCGATCTCAGATCT
	Reverse[b]	GATCTGAGATCGGAAGAGCGGTTCAGCAGGAATGCCGAG
ACTTGAT	Forward[a]	ACACTCTTTCCCTACACGACGCTCTTCCGATCTACTTGAT
	Reverse[b]	TCAAGTAGATCGGAAGAGCGGTTCAGCAGGAATGCCGAG
GATCAGT	Forward[a]	ACACTCTTTCCCTACACGACGCTCTTCCGATCTGATCAGT
	Reverse[b]	CTGATCAGATCGGAAGAGCGGTTCAGCAGGAATGCCGAG
TAGCTTT	Forward[a]	ACACTCTTTCCCTACACGACGCTCTTCCGATCTTAGCTTT
	Reverse[b]	AAGCTAAGATCGGAAGAGCGGTTCAGCAGGAATGCCGAG
GGCTACT	Forward[a]	ACACTCTTTCCCTACACGACGCTCTTCCGATCTGGCTACT
	Reverse[b]	GTAGCCAGATCGGAAGAGCGGTTCAGCAGGAATGCCGAG
CTTGTAT	Forward[a]	ACACTCTTTCCCTACACGACGCTCTTCCGATCTCTTGTAT
	Reverse[b]	TACAAGAGATCGGAAGAGCGGTTCAGCAGGAATGCCGAG

[a]Phosphorothioate bond between second-to-last and last (final "T" overhang) nucleotides
[b]5′ ends are phosphorylated

functional barcoded adapters: Synthesize HPLC-purified oligonucleotides (similar to those listed in Table 1) on the 0.05 μmol scale, resuspend each oligo in annealing buffer to a final concentration of 200 μM, mix the pair of forward and reverse barcoded oligos 1:1, denature in a wet heat block for

5 min at 95 °C, remove heat block to cool down at room temperature for 45 min, keep on ice for 30 min, and store annealed adapters at −20 °C. When ready to use, dilute adapters to the working concentration of Illumina genomic DNA adapters generating successful input DNA or ChIP DNA libraries. To determine adapter concentrations quickly, just compare DNA concentration measurements obtained on a Nanodrop spectrophotometer for Illumina genomic DNA adapter mix and custom-made adapters.

15. This gel extraction get rids of adapter-adapter dimers. These by-products are amplified by PCR preferentially and are visible on the gel as a strong intensity, compact band around 100–130 bp. PCR amplification of adapter-adapter dimers can compete with amplification of the regular adapter-ChIP DNA fragments, resulting in the partial or complete loss of the smear characterizing the generation of a successful sequencing library. Great care should be taken to avoid the isolation of DNA fragments around ~100 bp or lower at this step, hence the isolation in **step 10** of fragments between 150 and 500 bp. At this step, it commonly occurs that ChIP DNA is not visible, without incidence on the successful generation of a high-quality library. If adapter-adapter dimers still remain present after the library purification or are visible on the Bioanalyzer run, smaller DNA fragments, such as those dimers, can be discarded using AMPure beads (Agencourt) at a 0.8–1.0 bead-to-DNA ratio.

16. The amplification of adapter-ligated ChIP fragments must remain linear to prevent overrepresentation and underrepresentation of fragments in the final sequencing library that do not reflect biological phenomena. Illumina recommends a maximum of 18 PCR cycles at this step, with the most common range between 13 and 17 cycles. The number of sequence reads would likely be greater for overrepresented fragments due to PCR artifact, potentially leading to the identification of a false-positive binding site.

17. The final library should appear as a bright smear between 150 and 500 bp, of medium-to-high intensity. For ideal cluster generation and subsequent DNA sequencing of these clusters, DNA fragments between 150 and 350 bp should be isolated. According to Illumina's guidelines, the median size should be ~230 bp for optimal cluster generation. There should not be any band below 150 bp. If there is a faint band around 100–120 bp, it consists of adapter-adapter dimers and should be completely and carefully avoided. Presence of adapter-adapter dimers in the sequencing library will reduce the number of sequence reads passing quality filters and mapping to the reference genome.

18. Measure DNA concentration and $A_{260/280}$ ratio using Nanodrop. In our experience, concentrations of ChIP DNA sequencing libraries are at least above 8.0 ng/μL, and are at least above 15.0 ng/μL for input DNA. Good-quality libraries have $A_{260/280}$ ratios between 1.7 and 2.0. All libraries with lower DNA concentrations than 5.0 ng/μL are discarded, along with those with low $A_{260/280}$ ratios indicative of poor quality. We submit to the sequencing facility a minimum of 10 μL of sequencing library at a DNA concentration equal or greater than 5.0 ng/μL. When mixing libraries for multiplex sequencing, similar quantities of DNA should be mixed (equimolar ratios) to obtain a proper representation of all barcoded samples during sequencing. The PicoGreen dsDNA assay could be used to determine DNA concentrations more precisely. However, using Nanodrop, we rarely get more than a twofold difference between the number of sequence reads from the least and most abundant barcoded samples. The Qubit fluorometer tends to perform better than the Nanodrop at lower DNA concentrations and higher salt concentrations.
19. Illumina sequencing is a microfluidics-based sequencing-by-synthesis approach with two main steps: cluster generation on a cluster station and sequencing per se on the Genome Analyzer. During cluster generation, individual molecules from the sequencing library are attached onto a flowcell containing a lawn of complementary oligonucleotides. After initial bridge PCR amplification, the initial template is washed away and the flowcell-bound template replica is submitted to multiple rounds of bridge amplification. Each cluster contains ~1,000 molecules. Accurate size selection of the library is crucial for proper sequencing. Longer fragments would form fewer clusters of larger size, resulting in a lesser number of sequencing reads. On the other hand, shorter DNA fragments would generate too many clusters of smaller size that will not pass quality filters, hence decreasing the overall mapping values. Once the flowcell is installed on the Genome Analyzer, a sequencing primer is first annealed and four fluorescently labeled nucleotides with a reversible terminator are added. Four cycles of imaging, one for each nucleotide, are performed to determine the identity of the incorporated nucleotide. Next, the reversible terminator is cleaved off and followed by the addition of four modified nucleotides as previously described. This process is repeated up to the desired number of cycles. About 20–40 million short sequence reads are generated per lane of the Genome Analyzer IIx, while this number can easily reach 60–140 million reads per lane on the HiSeq sequencer.

20. Several statistics are used to determine if the sequencing run was successful, including the number of clusters passing quality filters, the error percentage, the total number of reads, various mapping values, and the cluster density. Multiplex sequencing runs have typically an elevated error rate compared to non-barcoded ones. Raw mapping values before barcode removal should not be used given the presence of the index. The percentage of uniquely mapping reads is usually reported as sequencing metric, along with the total number of reads. For yeast barcoded ChIP-Seq experiments, we observe around 60 % or more total uniquely mapping reads and around 10–15 % total multiple genome matches. Presence of adapter-adapter dimers in the sequencing library could reduce overall mapping to only 10–20 %.

21. Bowtie [31] and Burrows-Wheeler Aligner (BWA) [32] are fast and accurate short read aligners widely used in the high-throughput sequencing field. Output files are directly used by downstream applications from their software suite, or other computational programs, to perform specific bioinformatics tasks (RNA-Seq, isoform discovery, genome re-sequencing, ChIP-Seq, SNP calling, cloud computing, etc.). When using Bowtie for the yeast genome, we use the following parameters for read alignment: -n 2 -l 26 (2 mismatches on 26 bp).

22. Sometimes ChIP-Seq profiles for a particular chromosome or for a whole sample do not appear in the IGB browser. By scrolling down in the chromosome selection menu, other chromosome labels might appear and contain the actual profile. Rename files to use the correct chromosome label and be consistent. For chromosome 1, one might notice the following names: "chr1," "chr01," "chrI," "I," and "1."

23. Many peak scoring algorithms exist, including PeakFinder [8], FindPeaks [45], SISSRs [46], QuEST [47], PeakSeq [35], MACS [48], U-Seq [49], CisGenome [38], E-RANGE [50], spp R package [51], HPeak [52], SoleSearch [53], PeakRanger [54], and many others. CisGenome has a nice GUI and is very user friendly, requiring less programming skills [38]. First, the alignment file (ELAND or bowtie) is converted to a .aln file. Then, load the yeast genome and perform two-sample peak calling. Important parameters should be optimized, including the bin size B, the read extension length E, the window statistical cutoff C, and the half window size W. For W, point TF would have narrower peaks, thus a smaller W (<500 bp) than factors binding broader regions (RNA polymerase II, histone modifications). For E, in general, it should be determined with the following formula: E = fragment size—read length, to extend the fragment in the 3′ direction. Another popular algorithm is MACS, which uses a model-based approach [48].

It can be easily adapted to smaller genome contexts by modifying the effective genome size, and the bandwidth related to half of the estimated sonicated fragment size [48]. Higher confidence regions can be obtained by decreasing the P-value cutoff and modifying the model fold enrichment value (mfold) [24].

24. ChIP-Seq scoring samples in yeast include input DNA, ChIP on untagged strains, and, in the case of two-step IPs, agarose bead slurry only or ChIP on normal IgG (mock IP). Earlier ChIP-Seq efforts used mostly input DNA for scoring purposes, while recent yeast studies are using mock IPs as control samples. After pooling scoring replicates, an ~10 M reads scoring set is usually sufficient for proper algorithm performance.
25. It is critical to inspect binding regions in IGB to get a sense of the generated target list. The number of binding sites might be too high or too low. More importantly, the length of the binding site might be inaccurate and will constitute a problem, in particular if the fraction of very long binding regions is too high. We have seldom experienced cases in which target lists from point transcription factors contained many binding sites of length over 5 kb. Parameters of the peak scoring algorithm and/or filtering thresholds should be changed. When comparing the location of the called binding site to the actual peak from the ChIP-Seq signal, the significant region might include only the shoulders of the peak and not the center. If this phenomenon occurs regularly, the parameters of the scoring algorithm should be modified.

References

1. Costanzo MC, Hogan JD, Cusick ME et al (2000) The yeast proteome database (YPD) and *Caenorhabditis elegans* proteome database (WormPD): comprehensive resources for the organization and comparison of model organism protein information. Nucleic Acids Res 28:73–76
2. Prakash A, Tompa M (2009) Assessing the discordance of multiple sequence alignments. IEEE/ACM Trans Comput Biol Bioinform 6: 542–551
3. Borneman AR, Gianoulis TA, Zhang ZD et al (2007) Divergence of transcription factor binding sites across related yeast species. Science 317:815–819
4. Gilmour DS, Lis JT (1984) Detecting protein-DNA interactions *in vivo*: distribution of RNA polymerase on specific bacterial genes. Proc Natl Acad Sci U S A 81:4275–4279
5. Orlando V, Strutt H, Paro R (1997) Analysis of chromatin structure by *in vivo* formaldehyde cross-linking. Methods 11:205–214
6. Horak CE, Snyder M (2002) ChIP-chip: a genomic approach for identifying transcription factor binding sites. Methods Enzymol 350: 469–483
7. Harbison C, Gordon DB, Lee TI et al (2004) Transcriptional regulatory code of a eukaryotic genome. Nature 431:99–104
8. Johnson DS, Mortazavi A, Myers RM et al (2007) Genome-wide mapping of *in vivo* protein-DNA interactions. Science 316:1497–1502
9. Robertson G, Hirst M, Bainbridge M et al (2007) Genome-wide profiles of STAT1 DNA association using chromatin immunoprecipitation and massively parallel sequencing. Nat Methods 4:651–657
10. Euskirchen GM, Rozowsky JS, Wei CL et al (2007) Mapping of transcription factor binding regions in mammalian cells by ChIP: comparison of array- and sequencing-based technologies. Genome Res 17:898–909
11. Birney E, Stamatoyannopoulos JA, Dutta A et al (2007) Identification and analysis of functional

elements in 1% of the human genome by the ENCODE pilot project. Nature 447: 799–816

12. Celniker SE, Dillon LA, Gerstein MB et al (2009) Unlocking the secrets of the genome. Nature 459:927–930
13. Lefrancois P, Euskirchen GM, Auerbach RK et al (2009) Efficient yeast ChIP-Seq using multiplex short-read DNA sequencing. BMC Genomics 10:37
14. Teytelman L, Ozaydin B, Zill O et al (2009) Impact of chromatin structures on DNA processing for genomic analyses. PLoS One 4:e6700
15. Auerbach RK, Euskirchen G, Rozowsky J et al (2009) Mapping accessible chromatin regions using Sono-Seq. Proc Natl Acad Sci U S A 106:14926–14931
16. Preti M, Ribeyre C, Pascali C et al (2010) The telomere-binding protein Tbf1 demarcates snoRNA gene promoters in *Saccharomyces cerevisiae*. Mol Cell 38:614–620
17. Huber A, French SL, Tekotte H et al (2011) Sch9 regulates ribosome biogenesis via Stb3, Dot6 and Tod6 and the histone deacetylase complex RPD3L. EMBO J 30: 3052–3064
18. Zill OA, Scannell D, Teytelman L et al (2010) Co-evolution of transcriptional silencing proteins and the DNA elements specifying their assembly. PLoS Biol 8:e1000550
19. van Dijk EL, Chen CL, d'Aubenton-Carafa Y et al (2011) XUTs are a class of Xrn1-sensitive antisense regulatory non-coding RNA in yeast. Nature 475:114–117
20. Batta K, Zhang Z, Yen K et al (2011) Genome-wide function of H2B ubiquitylation in promoter and genic regions. Genes Dev 25: 2254–2265
21. Zhou X, O'Shea EK (2011) Integrated approaches reveal determinants of genome-wide binding and function of the transcription factor Pho4. Mol Cell 42:826–836
22. Cai L, Sutter BM, Li B et al (2011) Acetyl-CoA induces cell growth and proliferation by promoting the acetylation of histones at growth genes. Mol Cell 42:426–437
23. Eaton ML, Galani K, Kang S et al (2010) Conserved nucleosome positioning defines replication origins. Genes Dev 24:748–753
24. Zheng W, Zhao H, Mancera E et al (2010) Genetic analysis of variation in transcription factor binding in yeast. Nature 464: 1187–1191
25. Haynes BC, Skowyra ML, Spencer SJ et al (2011) Toward an integrated model of capsule regulation in *Cryptococcus neoformans*. PLoS Pathog 7:e1002411
26. Smith KM, Phatale PA, Sullivan CM et al (2011) Heterochromatin is required for normal distribution of *Neurospora crassa* CenH3. Mol Cell Biol 31:2528–2542
27. Venters BJ, Wachi S, Mavrich TN et al (2011) A comprehensive genomic binding map of gene and chromatin regulatory proteins in *Saccharomyces*. Mol Cell 41:480–492
28. Park PJ (2009) ChIP-seq: advantages and challenges of a maturing technology. Nat Rev Genet 10:669–680
29. Li H, Ruan J, Durbin R (2008) Mapping short DNA sequencing reads and calling variants using mapping quality scores. Genome Res 18:1851–1858
30. Li R, Li Y, Kristiansen K et al (2008) SOAP: short oligonucleotide alignment program. Bioinformatics 24:713–714
31. Langmead B, Trapnell C, Pop M et al (2009) Ultrafast and memory-efficient alignment of short DNA sequences to the human genome. Genome Biol 10:R25
32. Li H, Durbin R (2009) Fast and accurate short read alignment with Burrows-Wheeler transform. Bioinformatics 25:1754–1760
33. Nicol JW, Helt GA, Blanchard SG Jr et al (2009) The Integrated Genome Browser: free software for distribution and exploration of genome-scale datasets. Bioinformatics 25:2730–2731
34. Wilbanks EG, Facciotti MT (2010) Evaluation of algorithm performance in ChIP-seq peak detection. PLoS One 5:e11471
35. Rozowsky J, Euskirchen G, Auerbach RK et al (2009) PeakSeq enables systematic scoring of ChIP-seq experiments relative to controls. Nat Biotechnol 27:66–75
36. Euskirchen GM, Auerbach RK, Davidov E et al (2011) Diverse roles and interactions of the SWI/SNF chromatin remodeling complex revealed using global approaches. PLoS Genet 7:e1002008
37. Zhu LJ, Gazin C, Lawson ND et al (2010) ChIPpeakAnno: a Bioconductor package to annotate ChIP-seq and ChIP-chip data. BMC Bioinformatics 11:237
38. Ji H, Jiang H, Ma W et al (2008) An integrated software system for analyzing ChIP-chip and ChIP-seq data. Nat Biotechnol 26: 1293–1300
39. Janke C, Magiera MM, Rathfelder N et al (2004) A versatile toolbox for PCR-based tagging of yeast genes: new fluorescent proteins, more markers and promoter substitution cassettes. Yeast 21:947–962
40. Pfaffl M (2001) A new mathematical model for relative quantification in real-time RT-PCR. Nucleic Acids Res 29:e45

41. Quail MA, Kozarewa I, Smith F et al (2008) A large genome center's improvements to the Illumina sequencing system. Nat Methods 5:1005–1010
42. Cronn R, Liston A, Parks M et al (2008) Multiplex sequencing of plant chloroplast genomes using Solexa sequencing-by-synthesis technology. Nucleic Acids Res 36:e122
43. Craig DW, Pearson JV, Szelinger S et al (2008) Identification of genetic variants using bar-coded multiplexed sequencing. Nat Methods 5:887–893
44. Wong KH, Struhl K (2011) The Cyc8-Tup1 complex inhibits transcription primarily by masking the activation domain of the recruiting protein. Genes Dev 25:2525–2539
45. Fejes AP, Robertson G, Bilenky M et al (2008) FindPeaks 3.1: a tool for identifying areas of enrichment from massively parallel short-read sequencing technology. Bioinformatics 24: 1729–1730
46. Jothi R, Cuddapah S, Barski A et al (2008) Genome-wide identification of *in vivo* protein-DNA binding sites from ChIP-Seq data. Nucleic Acids Res 36:5221–5231
47. Valouev A, Johnson DS, Sundquist A et al (2008) Genome-wide analysis of transcription factor binding sites based on ChIP-Seq data. Nat Methods 5:829–834
48. Zhang Y, Liu T, Meyer CA et al (2008) Model-based analysis of ChIP-Seq (MACS). Genome Biol 9:R137
49. Nix DA, Courdy SJ, Boucher KM (2008) Empirical methods for controlling false positives and estimating confidence in ChIP-Seq peaks. BMC Bioinformatics 9:523
50. Mortazavi A, Williams BA, McCue K et al (2008) Mapping and quantifying mammalian transcriptomes by RNA-Seq. Nat Methods 5:621–628
51. Kharchenko PV, Tolstorukov MY, Park PJ (2008) Design and analysis of ChIP-seq experiments for DNA-binding proteins. Nat Biotechnol 26:1351–1359
52. Qin ZS, Yu J, Shen J et al (2010) HPeak: an HMM-based algorithm for defining read-enriched regions in ChIP-Seq data. BMC Bioinformatics 11:369
53. Blahnik KR, Dou L, O'Geen H et al (2010) Sole-Search: an integrated analysis program for peak detection and functional annotation using ChIP-seq data. Nucleic Acids Res 38:e13
54. Feng X, Grossman R, Stein L (2011) PeakRanger: a cloud-enabled peak caller for ChIP-seq data. BMC Bioinformatics 12:139

Chapter 16

High-Density Tiling Microarray Analysis of the Full Transcriptional Activity of Yeast

Lior David, Sandra Clauder-Münster, and Lars M. Steinmetz

Abstract

Understanding the relationship between DNA sequence variation and phenotypic variation in complex or quantitative traits is one of the major challenges in modern biology. We are witnessing a deluge of DNA sequence information and association studies of genetic polymorphisms with phenotypes of interest in families and populations. In addition, it has become clear that large portions of eukaryotic genomes beyond protein-coding genes are transcribed, generating numerous noncoding RNA (ncRNA) molecules whose functions remain mostly unknown.

DNA oligonucleotide microarrays constitute a powerful technology for studying the expression of genes in different organisms. The *Saccharomyces cerevisiae* tiling array presents a significant advance over previous array-based platforms. It has a high density of overlapping probes that start on average every 8 bp along each strand of the genome, enabling precise definition of transcript structure. Furthermore, the array includes probes specific for the polymorphic positions of another, distantly related yeast strain, allowing accurate measurement of allele-specific expression in a hybrid of the two strains. This technology thus allows high-resolution, quantitative, strand- and allele-specific measurements of transcription from a full eukaryotic genome. In this chapter, we describe the methods for extracting RNA, synthesizing first-strand cDNA, fragmenting, and labeling of samples for hybridization to the tiling array. Combining genome-wide information on variation in DNA sequence with variation in transcript structure and levels promises to increase our understanding of the genotype-to-phenotype relationship.

Key words Tiling microarray, Transcription, Gene expression, Gene structure, Noncoding RNA, Whole-genome microarray, Strand-specific transcription, Allele-specific expression, cDNA, Yeast

1 Introduction

A major challenge in biology is to understand how information encoded in genomes leads to phenotype [1]. Phenotypic diversity in many important traits arises from the variability in genetic information and regulation of its expression among individuals. Differences in DNA sequences of single genes have been shown to control differences in simple phenotypes. Most notably, studies of diseases with a Mendelian mode of inheritance have indicated

Jeffrey S. Smith and Daniel J. Burke (eds.), *Yeast Genetics: Methods and Protocols*, Methods in Molecular Biology, vol. 1205, DOI 10.1007/978-1-4939-1363-3_16, © Springer Science+Business Media New York 2014

mutations in protein-coding genes as the prevalent type of causative genetic variation. Quantitative traits, on the other hand, are conditioned by multiple loci, each with various alleles that interact among themselves and with the environment. These features significantly complicate the determination of the genetic basis of quantitative phenotypic variation. Although some methodologies have enabled the association of multiple genomic regions with a single quantitative phenotype, the genetic variability in these regions often explains only a small portion of the phenotypic variation of the trait [2, 3]. This underscores the complexity of the genotype-to-phenotype relationship and the need for quantitative technologies to unravel it.

Another discovery brought about by the increase in DNA sequence information is that although organisms like worm, fly, plant, fish, and human have significantly different numbers of cells and tissues, all share a similar set of protein-coding genes with respect to both number and functional identity. Traditionally, proteins were thought to be responsible for transmitting genetic information to bring about phenotype; recent work, however, suggests different classes of transcripts also contribute significantly [4, 5].

Transcripts can be roughly divided into two major groups: messenger RNAs (mRNAs), which code for proteins; and noncoding RNAs (ncRNAs), the majority of which still lack functional annotation [4, 6]. Functional roles have been assigned to some classes of ncRNAs, including regulation of transcription, determination of epigenetic states, and posttranscriptional regulation [7]. Thus, characterizing the transcriptome in various tissues, conditions, and genetic backgrounds will also enable the discovery and enumeration of ncRNAs, as well as the incorporation of transcriptional activity and regulation into the study of quantitative trait genetics [2, 5].

DNA microarrays have evolved over time to enable wider coverage for genome-wide analyses. By exploiting the accumulation of sequence information from model organisms, DNA microarray designs progressed from targeting only verified protein-coding genes to higher probe densities that cover other functional regions like promoters and exons of both verified and predicted genes [8–10]. Thereafter, technical developments allowed further reduction in feature size (the area used for synthesis of a DNA probe on the array); this along with the availability of full genome sequences enabled the development of tiling arrays [11–15].

Tiling arrays, on which probe arrangement is much denser than on previous platforms, systematically cover the entire genome sequence regardless of its annotation. These arrays therefore enable an unbiased interrogation of transcription as well as the delineation of all transcribed elements in the genome. Such an array design is clearly advantageous over previous designs, but also presents certain challenges. Tiling over most or all bases of a genome reduces the choice of optimal probe sequences, since the genome sequence determines the probe sequence. Furthermore, even with the most dense probe arrangement available, the genome coverage achievable by a single array is still a function of the genome size, leaving large genomes with less coverage or requiring more than one array per sample.

Our own applications of this method in the budding yeast *Saccharomyces cerevisiae* [11] have demonstrated the significant advantages of tiling arrays relative to other array-based technologies. The yeast genome is roughly 12 Mbp in size: thus, based on current feature size technology, the array could be designed with overlapping probes across the whole genome, rather than probes spaced at fixed intervals as necessitated by larger genomes. Moreover, tiling overlapping probes separately for each strand allows strand-specific measurements of transcription at 8 bp resolution and therefore a precise definition of transcript structure. The probe tiles are offset by 4 bp between opposite strands, allowing measurement at 4 bp resolution of double-stranded target samples such as genomic DNA or double stranded cDNA. Tiling probes on both strands with an offset compensates in part for the limited choice of probes, since targets (whether single- or double-stranded) can hybridize to probes with different nucleotide compositions that represent the same genomic region. Thus, using overlapping probes and different tiles for each strand to measure the same transcripts makes this array a highly sensitive and accurate platform for genome-wide transcriptome studies. Furthermore, due to the high-density design, this tiling array contains also probes with sequence specificity to the polymorphic positions of a second yeast strain, YJM789 [16]. The genome sequence of YJM789 is different by roughly 6 SNPs per kilobase compared to S288c, the reference strain sequence that was otherwise used for the array design. Allele specific expression at all of these polymorphic positions can therefore be measured with this array by hybridizing samples prepared from a hybrid of the two strains. Allele-specific expression coupled with genetic linkage mapping can identify variation in regulatory sequences that contributes to phenotypic variation [17].

The unique design of tiling arrays has necessitated modification of the standard protocols used for studies with traditional microarrays. Preparation of high quality target samples is essential for quantitatively measuring the levels and structures of all transcripts in a condition of interest using tiling arrays. This protocol covers RNA extraction, first-strand cDNA synthesis, sample fragmentation, and labeling of samples for hybridization to the tiling array. The cDNA synthesis step stops at preparation of first-strand cDNA to minimize the RNA processing steps and measure transcription in a strand-specific manner without introducing a bias in representation of transcripts. We have also incorporated the use of Actinomycin D, which prevents antisense artifacts by specifically inhibiting DNA-dependent (but not RNA-dependent) DNA synthesis.

Without this step, antisense artifacts due to spurious second-strand cDNA generation can occur during first-strand synthesis by the reverse transcriptase [18]. Altogether, the following protocol allows the generation of reliable and sensitive transcriptome maps of the full eukaryotic genome. This tiling array technology and design enables these transcriptome maps to be combined with other genetic methods, which will help clarify the molecular basis of phenotypic variation in complex or quantitative traits.

2 Materials

2.1 Extraction of Total RNA from Yeast Cells

1. Oak Ridge centrifugation tubes, PPCO (Nalgene).
2. Acid washed glass beads (Sigma).
3. Phase Lock Gel 50 mL tubes, Heavy (Eppendorf).
4. SYBR Green II RNA gels stain (Molecular Probes).
5. Bead buffer (75 mM NH_4OAc, 10 mM EDTA).
6. Phenol, acidic unbuffered water saturated (pH 4.3, Applichem).
7. Phenol–chloroform–isoamylalcohol; 25:24:1 (v/v/v) (Applichem).
8. Isopropanol solution (100 %, v/v).
9. Ethanol solution (100 %, v/v).
10. NH_4OAc solution (7.5 M).
11. SDS solution (10 %, w/v).
12. DEPC treated (RNase-free) water.

2.2 Purification of Poly(A) RNA

1. Oligotex mRNA maxi kit (Qiagen) (*see* **Note 1**).
2. DEPC water.

2.3 Residual Genomic DNA Digestion

1. TURBO DNA-free kit (Ambion).

2.4 Synthesis of First-Strand cDNA

1. Random Primers 3 μg/μL (Invitrogen).
2. Oligo(dT) 12–18 Primer, 0.5 μg/μL stock (Invitrogen).
3. dNTP + dUTP mix: dCTP, dATP, and dGTP each 10 mM; dTTP 8 mM; dUTP 2 mM.
4. SuperScript II Reverse Transcriptase (Invitrogen).
5. RNase H (Epicentre).
6. RNase cocktail (Ambion).
7. Actinomycin D stock solution, 1.25 mg/mL (Sigma).

2.5 Purification of First Strand cDNA

1. Affymetrix GeneChip Sample Cleanup module (Affymetrix). A kit for 10 reactions is also available. Alternatively, MinElute PCR Purification kit can be used (Qiagen).

2.6 Fragmentation of the cDNA and Labeling with Biotin

1. Affymetrix GeneChip WT Terminal labeling kit (Affymetrix). A kit for 10 reactions is also available.

2.7 Target Hybridization on the Tiling Microarray

1. *S. cerevisiae* yeast tiling array (Affymetrix PN 520055) (*see* **Note 2**).
2. GeneChip Hybridization, Wash and Stain kit (Affymetrix).
3. GeneChip control oligoB2 (Affymetrix).

3 Methods

Genome-wide analysis of the yeast transcriptome in a given condition: The following protocol includes extraction of RNA, synthesis of first-strand cDNA and processing it for hybridization on the high density tiling array. In our experience, the signal to noise ratio obtained from hybridizing cDNA derived from poly(A)-enriched RNA captures transcriptional information for most RNAs under standard (e.g., exponential) growth conditions (including mRNAs, ncRNAs, tRNAs, and rRNAs). The protocol described below is mainly structured around preparation of such samples. For some applications, researchers might want to look at total RNA directly, thus we also describe the preparation of first-strand cDNA from total RNA samples. The preparation of cDNA directly from total RNA essentially skips the enrichment of the poly(A) RNA step.

3.1 Extraction of Total RNA from Yeast Cells

1. Before starting: Warm a water bath up to 65 °C. Prepare a high-speed centrifuge (e.g., Sorvall), assemble the appropriate rotor (e.g., SS-34) to use with the Oak Ridge tubes and set its temperature to 10 °C. Also prepare a refrigerated centrifuge with a suitable rotor for 50 mL tubes to precipitate the cells and centrifuge the Phase lock Gel tubes. The extraction procedure is better carried out continuously from start to end and in a timely manner.
2. Per sample prepare:
   (a) Oak ridge tube with 5 g of glass beads, 1 mL of 10 % SDS and 12 mL of phenol. Place in the 65 °C water bath.
   (b) A 15 mL tube with 10 mL of bead buffer. Place in the 65 °C water bath as well.
   (c) A 50 mL Gel Phase lock tube with 10 mL of phenol–chloroform–isoamylalcohol.
   (d) Oak ridge tube for precipitation with 24 mL of isopropanol and 800 μL of 7.5 M NH_4OAc.

3. Harvest 100 mL culture with an OD_{600} of ~1 by centrifugation (1,200 g at room temperature (RT)) and discard the medium. This step can be easily carried out in two 50 mL plastic tubes per sample. Proceed directly to extract total RNA or snap freeze the cell pellets in liquid N_2 and store at −80 °C for later use (*see* **Note 3**). The number of cells to harvest from different conditions should be calculated to ensure a similar RNA yield (*see* **Note 1**). Lower the temperature of the centrifuge to 4 °C.
4. Resuspend cell pellets in 10 mL of 65 °C bead buffer (by vortex or pipette) and transfer cell suspension into the pre-warmed Oak ridge tube that contains the glass beads, SDS, and hot phenol. If the cells of one sample were collected in two separate 50 mL tubes, use the same 10 mL bead buffer to resuspend both pellets and join them together into one hot phenol Oak ridge tube.
5. Vortex vigorously the sample for 1 min at maximum vibration speed (~3,000 rpm) to break the cells' wall and incubate back at 65 °C for 1 min. Repeat vortex–incubation cycles 3 more times. Incubate for additional 5 min after which vortex for 1 min. Place tubes for 5 min on ice.
6. Centrifuge the Oak ridge tubes in an appropriate high-speed rotor (e.g., SS34 for Sorvall) for 15 min, 12,000 × *g* at 10 °C. Transfer the top aqueous phase into the Phase lock 50 mL tube by pipetting gently. Avoid the white interphase as much as possible. Lower the centrifuge temperature to 4 °C.
7. Shake the Phase lock gel tube vigorously to get a suspension of the aqueous sample with the organic phenol–chloroform–isoamylalcohol solution. Do not vortex. Centrifuge the tubes for 10 min, 1,750 × *g*, 4 °C.
8. Pour the top, gel separated, aqueous phase into the Oak ridge tube that contains isopropanol and NH_4OAc. Shake vigorously to mix. Centrifuge the tubes for 30 min, 17,210 × *g*, 4 °C to precipitate the RNA.
9. Decant the liquid leaving the pellet adhered to the bottom/side of the tube. Add 10 mL of 70 % ethanol, shake heavily to detach the pellet and wash it. Centrifuge again for 5 min, 17,210 × *g*, 4 °C to precipitate the RNA.
10. Decant the 70 % ethanol, leaving as few drops as possible. Air-dry the pellet on ice. Add 500 μL DEPC water to dissolve the RNA (leave on ice until pellet is dissolved). Transfer into a 1.5 mL Eppendorf tube.
11. Dilute a small aliquot 1:100–1:500 to determine the concentration and quality of the RNA sample (*see* **Notes 4** and **5**). You should get above 2 mg total RNA per sample. The absorbance ratio of 260/280 nm should be around 2.0 and that of

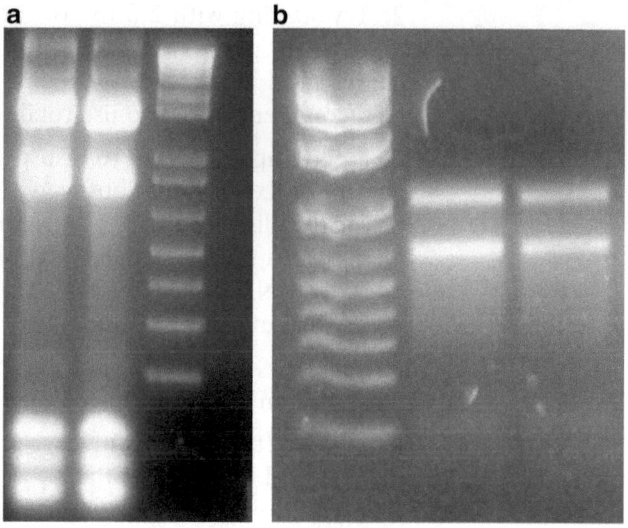

Fig. 1 Total RNA samples and poly(A) enrichment. (**a**) Yeast total RNA samples separated on 2 % (w/v) agarose gel. Note the bright distinct bands of the high molecular weight ribosomal RNA (rRNA) and of the low weight RNAs that indicate the integrity of the RNA sample. The poly(A) RNA molecules comprise the faint smear observed in the background of each sample. (**b**) Poly(A) RNA after the enrichment step. Note the integrity of the sample and the reduction in the rRNA quantity relative to the background smear that contains the rest of the transcripts. Both gels use a 1 kb DNA ladder as a size marker

260/230 around 2.0 (the lower ratio could indicate organic contaminations like phenol).

12. In addition, analyze 1 μL per sample on a 2 % (w/v) agarose gel (stained with SYBR Green II RNA gel stain or ethidium bromide). The gel should show a light smear with 2 bands for the high molecular weight ribosomal subunits and 3 bands for the low molecular weight RNAs (*see* Fig. 1a and **Note 6**). The quality of the extracted RNA can also be determined on a dedicated instrument such as the Bioanalyzer (Agilent technologies).
13. Store the total RNA samples at −80 °C or aliquot and continue to the next step.

3.2 Purification of Poly(A) RNA

In the case of preparing total RNA for hybridization move directly to Subheading 3.3. This step is carried out using the Oligotex mRNA maxi kit in accordance with the manufacturer's protocol (*see* **Note 1**).

1. Before starting with the purification of the poly(A) fraction, prepare the following: Set the heating block temperature to 70 °C. Place a 1.5 mL Eppendorf tube with DEPC water in the 70 °C heating block for the elution step.

2. Try starting with 2.5 mg of total RNA to ensure appropriate yield of poly(A)-enriched RNA. Make up the volume of the total RNA sample to 650 μL with DEPC water. It is recommended not to use total RNA older than 2 weeks.
3. Pipette up and down to resuspend the Oligotex beads in their tube. Mix the OBB buffer well before use. Precipitated salts can be dissolved at 37 °C. Add 650 μL of OBB buffer to the RNA sample and 135 μL of resuspended Oligotex beads. If less RNA is used, less Oligotex bead suspension can be used (*see* manufacture's manual).
4. Flick the tube to mix the sample and incubate at 70 °C for 7–10 min to denature secondary structures of the RNA.
5. Flick the tube again and leave at room temperature for 10 min to anneal the RNA to the poly(T)-coated beads.
6. Mark and place a spin column into a collection tube.
7. Spin the sample for 2 min at full speed in a tabletop microcentrifuge (14,000–18,000 × *g*) to precipitate the beads with the RNA. Discard the liquid and resuspend the beads in 600 μL of buffer OW2 by pipetting. Load the sample onto the spin column.
8. Centrifuge for 1 min at the full speed (>14,000 × *g*). Discard the flow-through from the tube and place the spin column back into the same collection tube.
9. Wash the sample again with 600 μL of buffer OW2 by pipetting inside the spin column. Be careful not to puncture the column membrane. Centrifuge for 1 min at full speed (>14,000 × *g*).
10. Transfer the column into a new and labeled RNase free tube that is placed in the 70 °C heating block. To elute the poly(A) RNA, add 50 μL of 70 °C DEPC water into the column and resuspend the resin by pipetting while in the heating block. Centrifuge for 1 min at full speed (>14,000 × *g*) and repeat with another 50 μL of 70 °C DEPC water. If multiple samples are processed in parallel, leave the resuspended ones in the 70 °C heating block until all are ready for centrifugation.
11. Usually the yield of poly(A)-enriched RNA is 1–1.5 % of the total RNA (*see* **Note 1**). Determine the concentration and purity of the sample by absorbance measurements.

3.3 Digestion of Residual Genomic DNA from the RNA sample

To avoid detection of residual genomic DNA in the hybridization step, a DNase treatment is applied prior to the cDNA synthesis step (*see* **Note 7**). This protocol employs the reliable and easy to use Turbo DNase kit that is quick and does not require heat inactivation or purification of the sample from the protein at the end.

1. Start the protocol with 10–30 μg of poly(A) RNA or 25 μg of total RNA. The exact amount to begin with depends on the level of residual DNA and the procedure should leave at least

9 µg of pure poly(A) RNA or 20 µg of total RNA to start the next step with. Before starting, set a heating block to 37 °C.

2. Perform the DNase reaction in up to 110 µL total. To the RNA sample, add 11 µL of (10×) buffer, 3 µL DNase and fill the reaction up to 110 µL with water. Incubate for 30 min at 37 °C. The volume and the reagent amounts can be scaled down according to the amount of residual DNA and the required volume for the next step.
3. Mix the inactivating suspension prior to use. Add 0.15 volumes of the inactivation reagent. Inactivate the DNase for 2 min at RT. Flick the tube occasionally to keep the inactivating reagent from precipitating.
4. Centrifuge for 2 min at full speed (>14,000 × *g*) in a tabletop microcentrifuge and collect the sample by carefully aspirating the liquid avoiding the pellet of the inactivating reagent.
5. Determine the concentration of the sample. It is recommended to load 0.1–0.2 µg on a 2 % (w/v) agarose gel to ensure the sample was not degraded prior to cDNA synthesis (*see* Fig. 1b and **Note 8**).

3.4 Synthesis of First Strand cDNA

This step is performed slightly differently for poly(A) and total RNA. Ideally start with 9 µg of poly(A) RNA or 20 µg of total RNA. Sometimes it may be necessary to perform two reactions per sample to ensure sufficient yield and this is generally recommended for total RNA (*see* **Note 1**). If the volume of the DNA free sample is too big for the cDNA synthesis reaction, mix the RNA with an equal volume of 100 % Ethanol and precipitate it by centrifugation (>14,000 × *g*). Dry and resuspend the pellet in the required volume of DEPC-water.

3.4.1 Synthesis of First Strand cDNA from Poly(A) RNA

(a) Set three heating blocks to 37 °C, 42 °C, and 70 °C. Alternatively, the reaction can be carried out in a thermal cycler.

(b) Dilute the Oligo(dT) primer to 0.05 µg/µL.

(c) Keep the RNA–random primer µg ratio at 1:0.5 and RNA–Oligo(dT) primer ratio at 1:0.01.

(d) Add 1.5 µL (4.5 µg) random primers and 1.8 µL (0.09 µg) diluted Oligo(dT) primer to 9 µg DNA free poly(A) RNA. Bring to 124 µL with water. Mix and incubate at 70 °C for 10 min for denaturation. Return to ice.

(e) Add 40 µL of (5×) first strand buffer, 20 µL of 0.1 M DTT, 5 µL of dNTP + dUTP mix, 1 µL of Actinomycin D (final conc. of 6.25 µg/mL), and 10 µL of Superscript II RT to a final volume of 200 µL. Incubate at 42 °C for 60 min.

(f) Inactivate the enzyme for 15 min at 70 °C.

(g) To digest the RNA template molecules, add 3 µL of RNase H and 3 µL of RNase cocktail. Incubate at 37 °C for 20 min.

3.4.2 Synthesis of First Strand cDNA from Total RNA

(a) The reaction is carried out in a thermal cycler in 0.2 mL PCR tubes.

(b) Dilute the Oligo(dT) primer to 0.05 μg/μL.

(c) Keep the RNA–random primer μg ratio at 1:0.086 and RNA–Oligo(dT) primer ratio at 1:0.0017.

(d) Take a 20 μg aliquot of DNA free total RNA. Add 0.6 μL (1.8 μg) random primers and 0.68 μL (0.034 μg) diluted Oligo(dT) primer. Bring to 58.32 μL with water. Mix and incubate at 70 °C for 10 min followed by 10 min at 25 °C and 4 °C on hold for denaturation and annealing. Alternatively, return to ice instead of the 4 °C holding step.

(e) Add 20 μL of (5×) first strand buffer, 10 μL of 0.1 M DTT, 5 μL of dNTP+dUTP mix, 1.68 μL of actinomycin D (final concentration of 20 μg/mL), and 10 μL of Superscript II RT to a final volume of 105 μL.

(f) Place the tube in a thermal cycler and run the following program: 25 °C for 10 min; 37 °C for 30 min; 42 °C for 30 min and 70 °C for 10 min to inactivate the enzyme.

(g) To digest the RNA template molecules, add 3 μL of RNase H and 3 μL of RNase cocktail. Incubate at 37 °C for 20 min.

3.5 Purification of First Strand cDNA

This step uses the cleanup kit from Affymetrix.

1. Add 500 μL of binding buffer to the reaction and mix by vortexing. The color should stay yellow similarly to the original binding buffer color. If the color changed to pink add 2 μL of 3 M sodium acetate to adjust the pH for optimal yield. Place a cleanup column in a 2 mL tube, load the sample and centrifuge for 1 min at 9,300 × *g* in a tabletop microcentrifuge. Discard the flow-through.
2. If the cDNA synthesis was carried out in two separate reactions per sample, combine each with 500 μL of binding buffer, load the first part on the column, centrifuge as above and discard the flow-through. Then load the other reaction on the same column and centrifuge again.
3. To wash the sample, add 750 μL of washing buffer and centrifuge for 1 min at 9,300 × *g*. Discard the flow-through. To dry the membrane after the wash, open the column cap and place it on the rotor in a direction that will not allow the cap to close during centrifugation. Spin with open caps for 5 min.
4. Transfer the dry column into a clean labeled 1.5 mL tube. To elute the clean cDNA apply 15 μL of elution buffer onto the membrane. Allow 1 min to release the sample and centrifuge for 1 min at maximum speed (>14,000 × *g*). Repeat the elution into the same tube with additional 15 μL of elution buffer.
5. Measure absorbance to estimate the quantity and concentration.

3.6 Fragmentation of the cDNA and Labeling with Biotin

This step uses the Affymetrix GeneChip WT Terminal labeling kit (*see* **Note 9**). Reactions are performed in 0.2 mL PCR tubes in a thermal cycler.

1. Start with an aliquot of 4.5 μg of clean cDNA obtained from poly(A) RNA or 5.5 μg cDNA obtained from total RNA. Add water to a total of 31.2 μL. In addition, aliquot 300 ng from the original cDNA sample and save it on ice for later use as a control for the fragmentation gel analysis.
2. Prepare a mix according to the number of reactions. Mix per reaction contains: 10 μL of water, 4.8 μL of cDNA fragmentation buffer, 1.0 μL of UDG (10 U/μL), 1.0 μL APE (1,000 U/μL). The total volume per reaction is 16.8 μL.
3. Combine 16.8 μL of the fragmentation mix with the 31.2 μL of the cDNA. Spin down. Place the reaction in a thermal cycler and run the following program: 37 °C for 60 min, 93 °C for 2 min, and 4 °C for at least 2 min. Store on ice if labeling will soon follow or in −20 °C if the labeling will be carried out later.
4. Check the quality of the fragmentation by loading 3 μL of the fragmented sample on a 2 % (w/v) agarose gel alongside with 300 ng of the saved untreated cDNA (*see* Fig. 2). Do not continue the protocol if the intact cDNA was degraded or if the sample was over- or under-fragmented.
5. For terminal labeling of the fragmented cDNA use the 45 μL sample. Add 12 μL of (5×) TdT buffer, 2 μL of TdT, and 1 μL of DNA labeling reagent (5 mM) to a total volume of 60 μL. Perform the following reaction on a thermal cycler: 37 °C for 60 min, 70 °C for 10 min, and 4 °C for at least 2 min.

3.7 Target Hybridization on the Tiling Microarray

These steps use the yeast tiling microarray (*see* **Note 2**), the GeneChip Hybridization, Wash and Stain kit, and the GeneChip control oligoB2 from Affymetrix. The hybridization, washing, staining, and scanning are carried out using Affymetrix equipment.

1. Set a heating block to 99 °C or boil water. Turn on the hybridization oven and set the temperature to 45 °C.
2. Take out the packed microarrays from the 4 °C storage and allow them to equilibrate to RT. Take the array out and mark it by the sample name and date on the front label. Place the microarray face down on a Kimwipe or a paper towel to avoid scratching the glass.
3. For hybridization, to each sample of 60 μL add 5 μL of 3 nM OligoB2 to a final concentration of 0.05 nM, 150 μL of (2×) Hybridization mix and 85 μL of water to a final volume of 300 μL per sample. If more than one sample is processed for hybridization, the hybridization mix should be prepared together for all and added individually to each sample.

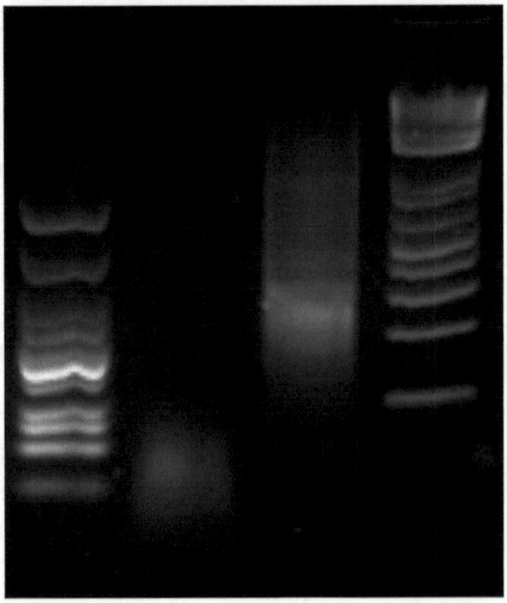

Fig. 2 Complementary DNA (cDNA) sample and fragmentation quality analysis. Separation of cDNA samples before (*third lane from the left*) and after fragmentation (*second lane*) on a 2 % (w/v) agarose gel. Equal amounts of samples from before and after fragmentation were loaded side by side. As a reference, the lowest band of the left DNA ladder is 25 bp and that of the right one is 100 bp in size. The cDNA sample was not degraded as can be seen from the size of the smear that is similar to that of the poly(A) RNA sample in Fig. 1. Note that the cDNA was fragmented enough since all of it is smaller than 100 bp, but not too digested since the intensity remained similar to that of the sample before the fragmentation

4. Place the sample in the 99 °C heating block (or in boiling water) for 10 min to denature the cDNA. Place denatured sample on ice until loading it on the microarray.
5. While denaturing, wet the microarray with 250 μL (1×) Pre-hybridization mix and place it in the 45 °C hybridization oven at 60 rpm for pre-hybridization until the sample is ready to load. The injection of liquid into the microarray is carried out from the bottom hole on the back side by puncturing the rubber septa with a small 10 μL tip. Before injecting liquid into the chamber, the upper septa must be punctured by another tip to allow a way out for the air. To inject the liquid place a larger tip inside a small tip or use a large tip with a narrow edge to avoid large punctures in the septa. Injection of liquid into the microarray should be done gently and slowly to avoid damage. Injecting 250 μL will leave a visible air bubble inside the chamber. Make sure this air bubble is free to rotate by turning and tapping gently on the side of the microarray.

6. While the microarray is pre-hybridizing, take the sample out from the ice and centrifuge it at maximum speed in a tabletop microcentrifuge for 5 min (>14,000 × *g*) to precipitate any particles.
7. While the sample is centrifuging, get back the microarray, place it on a Kimwipe and insert a small tip into the top septa. Take the pre-hybridization buffer out from the microarray. Take the sample after centrifugation, Aspirate 220 μL for poly(A) or 250 μL for total RNA derived cDNA slowly from the top part of the target sample and inject it slowly into the microarray. Avoid the bottom part of the sample and the possible pellet. Ensure that the air bubble inside the microarray is free to rotate. Cover the two septa with Tough Spot adhesive labels to avoid leaks during hybridization. Place the microarray in the 45 °C oven for 16 h at 60 rpm. Hybridizing 220 μL out of the 300 μL sample corresponds to 3.3 μg cDNA from poly(A) RNA. Hybridizing 250 μL out of the 300 μL sample corresponds to 4.58 μg cDNA from total RNA per microarray. Freeze the rest of the target sample in −20 °C.
8. Before the end of the hybridization step (16 h), turn the Affymetrix fluidics station on and prime the required number of positions according to the number of arrays used. Use Wash buffers A and B of the kit. Make sure to place enough buffers and water according to the number of arrays processed.
9. While running the priming protocol on the fluidics station, get the microarray from the hybridization oven. Peel off the Tough Spots, take the target sample out slowly from the microarray and return the sample into its original tube with the frozen remaining sample. The sample can be hybridized again so keep it at −80 °C at least until you see the scanning results are of sufficient quality or keep it longer if needed.
10. Immediately after thoroughly emptying the microarray from the target sample, refill the chamber slowly with Wash buffer A and perform the following manual washing steps: Inject 220 μL and rotate the microarray 180° by hand for 3 times allowing the bubble to rotate up and down. Draw slowly the Wash buffer A out and discard it. Repeat this wash two more times. Removing the sample and performing this manual wash step allow keeping the labeled sample for possible use in the future instead of washing it into the lines of the fluidics station. Notice: The microarray after the hybridization should not be kept without Wash buffer A so fill up the microarray chamber with Wash A buffer before attaching it to the fluidics station especially when processing several microarrays in parallel. To avoid trapping small air bubbles inside the chamber take an excess of buffer (~300 μL) and inject it slowly holding the microarray in an upright position allowing all the air to exit

until the buffer flows out from the upper tip. Check carefully that no tiny bubbles remained trapped. If there is a bubble, draw half of the volume back out and fill the chamber again in the same way.

11. Prepare 3 tubes per microarray with 600 μL of each Stain Cocktail 1, Stain Cocktail 2, and Array Holding Buffer. Place each tube in its corresponding position on the fluidics station. Upload and run the fluidics protocol FS450_0001 (if using fluidics station 450 or its matching protocol on fluidics station 400). The last step of the protocol leaves a filled microarray (with Wash buffer A) ready to scan. Take the microarray off the station and perform the shutdown protocol if required.
12. Turn on the scanner ahead of time to warm it up. Before scanning, clean the glass of the microarray using water and Kimwipe. Check that no tiny air bubbles were trapped behind the glass as those will obstruct the scanning. If you find a bubble, take it out manually by replacing the Wash buffer A as in step 10 above.
13. For scanning, the Tough Spots can be placed again on the back of the microarray to protect from leaks inside the scanner. Use the scanner according to the manufacturer instructions. For the yeast tiling array, at least an Affymetrix 3000 7G scanner is required to scan at the right resolution. The specific files that are supplied with the microarrays have to be installed prior to the scan for the program to recognize the array type (*see* **Note 2**). For a rough evaluation of the results *see* **Note 10**.

4 Notes

1. Amounts of RNA—Lower amounts of RNA transcripts are produced under some growth conditions or by certain mutant strains. Thus, we recommend using the Oligotex maxi kit that can process more than 1 mg of total RNA in order to obtain sufficient amounts of poly(A) RNA. Under more standard conditions the midi kit can be used to obtain 9 μg of poly(A) RNA. Similarly, the efficiency of cDNA synthesis can vary by conditions and in some cases two synthesis reactions per sample have to be carried out to obtain 4.5 μg of clean first-strand cDNA from poly(A) RNA or 5.5 μg cDNA from total RNA.
2. Yeast tiling microarray design—Affymetrix manufactures at least two designs of *S. cerevisiae* tiling arrays, which are commercially available. The authors use and recommend the design (PN 520055) that tile a probe every 8 bp on average separately for each strand on the same microarray. Other designs include a tile of 5 bp but of one strand only. The PN 520055 design is required if allele-specific expression is to be measured from

hybrid strains as other designs do not include probes for polymorphic positions.

3. Although RNA samples can be stored at −80 °C after the extraction, for quality reasons it is recommended to continue the poly(A) RNA enrichment protocol or DNase treatment soon after the extraction and within a couple of weeks at most.
4. RNA and cDNA quality—The key for accurate transcriptome measurement is obtaining high quality RNA. The RNA extraction and poly(A) RNA enrichment steps (Subheadings 3.1 and 3.2) should be carried out promptly and in an RNase-free environment including the reagents and equipment (e.g., gel buffer and electrophoresis box). Extracting RNA from freshly harvested cells is recommended over using frozen cell pellets.
5. It is recommended that the time from harvesting of the cells to the first vortexing in the hot phenol tube should not exceed 8 min to capture reliably the RNA repertoire. Therefore, because of the timely protocol, avoid processing too many samples in parallel.
6. Gel analysis of the total RNA—This analysis should indicate the quality of the sample. A heavy smear down the lane or very bright bands at low molecular weight may be indicative of RNA degradation or RNase contamination. Ideally the high-molecular weight ribosomal RNA bands should be strong and distinct as they comprise most of the RNA in the sample. Faint bands and brighter smear at a size lower than the ribosomal RNA bands, indicate degradation and require repeating the extraction of the RNA sample.
7. Genomic DNA contamination—Some genomic DNA could be extracted along with the RNA sample. To avoid reading signals from the genomic DNA while measuring the transcriptome, a DNase treatment is applied. The DNase treatment could be carried out either after the poly(A) RNA enrichment step (as described above) or alternatively right after the total RNA extraction. The later is necessary in the case of measuring the transcriptome from total RNA samples or recommended if multiple poly(A) RNA samples will be prepared from the same total RNA sample. In any case, the above protocol works well also if the poly(A) enrichment step follows the DNase treatment rather than preceding it.
8. Gel analysis before cDNA synthesis Regardless of whether poly(A) enriched or total RNA samples are used for cDNA synthesis, the quality of the sample should be confirmed on a 2 % (w/v) agarose gel right before the cDNA synthesis step. In the case of DNase treatment after the poly(A) RNA enrichment (as described above), it is recommended to run an analysis gel only once after the DNase treatment. But if the order of

poly(A) enrichment and DNase treatment is swapped the gel analysis should be carried out after the poly(A) enrichment step before the cDNA synthesis. As indicated in Fig. 1b, cDNA synthesis should be carried out only if the RNA sample is not degraded.

9. Target labeling—If problems with the target cDNA labeling are suspected, the labeling efficiency can be tested by a gel shift assay according to the GeneChip® Whole Transcript (WT) Sense Target Labeling Assay Manual from Affymetrix.
10. A rough sanity check of the scan image—A quick quality control check of the scan image can indicate the quality of the data. Check for the checkerboard pattern of the borders that was created by hybridizations of the synthetic oligoB2. The intensity of the off (dark) features will indicate the background signal. High contrast and sharp borders of the checkerboard pattern with high intensity of the oligoB2 probes indicate that the labeling, hybridization, washing, staining, and scanning were good. If very low intensity signal is found for the genomic probes compared to the oligoB2 synthetic checkerboard probes, the sample preparation was poor suggesting over- or under-fragmentation or too little material to begin with. Check also the overall picture for dim areas that indicate air bubbles were trapped, dust particles or manufacturing defects. Some of the analysis algorithms will discard these dim regions automatically. Zoom in to check for the decrease between the perfect match and mismatch probes that reflects the specificity of the hybridization signal.

References

1. Mackay TFC (2001) The genetic architecture of quantitative traits. Annu Rev Genet 35:303–339
2. Mackay TFC, Stone EA, Ayroles JF (2009) The genetics of quantitative traits: challenges and prospects. Nat Rev Genet 10:565–577
3. Manolio TA, Collins FS, Cox NJ et al (2009) Finding the missing heritability of complex diseases. Nature 461:747–753
4. Dinger ME, Amaral PP, Mercer TR et al (2009) Pervasive transcription of the eukaryotic genome: functional indices and conceptual implications. Brief Funct Genomic Proteomic 8:407–423
5. Sieberts S, Schadt E (2007) Moving toward a system genetics view of disease. Mamm Genome 18:389–401
6. Amaral PP, Dinger ME, Mercer TR et al (2008) The eukaryotic genome as an RNA machine. Science 319:1787–1789
7. Mattick JS (2011) The central role of RNA in human development and cognition. FEBS Lett 585:1600–1616
8. Lockhart DJ, Dong H, Byrne MC et al (1996) Expression monitoring by hybridization to high-density oligonucleotide arrays. Nat Biotechnol 14:1675–1680
9. Schena M, Shalon D, Davis RW et al (1995) Quantitative monitoring of gene expression patterns with a complementary DNA microarray. Science 270:467–470
10. Wodicka L, Dong H, Mittmann M et al (1997) Genome-wide expression monitoring in *Saccharomyces cerevisiae*. Nat Biotechnol 15: 1359–1367
11. David L, Huber W, Granovskaia M et al (2006) A high-resolution map of transcription in the yeast genome. Proc Natl Acad Sci U S A 103: 5320–5325

12. Kapranov P, Cawley SE, Drenkow J et al (2002) Large-scale transcriptional activity in chromosomes 21 and 22. Science 296:916–919
13. Selinger DW, Cheung KJ, Mei R et al (2000) RNA expression analysis using a 30 base pair resolution *Escherichia coli* genome array. Nat Biotechnol 18:1262–1268
14. Tjaden B, Saxena RM, Stolyar S et al (2002) Transcriptome analysis of *Escherichia coli* using high-density oligonucleotide probe arrays. Nucleic Acids Res 30:3732–3738
15. Yamada K, Lim J, Dale JM et al (2003) Empirical analysis of transcriptional activity in the *Arabidopsis* genome. Science 302: 842–846
16. Wei W, McCusker JH, Hyman RW et al (2007) Genome sequencing and comparative analysis of *Saccharomyces cerevisiae* strain YJM789. Proc Natl Acad Sci U S A 104:12825–12830
17. Gagneur J, Sinha H, Perocchi F et al (2009) Genome-wide allele- and strand-specific expression profiling. Mol Syst Biol 5:274
18. Perocchi F, Xu Z, Clauder-Munster S et al (2007) Antisense artifacts in transcriptome microarray experiments are resolved by actinomycin D. Nucleic Acids Res 35:e128

Chapter 17

Analysis of Silencing in *Saccharomyces cerevisiae*

Andrew Miller and Ann L. Kirchmaier

Abstract

Silencing assays have proven to be powerful tools not only for understanding how epigenetic processes function and defining the structural components of silent chromatin, but also for a useful readout for characterizing the functions of proteins involved in chromatin biology that influence epigenetic processes directly or indirectly. This chapter describes a collection of assays for monitoring silencing in *Saccharomyces cerevisiae*, including qualitative and quantitative methods as well as protocols that provide either indirect or direct measurements of the transcriptional state of loci regulated by silent chromatin.

Key words *Saccharomyces cerevisiae*, Silencing, Telomere position effect, *SIR1*, *SIR2*, *SIR3*, *SIR4*

1 Introduction

1.1 Silent Chromatin in Saccharomyces cerevisiae

Silent chromatin is found in the *Saccharomyces cerevisiae* genome at *HML*, *HMR*, the rDNA locus, and telomeres [1–3] (Fig. 1). In contrast to transcriptional repression, which occurs in a promoter-dependent fashion, silencing inactivates transcription in a regional manner [1]. Thus, in addition to endogenous genes present at silenced loci, reporter genes also become inactivated when inserted into these regions. The structure and composition of chromatin at silenced loci are distinct from actively transcribed regions of the genome. Silent chromatin has a more compact and ordered structure than euchromatin [4–7] and the DNA in silent chromatin shows greater resistance to digestion by restriction enzymes or modification by DNA methyltransferases than does DNA in euchromatic regions [8–10].

Silencing in *S. cerevisiae* is mediated via a complex process involving underlying DNA sequence elements and associated DNA binding factors that facilitate recruitment of silent chromatin components, SIR (Silent Information Regulator) proteins, and altered histone modifications [1]. The formation and composition of silent

Jeffrey S. Smith and Daniel J. Burke (eds.), *Yeast Genetics: Methods and Protocols*, Methods in Molecular Biology, vol. 1205, DOI 10.1007/978-1-4939-1363-3_17, © Springer Science+Business Media New York 2014

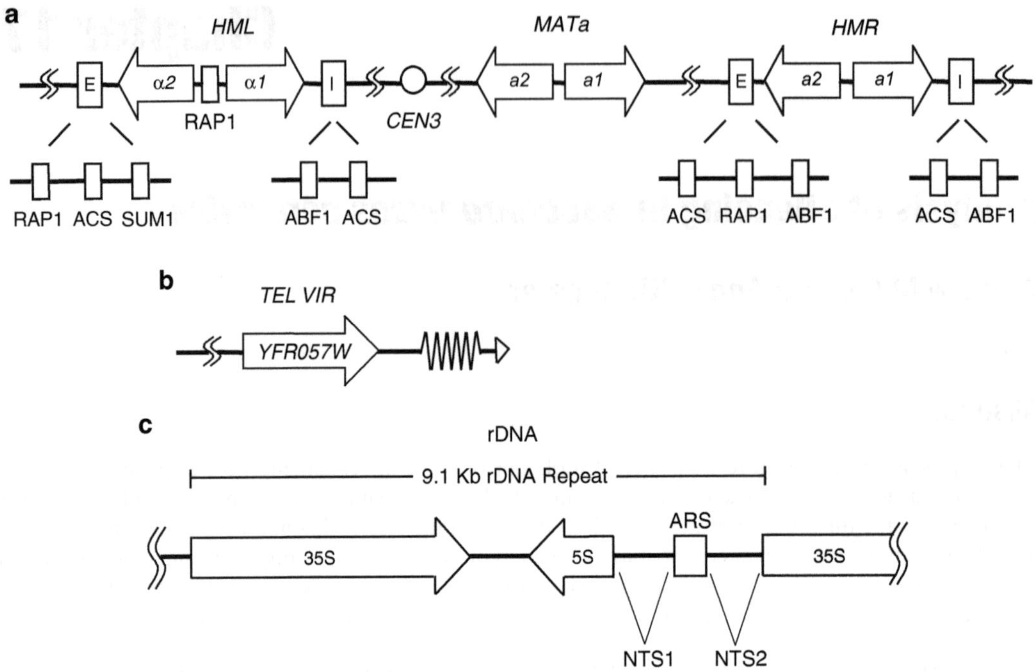

Fig. 1 Silenced loci in *S. cerevisiae*. (**a**) Mating-type loci *HML*, *MAT*, and *HMR* loci on Chromosome III. At the transcriptionally active *MAT* locus, the ***a1*** and ***a2*** genes are present in ***MATa*** cells (shown) and *MAT* contains the *α1* and *α2* genes in *MATα* cells. The silent mating-type loci, *HML* and *HMR*, are flanked by *E* and *I* silencers, which contain the indicated ORC, Abf1p, Rap1p, and Sum1p binding sites. The *α1 α2* promoter region at *HML* also contains a Rap1p binding site that acts as a proto-silencer and promotes the maintenance of silencing at *HML* [52]. (**b**) Telomere VIR. The endogenous subtelomeric gene *YFR057W* is located approximately 5 kbp from the right end of Chromosome VI and is silenced in wild-type cells. (**c**) The rDNA repeat. The rDNA locus consists of 100–200 copies of a 9.1 kb repeat containing the 5S and 35S rRNA genes plus a "nontranscribed spacer" regions (NTS1 and NTS2) and an origin of DNA replication (ARS)

chromatin in *S. cerevisiae* have been best characterized at the silent mating-type loci *HML* and *HMR*. The *HM* loci contain transcriptionally inactive copies of genes encoding α and **a** mating-type information, whereas the *MAT* locus is actively transcribed, and can contain either α or **a** information, which defines the mating-type of the cell. Unlike the *MAT* locus, each *HM* locus is flanked by DNA sequences termed the *E* and *I* silencers, which are bound by proteins that act to facilitate or stabilize SIR protein association with the *HM* loci. *HMR-E*, for example, contains an ARS sequence bound by ORC (Origin Recognition Complex) and two binding sites for the transcription factors Abf1p and Rap1p. Silencing at the *HM* loci also requires the SIR proteins 1, 3, and 4 as well as a NAD$^+$-dependent histone deacetylase Sir2p [1]. During silent chromatin formation at *HMR*, the proteins bound to *HMR-E* initially recruit the Sir proteins to this locus. Sir2-4p then propagates through the locus by their association with hypoacetylated nucleosomes. Spreading of

Sirs throughout a silenced region occurs upon changes in the nature of the chromatin surface mediated by deacetylation of histones by Sir2p, which enhances the association of Sir3-4p with nucleosomes [1, 11–16]. Silent chromatin formation also depends upon the loss of other histone modifications from chromatin at silenced loci [1, 16, 17]. Silent chromatin is thought to form similarly at telomeres, except that multiple Rap1p binding sites near telomere ends largely facilitate recruitment of SIR proteins [18]. Once formed, silent chromatin is maintained throughout the cell cycle and then is inherited upon DNA replication [19].

Differences in silencing exist between the *HM* loci, telomeres, and the rDNA locus. The *HM* loci are generally silenced in all cells in a wild-type population. In contrast, genes in sub-telomeric regions are silenced in a subset of cells within a population [20, 21], a phenomenon termed telomere position effect, TPE. Studies investigating natural telomeres in *S. cerevisiae* also suggest repression of open reading frames in subtelomeric regions is not ubiquitous at all chromosome ends [22]. Similar to at telomeres, reporter genes integrated at the rDNA locus can display position effect variegation [3]. Differences in the requirements for silent chromatin components also exist between the *HM* loci, telomeres, and the rDNA locus. For example, both *HM* and telomeric silencing are completely abolished in *sir2Δ*, *sir3Δ*, and *sir4Δ* strains, whereas although *SIR1* is important for establishing silencing at the *HM* loci, *SIR1* is not required for silencing of telomeric reporter genes under most conditions [21]. In contrast, rDNA silencing of reporter genes and endogenous Pol II-dependent NTS transcripts requires Sir2p, but not the other SIR proteins [1, 23]. In addition, mutations in modifiers of silencing at one locus do not always affect silencing similarly at other loci (e.g., [24]).

Silencing assays have been predominantly developed to monitor silent chromatin at *HML*, *HMR*, and telomeres, although reporter genes have also been incorporated at the rDNA locus (e.g., [3, 25–29]). The methods outlined below have been developed to assess both mating-type and telomeric silencing. Strategies for evaluating rDNA silencing are described elsewhere (e.g., [3, 23]). Silencing assays capitalize on evaluating phenotypic differences that are dependent on either gene silencing or derepression of transcription at one or more of the silenced loci. The phenotypes monitored are typically response to pheromones, mating ability, colony color, growth under selective conditions, or measurement of transcripts encoded by genes at silenced loci. Both qualitative and quantitative assays have been designed to assess silencing in either a direct or indirect manner. And, depending on assay design, phenotypic readouts can be observed as either a loss or a gain of silencing. Below we describe several qualitative (Patch Mating Assay, Colony Color Assay) and quantitative (Quantitative Mating Assay, α-Factor Confrontation Assay, Telomeric Silencing Assay, and Quantitative

Real-Time PCR Assay) for evaluating silencing. All of these assays provide indirect measurements of silencing except for the Quantitative Real-Time PCR assay, which directly measures mRNA transcripts originating from test loci.

1.2 Evaluating Silencing with Mating Assays

S. cerevisiae has two haploid mating-types, **a** or α, which are defined by the information expressed from the mating-type locus, *MAT*, on Chromosome III [30] (Fig. 1). *MAT**a*** yeast mate with *MATα* yeast by fusing to form an **a**/α diploid cell incapable of mating. Haploid cells divide mitotically, whereas diploid cells can either undergo mitotic division or, under conditions of starvation, undergo meiosis, or sporulation, resulting in the formation of tetrads consisting of two **a** spores and two α spores [30, 31]. In addition to *MAT*, the left and right arms of Chromosome III contain the silent mating-type loci, *HML* and *HMR*, that contain copies of mating-type information identical to that which can be present at *MAT*, except silenced. In the wild, haploid yeast may undergo mating-type conversion where either the α or **a** information stored at *HML* or *HMR*, respectively, is transferred to the actively transcribed *MAT* locus via homologous recombination initiated by the action of the HO endonuclease [32]. However, common laboratory strains, including S288C and W303, are genetically engineered to lack the HO endonuclease and are therefore genetically stable at *MAT*. Loss of silencing at the *HM* loci in a haploid cell creates what is known as a "pseudo-diploid" cell that expresses both **a** and α mating-type information and is therefore incapable of mating. Mating assays focused on silencing take advantage of mutations that create either a loss of function, i.e., lose mating in a previously silencing-competent strain, or a gain of function, i.e., restore mating in silencing-defective strain.

One genotype commonly used to identify mutants that affect silencing includes *MATα* plus an *HMR* locus with a crippled silencer, *HMR**ae*****. *HMR**ae***** contains point mutations in both the Abf1p and Rap1p binding sites at *HMR-E*, resulting in derepression at *HMR**ae***** due to a failure to recruit Sir proteins [26, 33, 34]. *MATα HMR**ae***** yeast express both α and **a** mating-type information simultaneously, creating a "pseudo-diploid" strain that is incapable of mating. *HMR**ae***** was originally constructed to define essential elements of the silencer itself [33]. Its role has been expanded to identify genes that influence silencing by screening for suppression of silencing defects at this locus (e.g., [26, 27, 34–36]). In other words, when a secondary mutation or deletion results in the restoration of silencing at *HMR**ae*****, the yeast will express only α information from *MAT* and the defect in mating will be suppressed (*see* Fig. 2).

1.2.1 Qualitative Mating Assay: Patch Mating

Mating assays, in general, rely on complementation of auxotrophic markers between two haploid strains. Growth on minimal medium indicates successful mating between strains of opposite mating types.

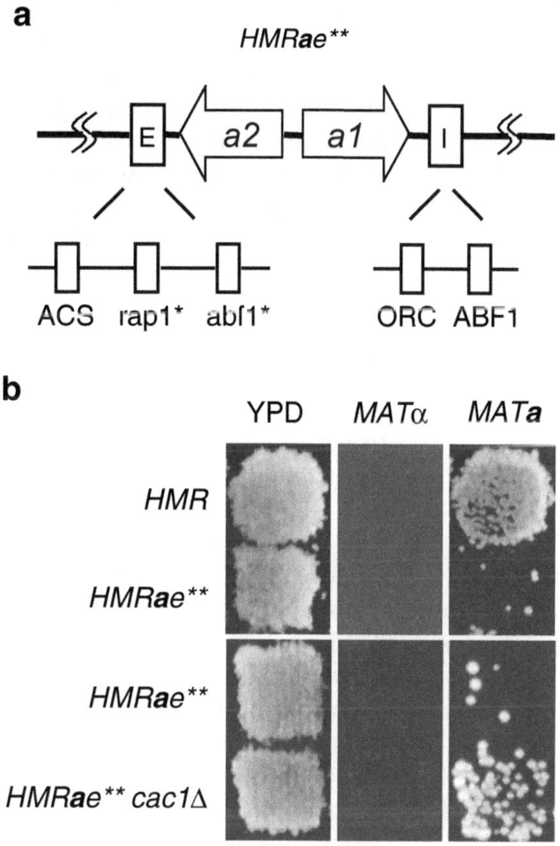

Fig. 2 Monitoring silencing at *HMR* and *HMRae*** by patch mating assays. (**a**) Schematic of *HMRae***. *HMRae*** contains point mutations at the Rap1p and Abf1p binding sites at the *E* silencer (*see* Table 2). Not to scale. (**b**) Patch mating assays. *MATα HMR* or *HMRae*** strains with the indicated genotypes were grown on minimal medium (YM) plates with supplements for 24 h at 30 °C, then replicated to rich medium (YPD) plates or to *MATα* or *MATa* lawns on YM plates lacking supplements and incubated for 2 days at 30 °C prior to collecting images with an Alpha Innotech imager (San Leandro, CA) and ChemiImager 5500 v.2.02 software

Here, the yeast strain being analyzed is termed the "experimental strain." The strain of the opposite mating type to the experimental strain is called the "tester strain." The tester contains complements to all the auxotrophic markers in the "experimental strain," yet lacks at least one gene encoding an auxotrophic marker that is present in the experimental strain. Patch Mating Assays such as the one shown in Fig. 2 can be used to evaluate silencing defects qualitatively. Numerous genetic backgrounds can be assayed, including *HML* in *MAT**a*** strains and *HMR* in *MATα* strains, under a variety of conditions. This method can also be used to identify suppressors of silencing defects in sensitized genetic backgrounds, such as in *MATα HMR**ae***** cells noted above.

1.2.2 Quantitative Mating Assays

Quantitative Mating Assays measure the efficiency of mating between an experimental strain and a tester strain at the level of a single cell. Like Patch Mating Assays, Quantitative Mating Assays provide an indirect measurement of silencing because a cell's ability to mate correlates with the expression state of *HML* or *HMR*. Quantitative Mating Assays can provide up to a ~10^5-fold range of linearity in which to measure the relative efficiency of silencing of two different strains. This range will vary somewhat, depending on the genotype of the parental "experimental" strain being evaluated. This assay is particularly useful for testing for synthetic interactions between two mutants that affect silencing as well as for detecting small differences in mating efficiencies between two strains that cannot be assessed by Patch Mating Assays alone. How to perform Quantitative Mating Assays and the determination of mating efficiencies are discussed below.

1.3 Examining Silencing at HML with α-Factor Confrontation Assays

The α-Factor Confrontation Assay provides an indirect measurement of silencing by capitalizing on the yeast cell's response to the mating pheromone α-factor. α-factor is a 13 aa peptide secreted by *MATα* cells that binds to the α-factor pheromone receptor Ste2p, a seven transmembrane domain receptor, on ***MATa*** cells. ***MATa*** haploids respond to α factor by arresting in G1 phase and undergoing distinct morphological changes in which the cell grows projection(s) called "shmoos" in preparation for fusion with a *MATα* cell [37]. In contrast, *MATα* and **a**/*α* diploid cells do not respond to α factor and will instead continue to progress through the mitotic cell cycle, characterized by budding and the formation of daughter cells [37]. This morphological difference between the arrested and cycling states is monitored in α-Factor Confrontation Assays.

α-Factor Confrontation Assays can be used to monitor silencing in ***MATa*** cells because mutations or growth conditions that result in defects in silencing at *HML* will cause haploid yeast to express both **a** and α information simultaneously. Such pseudo-diploid cells, will continue to progress through the cell cycle, will not arrest in G1 phase, will not form shmoos, and will not mate. Silencing assays based on α-factor sensitivity may be performed using either solid or liquid media.

1.4 Assessing Silencing with Colony Color Assays: The HMRΔ::ADE2 Reporter

A colorimetric assay based on the expression of *ADE2* can be used to evaluate the nature of silencing defects. In yeast, the *ADE2* gene product, phosphoribosylamino-imidazole-carboxylase, functions at the sixth step of de novo purine nucleotide biosynthesis. Ade2p adds a carboxyl group to the C4 carbon on the imidazole ring of 5′-phosphoribosyl-5-aminoimidazole. Deletion, mutation, or silencing of *ADE2* results in the accumulation of 5′-phosphoribosyl-5-aminoimidazole in vacuoles, which results in the formation of red pigmentation [38]. In contrast, when *ADE2* is fully expressed,

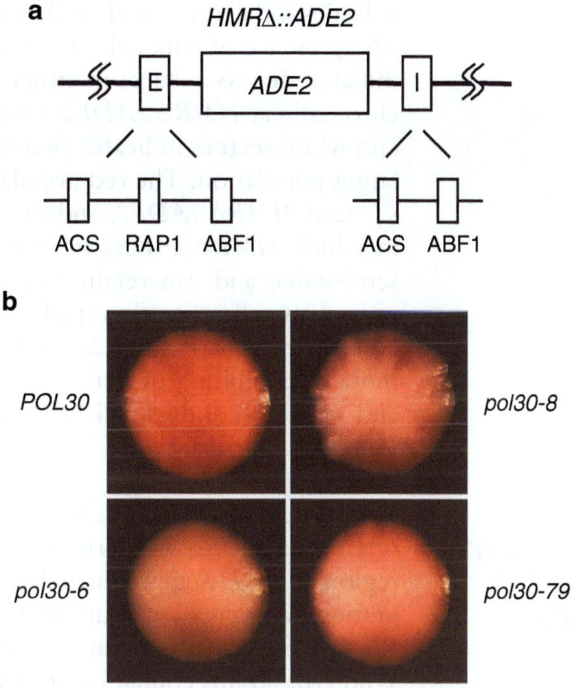

Fig. 3 Silencing of *ADE2* reporter at *HMR*. (**a**) Schematic of *HMRΔ::ADE2*. The ***a1*** and ***a2*** genes at *HMR* have been replaced with *ADE2* (*see* Table 1). Not to scale. (**b**) Colony Color Assay. *MAT**a*** *HMRΔ::ADE2* yeast with the indicated genotypes were grown logarithmically, plated onto YPD at a density of 100–300 cells per plate and incubated at 30 °C for 3 days. The plates were then transferred to 4 °C and incubated for an additional 3 days to allow for the accumulation of pigment prior to collecting images with a Leica MZ125 microscope and SPOT 4.1 imaging software

this metabolic intermediate does not accumulate and yeast will grow as white colonies. Thus, colony color can be employed as an assay to monitor silencing when *ADE2* is integrated, for example, between the *E* and *I* silencers at *HMR*, *HMRΔ::ADE2* (Fig. 3a) (e.g., [34, 39, 40]).

Four different colony phenotypes may be observed in Colony Color Assays; red, white, pink, or sectored colonies. Based on these phenotypes, this assay can provide information on the nature of the defects in silencing. Wild-type *HMRΔ::ADE2* yeast grow as red colonies on rich medium (YPD plates) due to the accumulation of the substrate for the *ADE2* gene product because *ADE2* is silenced. In contrast, the lack of red pigment accumulation indicates *ADE2* is expressed. Therefore, growth of white colonies indicates complete derepression at *HMR* has occurred. The presence of pink colonies implies that less pigment has accumulated due to a defect in either maintaining or inheriting silent chromatin at *HMRΔ::ADE2*. Finally, growth of colonies with red and white sectors indicates two sub-populations exist within the larger population. The red population is derived from cells with a silenced *HMRΔ::ADE2*, and the white population contains cells in which *HMRΔ::ADE2* is derepressed. These populations are semi-stable and can retain their transcriptional states at *ADE2* over many cell generations [20]. Examples of a wild-type (*POL30*) colony in which *HMRΔ::ADE2* is silenced and colonies of mutants exhibiting defects in maintaining or inheriting (*pol30-6* and *pol30-79*) and establishing (*pol30-8*) silencing are shown in Fig. 3b.

1.5 Evaluating Telomeric Silencing with the URA3-TEL VII-L Reporter

Studies demonstrating position effect variegation at telomeres have often used a *URA3* reporter gene integrated in the subtelomeric region at *ADH4*, approximately 1.1 kb from *Tel VII-L*. This integration event creates a truncated sub-telomeric region lacking the X and Y elements found at most natural telomeres (Fig. 4a) [20]. Wild-type strains containing *URA3-TEL VII-L* can grow on synthetic complete media lacking uracil as well as on media containing 5′fluoro-orotic acid, 5-FOA. Growth on media lacking uracil indicates *URA3* is not silenced in some cells in the population. Growth in the presence of 5-FOA, which is toxic to cells expressing the *URA3* gene product, orotidine 5′ phosphate decarboxylase, indicates that *URA3* is silenced in other cells in the population. Both transcriptional states are semi-stable within the population (e.g., [20, 41]). An example of a Telomeric Silencing Assay using *URA3-TEL VII-L* is shown in Fig. 4b, and this assay is described below. Several additional reporter genes, including *ADE2*, *HIS3*, and *TRP1*, are also available for integration at *Tel VII-L* or *Tel V-R* for the evaluation of telomeric silencing and are described elsewhere [20, 28, 42].

1.6 Monitoring HM and Telomeric Silencing by Quantitative Real-Time PCR

The Patch Mating, Quantitative Mating, Colony Color, α-Factor Confrontation, and Telomeric Silencing Assays monitor phenotypes that provide indirect measurements of silencing. For a more direct measurement of the transcriptional state of loci of interest, other strategies designed to measure transcripts from these loci can be employed, including RNA blots (e.g., [15, 23]), primer extension (e.g., [23]), or Quantitative Real-Time PCR as described below (e.g., [15]). As ***a1*** mRNA has a half-life of less than 3 min [43], changes in steady state levels of transcripts can provide an accurate measurement of the transcription initiation from *HMR*.

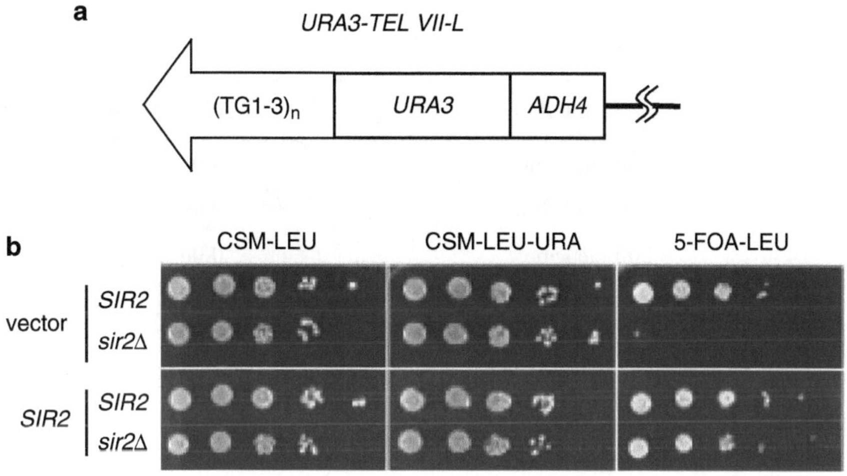

Fig. 4 Silencing of a *URA3* reporter at *Tel VII-L*. (**a**) Schematic of *URA3-TEL VII-L*. A *SalI-EcoRI* fragment containing the *URA3* reporter from pVII-L URA3-TEL was integrated at *ADH4* adjacent to *Tel VII-L* by homologous recombination [20, 41] (*see* Table 2). Not to scale. (**b**) Telomeric Silencing Assay. Tenfold serial dilutions of logarithmically growing yeast with the indicated genotypes were plated onto the indicated media and incubated for 2 days at 30 °C prior to collecting images with an Alpha Innotech imager (San Leandro, CA) and ChemiImager 5500 v. 2.02 software. The strains shown maintain a 2 μm vector (Yep351) [53] or a 2 μm plasmid expressing *SIR2* (pLP349) [54]. For *SIR2* cells, telomere position effect is demonstrated by variegated growth on plates lacking uracil (colonies are derived from cells in the population in which *URA3* is expressed) and on plates containing 5-FOA (colonies are derived from cells in the population in which *URA3* is silenced). Lack of growth of *sir2Δ* cells on plates containing 5-FOA indicates *URA3* was derepressed in all cells in the population

2 Materials

2.1 Evaluating Silencing with Mating Assays

2.1.1 Qualitative Mating Assay: Patch Mating

Prepare all media (*see* **Note 1**) with distilled deionized water and autoclave to sterilize.

1. Yeast Peptone Dextrose (YPD) liquid medium; 1 % w/v Bacto™-Yeast Extract, 2 % w/v Bacto™-Yeast Peptone, 2 % w/v D-glucose.
2. YPD solid medium; 1 % w/v Bacto™-Yeast Extract, 2 % w/v Bacto™-Yeast Peptone, 2 % w/v D-glucose, 2 % w/v Difco™ agar, granulated.
3. Yeast Minimal (YM) liquid medium; 0.7 % w/v Difco™ Yeast Nitrogen Base w/o Amino Acids, 2 % w/v D-glucose (*see* **Note 1** and Table 1 for supplements).
4. YM solid medium; 0.7 % w/v Difco™ Yeast Nitrogen Base w/o Amino Acids, 2 % w/v D-glucose, 2 % w/v Difco™ Agar (*see* **Note 1** and Table 1 for supplements).
5. Tester strains *MAT***a** *his4* or *MATα his4* or equivalent and experimental strains of interest (*see* **Notes 2** and **3** and Table 2).

Table 1
Supplemental stock solutions for YM plates

Stock solution	Working concentration (g/mL)
200× adenine	0.04
200× histidine	0.04
200× leucine	0.06
200× lysine	0.06
200× methionine	0.04
200× tryptophan	0.04
100× uracil	0.02

6. Sterile fine pile velvet, 150 × 150 mm (Sigma cat. no. Z363405-36EA or equivalent) (*see* **Note 4**).
7. Replica plater, 100 mm diameter (Sigma, cat. no. Z363391-1EA, BioWorld, cat. no. 43170004-1 or equivalent).
8. Alpha Innotech imager (San Leandro, CA) and ChemiImager 5500 Version 2.02 software or equivalent.

2.1.2 Quantitative Mating Assays

Prepare all media with distilled deionized water and autoclave to sterilize.

1. Yeast Peptone Dextrose (YPD) liquid medium; 1 % w/v Bacto™-Yeast Extract, 2 % w/v Bacto™-Yeast Peptone, 2 % w/v D-glucose.
2. YPD solid medium; 1 % w/v Bacto™-Yeast Extract, 2 % w/v Bacto™-Yeast Peptone, 2 % w/v D-glucose, 2 % w/v Difco™ agar, granulated.
3. Yeast Minimal (YM) liquid medium; 0.7 % w/v Difco™ Yeast Nitrogen Base w/o Amino Acids, 2 % w/v D-glucose (*see* **Note 1** and Table 1 for supplements).
4. YM solid medium; 0.7 % w/v Difco™ Yeast Nitrogen Base w/o Amino Acids, 2 % w/v D-glucose, 2 % w/v Difco™ agar, granulated (*see* **Note 1** and Table 1 for supplements).
5. Mating tester strains *MAT**a** his4* or *MATα his4* or equivalent (*see* **Note 3** and Table 2) and experimental strains.
6. 3 mL sterile syringe and Swinnex Filter Holder, 25 mm (Millipore cat. N. SX0002500).
7. Cellulose Nitrate Membrane Filters 0.45 μm pore size, 25 mm diameter (Whatman cat. no. 7184-002). Assemble filters in filter holders using forceps, wrap in aluminum foil and autoclave prior to use.

Table 2
Yeast strains for silencing assays

Strain	Genotype	Source
JRY2726	*MATa his4*	P. Schatz
JRY2728	*MATα his4*	P. Schatz
BY4741	*MATa his3Δ1 leu2Δ200 met15Δ0 ura3Δ0*	
BY4742	*MATα his3Δ1 leu2Δ200 lys2Δ0 ura3Δ0*	
YSC1021-553985	BY4741 *MATa his3Δ1 leu2Δ200 met15Δ0 ura3Δ0 bar1Δ::KanMX*	Open Biosystems
W303	*MATa* or *α ade2-1 his3-11,15 leu2-3,112 trp1-1 ura3-1 can1-100*	R. Rothstein
JRY4012	*W303 MATa ADE2 lys2Δ::hisG*	J. Rine
JRY4013	*W303 MATα ADE2 lys2Δ::hisG*	J. Rine
AKY1968	W303 *MATα hht1-hhf1Δ::LEU2 hht2 hhf2Δ::HIS3* plus PK189 (*HHT2 HHF2 ARS/CEN/URA3*)	[15][a]
AKY944	W303 *MATa hht1-hhf1Δ::LEU2 hht2 hhf2Δ::HIS3* plus PK189 (*HHT2 HHF2 ARS/CEN/URA3*)	[15][a]
AKY1785	W303 *MATa HMRΔ::ADE2 POL30Δ::hisG* plus PK169 (*POL30 ARS/CEN/URA3*)	[34][a]
AKY1692	W303 *MATa HMRΔ::ADE2 hht1-hhf1Δ::LEU2 hht2 hhf2Δ::HIS3* plus PK189 (*HHT2 HHF2 ARS/CEN/URA3*)	[34][a]
JRY5273	W303 *MATα HMRae***	[35]
AKY1677	W303 *MATα HMRae***	[36]
AKY1744	W303 *MATα HMRae** hht1-hhf1Δ::LEU2 hht2 hhf2Δ::HIS3* plus PK189 (*HHT2 HHF2 ARS/CEN/URA3*)	[34][a]
AK1785	W303 *MATα HMRae**pol30Δ::hisG* plus PK169	[34][a]
AKY4454	W303 *MATα URA3-TELVII-L*	[41]
AKY4460	W303 *MATα URA3-TRP1-TELVII-L*	[41]

[a]Parental strains for screening PCNA or histone H3 and H4 mutants in silencing assays

2.2 Examining Silencing at *HML* with α-Factor Confrontation Assays

Prepare all media with distilled deionized water and autoclave to sterilize.

1. Yeast Peptone Dextrose (YPD) liquid medium; 1 % w/v Bacto™-Yeast Extract, 2 % w/v Bacto™-Yeast Peptone, 2 % w/v D-glucose.
2. YPD solid medium; 1 % w/v Bacto™-Yeast Extract, 2 % w/v Bacto™-Yeast Peptone, 2 % w/v D-glucose, 2 % w/v Difco™ agar, granulated.
3. Yeast α-factor mating pheromone (*see* **Note 5**).
4. ***MATa*** or ***MATa** bar1*Δ yeast strains (*see* **Note 6**).
5. Microscope (Nikon Alpaphot-2 YS2 or equivalent), slides, coverslip.
6. Cell counter.

2.3 Assessing Silencing with Colony Color Assays: The *HMRΔ::ADE2* Reporter

Prepare all media with distilled deionized water and autoclave to sterilize.

1. Yeast Peptone Dextrose (YPD) liquid medium; 1 % w/v Bacto™-Yeast Extract, 2 % w/v Bacto™-Yeast Peptone, 2 % w/v D-glucose.
2. YPD solid medium; 1 % w/v Bacto™-Yeast Extract, 2 % w/v Bacto™-Yeast Peptone, 2 % w/v D-glucose, 2 % w/v Difco™ agar, granulated.
3. Wild-type yeast containing *HMRΔ::ADE2* and isogenic experimental strains (Table 2; *see* **Note 7**).
4. Velvet, 150 × 150 mm (Sigma cat. no. Z363405-36EA or equivalent).
5. Microscope and imaging software such as a Leica MZ125 microscope and SPOT 4.1 imaging software, or equivalent.

2.4 Evaluating Telomeric Silencing with the *URA3-TEL VII-L* Reporter

Prepare all media with distilled deionized water and autoclave to sterilize.

1. Yeast Peptone Dextrose (YPD) liquid medium; 1 % w/v Bacto™-Yeast Extract, 2 % w/v Bacto™-Yeast Peptone, 2 % w/v D-glucose.
2. YPD solid medium; 1 % w/v Bacto™-Yeast Extract, 2 % w/v Bacto™-Yeast Peptone, 2 % w/v D-glucose, 2 % w/v Difco™ agar, granulated.
3. 5-FOA plates; 0.7 % w/v Difco™ Yeast Nitrogen Base w/o Amino Acids, 2 % w/v D-glucose, 0.1 % w/v 5-FOA, 0.005 % uracil w/v, 2 % w/v Difco™ agar, granulated (*see* **Note 8**).
4. Complete Synthetic Medium lacking Uracil (CSM-URA) agar plates, 0.077 % w/v CSM-URA (Sunrise Science Products, cat. no. 1004-10); 0.7 % w/v Difco™ Yeast Nitrogen Base

w/o Amino Acids, 2 % D-glucose, 2 % w/v Difco™ agar, granulated.

5. Complete Synthetic Medium (CSM) agar plates, 0.079 % w/v CSM (Sunrise Science Products, cat. no. 1001-10); 0.7 % w/v Difco™ Yeast Nitrogen Base w/o Amino Acids, 2 % D-glucose, 2 % w/v Difco™ agar, granulated.
6. Yeast containing *URA3-TEL VII-L* (Table 2).
7. Multichannel pipettes (P20 and P200, or equivalent), 96-well round bottom tissue culture plate with lid (Costar cat. no. 3799), Pipetting Reservoir (VWR cat. no. 82026-356).
8. Velvet, 150×150 mm (Sigma cat. no. Z363405-36EA or equivalent).
9. Alpha Innotech imager (San Leandro, CA) and ChemiImager 5500 Version 2.02 software or equivalent.

2.5 Monitoring HM and Telomeric Silencing by Quantitative Real-Time PCR

Prepare all media with distilled deionized water and autoclave to sterilize.

1. Yeast Peptone Dextrose (YPD) liquid medium; 1 % w/v Bacto™-Yeast Extract, 2 % w/v Bacto™-Yeast Peptone, 2 % w/v D-glucose.
2. YPD solid medium; 1 % w/v Bacto™-Yeast Extract, 2 % w/v Bacto™-Yeast Peptone, 2 % w/v D-glucose, 2 % w/v Difco™ agar, granulated.
3. Yeast Minimal (YM) liquid medium; 0.7 % w/v Difco™ Yeast Nitrogen Base w/o Amino Acids, 2 % w/v D-glucose (*see* **Note 1** and Table 1 for supplements).
4. YM solid medium; 0.7 % w/v Difco™ Yeast Nitrogen Base w/o Amino Acids, 2 % w/v D-glucose, 2 % w/v Difco™ Agar (*see* Table 1 for supplements).
5. AE buffer; 5 mM NaOAc, pH 5.3, 10 mM EDTA, pH 8.0 prepared with DEPC-treated distilled H_2O (*see* **Note 9**).
6. 10 % SDS; 50 g sodium dodecyl sulfate per 500 mL distilled H_2O. Filter-sterilize through a 45 μm vacuum filter.
7. 10× Reaction buffer; 200 mM Tris–HCl, pH 7.5, 10 mM $CaCl_2$, 10 mM $MgCl_2$.
8. DNase I (Sigma cat. no. D-5307).
9. ReadyMade Primer Random Hexamer (IDT, cat. no. 51-0118-01 or equivalent).
10. dATP, dTTP, dCTP, dGTP (Promega, cat. nos. U120A, U123A, U122A, U121A, or equivalent).
11. RNAseOUT Ribonuclease Inhibitor (Invitrogen, cat. no. 10777-019).

12. M-MLV reverse transcriptase, 200 U/μL (Invitrogen, cat. no. 28025-013) or SuperScript II RNase H-Reverse Transcriptase (Invitrogen, cat. No. 18064-014).
13. 5× First Strand Buffer; 250 mM Tris–HCl, pH 8.3, 375 mM KCl, 15 mM $MgCl_2$. (or Invitrogen, cat. no. Y00146).
14. Thermocycler.
15. SYBR Green PCR Master Mix (Applied Biosystems, cat. no. 4309155) or equivalent.
16. StepOne Plus Real-Time PCR System (Applied Biosystems) or equivalent.

3 Methods

3.1 Evaluating Silencing with Mating Assays

3.1.1 Qualitative Mating Assay: Patch Mating

Day 1

1. Use a colony of the appropriate tester strain (e.g., *MAT**a** his4*) from a recently streaked YPD plate to inoculate 3 mL of liquid YPD medium. Grow overnight with aeration at 30 °C (*see* **Notes 2** and **3**).
2. Patch experimental strains (e.g., *MATα HMR**a**e*** and derivatives) to be tested onto a YPD plate or a YM plate plus appropriate supplements (*see* **Notes 2** and **10**).

Day 2

1. Spread 0.3–0.5 mL of the overnight culture of tester strain (e.g., *MAT**a** his4*) onto a YM plate lacking supplements.
2. Using a sterile velvet and replica plater, replica plate the patched experimental strains from **step 2** Day 1 onto a YPD plate that will serve as a control for growth.
3. Using a fresh velvet, replica plate the experimental strains onto the tester strain lawn of the opposite mating-type on the YM plate lacking supplements. Incubate approximately 24–48 h at 30 °C (*see* **Notes 11** and **12**).

Day 3 or 4

1. Collect images of the strains on the plates. Remove the lid from the plate and place the plate on top of a velvet. Take images of all plates at the same magnification using an Alpha Innotech imager (San Leandro, CA) and ChemiImager 5500 Version 2.02 software or equivalent (*see* **Note 12**).

3.1.2 Quantitative Mating Assays

1. Grow experimental wild-type and mutant strains plus a tester strain of the opposite mating type logarithmically in liquid YPD or selective medium (YM plus supplements). Grow three biological replicates for each genotype to be evaluated and

determine the OD_{600} of the cultures using a spectrophotometer. Dilute each culture to an OD_{600} of 1.0 into YPD medium (*see* **Note 13**).

2. For each strain, mix 200 μL (2×10^6 cells) of the experimental strain with 1 mL (1×10^7 cells) of the tester strain and load into a syringe attached to a sterile Swinnex Filter Holder. Collect the mixture of cells onto a sterile 0.45 μm pore, 25 mm nitrocellulose filter disk (*see* **Note 14**).
3. Disassemble the filter holder and, using sterile forceps, place the filter disk, cells side up, onto a YPD plate. Incubate at 30 °C for approximately 5 h.
4. Using sterile forceps, transfer the filter covered with cells to 1 mL of sterile water in a sterile 12×75 mm culture tube or equivalent. Resuspend the cell mixture by using a vortexer to dissociate cells from the filter and break up cell clumps.
5. Perform serial dilutions of the cell mixture into YM medium lacking supplements based on the initial concentration of experimental cells present in the cell mixture (2×10^6 cells/mL) as the starting concentration. The dilution series should contain the following concentrations: 1×10^3, 1×10^4, and 1×10^5 and 1×10^6 experimental cells/mL.
6. For each strain, plate 100 and 300 cells onto separate YM plates plus supplements to select for growth of the haploid experimental strain as well as any diploids resulting from mating events. These plates should contain medium that will not support growth of the haploid mating tester strain. Incubate the plates at 30 °C for 2 days.
7. For each strain, plate an appropriate number of cells onto two separate YM plates lacking supplements to select for growth of diploids (*see* **Note 15**). Incubate the plates at 30 °C for 2 days.
8. For each strain, separately count the number of colonies present on each YM plate plus supplements. For each biological replicate of each strain, determine the average number of colonies from both plates. This number represents the Plating Efficiency and is used for all subsequent calculations. Calculate the Plating Efficiency for each biological replicate of each experimental strain in the following manner (*see* **Note 16**):

   Plating Efficiency = {(colonies of experimental strain on YM plates containing supplements plated with "100" cells per plate) + [(colonies on YM plates containing supplements plated with "300" cells per plate)/3]}/2
9. For each strain, separately count the number of colonies present on each YM plate without supplements. For each biological replicate of each strain, determine the average number of colonies from both plates. Calculate the Mating Efficiency for each biological replicate of each experimental strain as follows (*see* **Note 17**):

Mating Efficiency = [(**a**/α diploids)$_{\text{YM without supplements}}$/(**a**/α diploids + haploids of the experimental strain)$_{\text{YM with supplements}}$]

For example, for a strain that was plated onto YM plates without supplements at 100 and 1×10^3 experimental cells/plate, the Mating Efficiency is calculated as follows:

Mating Efficiency = {(**a**/α diploid colonies on the YM plate plated with "100" cells per plate) + [(**a**/α diploid colonies on the YM plate plated with "1×10^3" cells per plate)/10]/2}/(Plating Efficiency calculated in **step 8**).

10. For each biological replicate, calculate the Relative Mating Efficiency of an experimental strain relative to wild-type. The wild-type strain's mating efficiency is set to 1 and is calculated as follows:

    Wild-Type Mating Efficiency = Mating Efficiency$_{\text{Wild-Type}}$/Mating Efficiency$_{\text{Wild-Type}}$ = 1

    The Relative Mating Efficiency, or the mating efficiency of each experimental strain relative to wild-type, is calculated as follows:

    Relative Mating Efficiency = Mating Efficiency$_{\text{Mutant}}$/Mating Efficiency$_{\text{Wild-Type}}$

11. The calculations above describe how to calculate the relative mating efficiency of a single biological replicate. Using all biological replicates for each strain, next determine the average ± standard deviation of the Relative Mating Efficiency for each experimental strain compared to wild-type, which is set to 1. Statistical analyses of the data should be conducted using an appropriate nonparametric statistical test such as the Wilcoxon rank sum test.

3.2 Examining Silencing at HML with α-Factor Confrontation Assays

3.2.1 Method 1: Liquid Cultures

1. Pick colonies of freshly streaked yeast strains and inoculate 3 mL liquid YPD medium and grow cultures logarithmically for several hours to overnight at 30 °C with aeration (*see* **Note 2**).
2. Determine the OD_{600} of the cultures. If the cells are no longer in log phase growth, dilute the cultures to an $OD_{600} \approx 0.2$ (2×10^6 cells/mL) in fresh YPD medium and grow for another ~1–2 h at 30 °C with aeration.
3. Pellet the cells by low-speed centrifugation for 5 min. Remove medium and wash cells with fresh YPD medium. This wash step is critical for *BAR1* yeast to remove any Bar1p protease that has accumulated in the medium. Pellet the cells by low-speed centrifugation for 5 min and resuspend the strains to an $OD_{600} \approx 0.2$–0.5 (2–5×10^6 cells/mL) in fresh YPD medium. Observe the cells under microscope to confirm that all strains are growing logarithmically.

4. Add α-factor to liquid cultures at a final concentration of approximately 10 μg/mL for *BAR1* cells or 0.1 μg/mL for *bar1Δ* cells. Incubate the cells at 30 °C with aeration for up to 5 h (*see* **Note 18**).
5. After 3, 4, and 5 h, observe at least 100 cells under microscope and count the number of cells that have formed shmoos, are unbudded, have small or large buds, or that have a "dumbbell" morphology in which a bud has emerged from the tip of a schmoo (*see* **Notes 18–20**).

3.2.2 Method 2: Solid Media

1. Spot 5 μL of 200 μg/mL α-factor onto a 5 mm diameter area of an YPD plate and let dry.
2. Streak a *MAT**a*** strain across the spot of α-factor.
3. Immediately after streaking, observe cells streaked on the α-factor spot using a microscope and identify and mark well-separated cells.
4. Incubate the cells at 23 °C and score at least 100 cells over time (~3–6 h) for the formation of schmoos as above.

3.3 Assessing Silencing with Colony Color Assays: The HMRΔ::ADE2 Reporter

1. Pick single colonies from freshly streaked wild-type and mutant yeast strains and inoculate 3 mL liquid YPD medium for each strain, incubate overnight with aeration at 30 °C (*see* **Note 2**).
2. The following morning, determine the optical density, OD_{600}, of the liquid cultures.
3. Initially dilute each culture in liquid YPD medium to an $OD_{600} = 1.0$ ($\approx 1 \times 10^7$ cells/mL).
4. Perform serial (1:10) dilutions of each strain using liquid YPD medium to achieve a final concentration of 1×10^3 cells/mL for each strain to be tested (*see* **Note 21**).
5. Using the 1 mL sample at a concentration of 1×10^3 cells/mL for each strain to be tested, plate 100–300 cells of each strain onto separate YPD plates (*see* **Note 22**).
6. Incubate the plates at 30 °C for 2–3 days and monitor growth for sufficient colony size (~2 mm) to visualize color.
7. Transfer the plates to 4 °C for approximately 3–4 days to enhance color development (*see* **Note 23**).
8. Collect color images of colonies of all control and experimental yeast strains. Remove the lid from the plate, place the plate on top of a velvet on a microscope stage of a Leica MZ125 microscope or equivalent, and collect images using SPOT 4.1 imaging software or equivalent (*see* **Notes 24** and **25**).

3.4 Evaluating Telomeric Silencing with the URA3-TEL VII-L Reporter

1. Pick fresh colonies of yeast and inoculate 3 mL of YPD for liquid cultures. Incubate for several hours or overnight at 30 °C with aeration to grow cells logarithmically.
2. The next morning, determine the OD_{600} of the cultures using a spectrophotometer and dilute all cultures in liquid YPD medium to a final cell density of 1×10^7 cells/mL.
3. Perform four 1:10 serial dilutions so that the final sample has a cell density of 1×10^3 cells/mL. The dilution series should include (1×10^7, 10^6, 10^5, 10^4, and 10^3 cells/mL) (*see* **Note 26**).
4. Using a P20 multichannel pipette, mix the samples in the first dilution (1×10^7 cells/mL) and spot 3 μL in a column onto YPD or CSM, CSM-URA, and 5-FOA plates, using a grid with 10×10 mm squares placed underneath and visible through the agar plates as a guide. Change pipette tips between each dilution and repeat spotting the dilution series in columns from left to right across the plate (*see* **Note 27**).
5. Incubate the plates at 30 °C and monitor growth at 24 and 48 h.
6. Collect images of cell growth on plates. Remove the lid from the plate and place the plate on top of a velvet, cell side up. Take images of all plates at the same magnification using an Alpha Innotech imager (San Leandro, CA) and ChemiImager 5500 Version 2.02 software or equivalent (*see* **Note 28**).

3.5 Monitoring HM and Telomeric Silencing by Quantitative Real-Time PCR

3.5.1 RNA Isolation from Cells (See Note 29)

1. Harvest $1\text{–}2 \times 10^8$ cells (10 mL of cells at 1–2 OD_{600} in liquid YPD or YM medium) (*see* **Note 30**).
2. Resuspend cells in 40 μL AE buffer, transfer to microfuge tube, and add 40 μL 10 % SDS. Use a vortexer to mix.
3. Add 40 μL phenol (*see* **Note 31**), use a vortexer to mix sample. Incubate at 65 °C for 5 min. Freeze on dry ice and then separate phases by centrifugation in a microfuge at maximum speed (~$16{,}000 \times g$) for 10 min (*see* **Note 32**).
4. Transfer aqueous phase to a fresh microfuge tube, add an equal volume of a 1:1 mixture of phenol–chloroform, mix by vortexing, place samples on a rotator and rotate for 5 min to mix samples. Separate phases by centrifugation in a microfuge at maximum speed (~$16{,}000 \times g$) for 5 min.
5. Transfer aqueous phase to fresh microfuge tube, add 40 μL 3 M NaOAc, pH 5.3 and 2.5 vol (~1 mL) 100 % ethanol. Mix by rocking, and then pellet RNA by centrifugation in a microfuge at maximum speed (~$16{,}000 \times g$) for 5 min.
6. Wash RNA pellet with 80 % ethanol, pellet RNA by centrifugation in a microfuge at maximum speed (~$16{,}000 \times g$) for 5 min and allow RNA pellet to dry.
7. Resuspend pellet in 15 mL 0.2 mM EDTA, pH 8.0. Determine the concentration of the RNA sample at $OD_{260/280}$ using a spectrophotometer.

3.5.2 cDNA Preparation for Real-Time PCR

cDNA can be generated from the isolated RNA using a commercially available kit as per manufacturer's instructions, e.g., QuantiTect Rev. Transcription Kit (Qiagen cat. no. 205311) or equivalent. An alternative protocol for the synthesis of cDNA is outlined below.

DNAse I Treatment

1. In a thin-walled reaction tube suitable for a thermocycler, mix 4 μg RNA with 1 μL 10× Reaction Buffer, 1 μL DNAse I (1 U) plus H_2O in a total volume of 10 μL. Using a thermocycler, incubate for 15 min at 25 °C. Add 1 μL 50 mM EDTA, pH 8.0 and incubate for 10–15 min at 70 °C to inactivate DNAse I, then cool sample to 4 °C.

cDNA Synthesis

1. In a thin-walled reaction tube suitable for a thermocycler, mix 5 μL of the DNase-treated RNA, 5 μL H_2O, 1 μL (100 ng) random hexamer primers plus 1 μL 10 mM dNTPs. Incubate at 65 °C for 5 min, place the tube on ice. Pellet sample by centrifugation for 30 s at low speed (~1,000–4,000 × *g*).
2. Add 4 μL 5× First Strand Buffer, 2 μL 0.1 M dithiothreitol, 1 μL RNaseOUT (40 U/μL). Mix by pipetting and incubate at 37 °C for 2 min.
3. Add 1 μL M-MLV Reverse Transcriptase or SupersScript II Reverse Transcriptase. Mix by pipetting and incubate at 25 °C for 10 min. Incubate at 37 °C for 50 min.
4. Incubate the sample at 70 °C for 15 min to inactivate the enzyme.
5. Add 10 μL of cDNA to 740 μL H_2O. Store the sample at −20 °C prior to use.

3.5.3 Real-Time PCR Reaction

1. For each biological replicate of each sample, prepare three technical replicates of PCR reactions as follows: In a thin-walled PCR reaction tube suitable for Real-Time PCR, for a Real-Time PCR reaction volume of 10 μL, mix 2.5 μL (or determine empirically if using a cDNA synthesis kit) of the diluted cDNA, 2.5 μL of the appropriate Primer Pair stock (e.g., *see* Table 3 and **Note 33**) plus 5 μL SYBR Green PCR Master Mix. Analyze samples by Real-Time PCR in appropriate instrumentation (e.g., StepOne Plus Real-Time PCR System, Applied Biosystems) according to manufacturer's instructions.
2. For each biological replicate, determine the average the technical replicates and determine the levels of *α1*, ***a1***, or *YFR057w* mRNA relative to an internal control (*SCR1* or *ACT1*) for each strain and compare to that observed in control yeast in

Table 3
Oligonucleotides for evaluating silencing by Quantitative Real-Time PCR

Region	Oligonucleotide pair
a1	5′TTTAGAAGAAAGCAAAGCCTTAATTCC 5′CTTGAAGTGGAGTAATGCCACATT
α1	5′CACAGTTTGGCTCCGGTGTA 5′CCGCGTGCCATTCTTCAG
YFR057W	5′GCCAAGCTTCCAATATCACGA 5′GGAATGATCTTGGAAATCGATCA
SCR1	5′CGCGGCTAGACACGGATT 5′GCACGGTGCGGAATAGAGAA
ACT1	5′GACGCTCCTCGTGCTGTCTT 5′GTCTTTTTGACCCATACCGACC

The optimized concentrations of each of the above primer pairs for quantitative Real-Time PCR are 900 nm/900 nm

which the locus of interest is derepressed (e.g., in *sir2Δ* cells), which has been set to 100 %. Data is calculated according to manufacturer's instructions and as follows:

$$2^{[(\text{locus } C_T - SCR1 \text{ or } ACT1\, C_T)_{sir2\Delta \text{ control strain}} - (\text{locus } C_T - SCR1\, C_T)_a]} \times 100$$

where a is the genotype of interest.

For each sample, express the data as the average ± standard deviation of all biological replicates. Statistical analyses of the data should be conducted using an appropriate nonparametric statistical test such as the Wilcoxon rank sum test.

4 Notes

1. Supplemental stock solutions should be spread onto YM plates using a sterile glass spreader as needed and allowed to dry prior to use. Stocks commonly used for complementing auxotrophies found in W303 and BY4741 strains are listed in Table 1. Mild heating will aid in dissolving of uracil while making a stock solution. Prior to use, all stocks should be sterilized using either a 0.45 μm filtration apparatus or an autoclave. Tryptophan stocks should be stored in the dark by wrapping the stock bottle in aluminum foil to prolong their stability. 0.15 mL of 200× stock solutions and 0.3 mL of 100× stock solutions should be spread onto YM plates of approximately

30 mL total volume. Alternatively, yeast can be grown in appropriate Complete Synthetic ("drop out") Medium, CSM (Sunrise Science Products).

2. The growth and manipulation of yeast has been covered in depth elsewhere [44]. Throughout this chapter, 30 °C has been listed as the standard temperature for growing yeast. However, certain yeast mutants are temperature sensitive and should instead be grown at permissive temperature while performing the assays discussed in this chapter. Here, it is assumed the reader will use appropriate growth conditions for their strains of interest.
3. Commonly used tester strains for mating assays include *MAT**a** his4* for determining the mating efficiency of most *MATα* strains or *MATα his4* for most *MAT**a*** strains (Table 2). Each tester strain complements the auxotrophic markers present in most laboratory strains (e.g., W303, S288C). This quality allows for the selection of diploids on medium lacking any supplements. Alternatively, any haploid strain with appropriate auxotrophic markers can be used as the tester strain. High background growth of some tester strains on minimal medium can sometimes occur due to reversion events if the tester strain chosen has a point mutation instead of a deletion at a locus being used for selection, e.g., *ade2-1* versus *ade2Δ*. Several parental strains (e.g., *MATα HMR**a**e***) available for creating experimental strains for evaluating silencing by Patch or Quantitative Mating Assays are listed in Table 2.
4. Homemade or commercial velvets should be wrapped in aluminum foil and sterilized by autoclaving prior to replica plating. Alternatively, sterile gauze can be used for replica plating.
5. The peptide α-factor, TRP-HIS-TRP-LEU-GLN-LEU-LYS-PRO-GLY-GLN-PRO-MET-TYR, can be synthesized in house or obtained commercially (e.g., Sigma-Aldrich cat. no. T-6901, or GeneScript cat. no. RP01002-10 mg). To generate a stock solution, resuspend the peptide in methanol at a final concentration of 1 mg/mL and store at −20 °C. The potency of α-factor can vary, depending on the purity of the synthesized peptide and the strain background being analyzed. Therefore, the concentration of α-factor required to stably arrest wild-type *BAR1* or *bar1* cells for 4–5 h should be determined empirically in an experiment analogous to the one described in Subheading 3.2.1 prior to routine use of a new stock of α factor.
6. *BAR1/SST1* encodes a protease, Barrier protease, which is secreted into the periplasmic space and degrades α-factor [45]. Therefore, yeast lacking *BAR1* require significantly less (~1/100) α-factor than *BAR1* strains to induce an arrest and

bar1 strains generally remain arrested for longer periods of time. Strains lacking *BAR1* can be generated by one step gene replacement or by genetic crosses [46] using for example *MATa bar1Δ::KANMX* yeast in the BY4741 genetic background (Table 2).

7. *HMRΔ::ADE2 sir2Δ*, *sir3Δ*, or *sir4Δ* yeast can be used as negative control strains that will grow as white colonies due to the loss of silencing at *HMR*.
8. For 1 L of medium for pouring 5-FOA plates, add 20 g Difco granulated agar to 500 mL dH_2O and autoclave to sterilize. After autoclaving, place in a 65 °C water bath to keep agar from hardening. Heat 400 mL H_2O to 65 °C, add 1 g 5-FOA, 7 g Yeast Nitrogen Base without Amino Acids, 20 g glucose, and 50 mg Uracil. Mix to dissolve, bring final volume to 500 mL, and sterilize using a 0.45 μm filter. Add mixture to the 500 mL of autoclaved agar. Avoid contamination by spraying 95 % Ethanol on the opening of each flask prior to mixing the solutions together. Pour plates immediately either in a hood or next to an open flame.
9. Use H_2O that has been treated with Diethyl pyrocarbonate, DEPC, to inactivate any RNases present. Add 1 mL DEPC to 1 L distilled H_2O, shake thoroughly, incubate 1 h at room temperature in a fume hood, then autoclave to sterilize and inactivate DEPC.
10. Streak strains from glycerol stocks stored at −80 °C onto either a YPD plate or a YM plate plus supplements. Incubate yeast at 30 °C for approximately 2 days prior to picking fresh colonies for use in analysis. Patch fresh colonies in 10 × 10 mm squares onto a YPD or a selective medium plate (YM plus supplements). A paper template containing a grid of 10 × 10 mm squares can be placed under the plate and viewed through the agar to facilitate aligning each strain while creating patches.
11. Growth on YM plates lacking supplements indicates mating of the "experimental" and required for growth, whereas neither haploid strain will be able to grow. Silencing is indicated by growth of diploids on minimal medium (YM plates lacking supplements). The YPD plate serves as a control for growth to ensure a similar amount of each strain had been plated initially and then transferred during replica plating.
12. Mating assays can also identify mutants with silencing-independent defects, e.g., mutations in the sterile, *STE*, genes [47]. Therefore, independent assays (e.g., Colony Color Assay or Quantitative Real-Time PCR Assay) should be conducted to confirm that any differences in the mating efficiency between mutant and wild-type strains are associated with silencing.

13. Quantitative mating assays should be performed in triplicate for each genotype. For strains grown in YPD medium, assume an OD_{600} of 1.0 is approximately 1×10^7 cells/mL. For strains grown in YM medium, assume an OD_{600} of 1.0 is approximately 2×10^7 cells/mL.
14. The genotypes of the experimental strain and the mating tester strain should be complementary for all autotrophic markers to permit growth of the diploid on a YM plate lacking supplements.
15. Ideally, a sufficient number of experimental cells should be plated that will result in the growth of 25–300 colonies of diploid cells per YM plate to facilitate counting colonies at the end of the experiment. Wild-type cells, for example, will typically exhibit mating efficiencies between 15 and 90 %, depending on the mating-type of the cell and the strain background being evaluated and, therefore, should be plated onto YM plates at concentrations of 100 and 300 cells/plate. In contrast, $<1 \times 10^6$ *sir2Δ* cells will mate. A rough estimate of the appropriate number of cells to plate for Quantitative Mating Assays can be determined in advance by comparing the mating efficiencies of the strains being evaluated to wild-type cells in preliminary Patch Mating Assays. Alternate concentrations of experimental cells to plate may include the following: 100, 300, 1×10^3, 3×10^3, 1×10^4, 3×10^4, or 1×10^5 cells/plate.
16. The number of colonies on the Plating Efficiency plates provides the number of cells that were available for mating in each culture. The Plating Efficiency may vary between the strains being evaluated. Counting colonies from plates containing different dilutions of each strain and determining the average the number of colonies present prior to conducting further calculations will control for assay variability.
17. Only diploids will be viable on the YM plates without supplements. The colonies present on these plates represent successful mating events between the tester and experimental strain. In contrast, the number of colonies on the YM plates plus supplements determined in **step 8** represents all viable cells present (both the experimental strain and diploids derived from the experimental strain). In other words, the mating efficiency is the ratio of diploids (**a**/α) to the total number of cells available for mating.
18. When silencing is lost at *HML* in *MAT**a*** cells, the **a**/α pseudodiploids will remain in log phase in the presence of α-factor and cultures will contain cells in various stages of the cell cycle, including unbudded cells and cells with small or large buds. In contrast, α-factor-arrested cells will contain a mixture of unbudded cells and shmoos at early time periods, mostly

shmoos at intermediate time periods and shmoos with multiple projections at late time periods. Cells may also initially arrest in G1 phase, then break out of the arrest. This behavior can be caused by an insufficient concentration of α factor due to its degradation by Bar1p or defects in maintaining silencing at *HML*. In either case, the cells that break out of the α-factor arrest will often bud from the tip of a projection, and have a morphology that is similar to a "dumbbell".

19. The varied incubation times at this step is to ensure that cells are counted once all the cells in the culture have completed the cell cycle and those cells with a silenced *HML* locus will have had sufficient time accumulate in G1 phase after the addition of α-factor. To synchronize a log phase culture, cells should be incubated in α-factor for at least two times the duration of a single cell cycle. For example, in liquid YPD medium at 30 °C, a wild-type cell will double approximately every 90 min. Therefore, to arrest an unsynchronized culture grown in YPD medium, approximately 3 h will be required for all the cells in the culture to achieve a G1 arrest, and often an additional hour for most cells to form shmoos.

20. α-Factor Confrontation Assays can also be applied to evaluate mutations in pathways unrelated to silencing. For example, mutations in the signal transduction pathway activated by α factor binding to Ste2p can also lead to defects in G1 arrest and shmoo formation in the presence of α-factor. Thus, to ensure that failure of a *MAT**a*** cell to arrest in the presence of α-factor is related to defects silencing at *HML*, any results from α-factor confrontation assays should be confirmed by an independent method such as the Quantitative Real-Time PCR Assay.

21. When performing serial dilutions, for each 1:10 dilution in a total volume of 1 mL, add 100 μL of sample to 900 μL of H_2O or liquid YPD medium. Mix using a vortexer immediately prior to performing each of the dilutions in the series as yeast will settle to the bottom of a liquid culture rapidly. The use of larger volumes when performing dilutions helps to reduce variability by ensuring an accurate number of cells are transferred between each dilution.

22. When plating small volumes, i.e., 100 μL, add the 100 μL of cells to 200 μL of sterile H_2O or liquid YPD medium prior to plating. Spreading cells evenly across the plate is easier when using a larger volume, especially if the plate is somewhat dry.

23. Storage at 4 °C will enhance accumulation of the red pigment in strains in which *ADE2* is partially or fully silenced. The accumulating pigment in W303-based strains will be red as in Fig. 3b, whereas colonies of BY series strains in the S288C

genetic background tend to develop a red-brown color when *ADE2* is silenced [48].

24. Take images of wild-type and experimental strains at the same time and magnification as well as under identical lighting conditions.
25. A recent study has provided evidence that silencing-independent co-regulation of *ADE2* and *URA3* (purine and pyrimidine biosynthesis pathways) may occur under certain conditions [49]. Thus, results obtained using the *ADE2* reporter should be also compared to those observed in a second silencing assay that is not based on *URA3* expression (e.g., a silencing assay based on mating or direct measurement of transcripts from the locus of interest as in the Patch Mating Assay, Quantitative Mating Assay or Quantitative Real-Time PCR Assay).
26. When screening multiple strains simultaneously, serial dilutions can be performed by using a multichannel pipette and sterile 96-well round bottom tissue culture plates with a lid. Up to eight strains can be spotted onto a single 100 mm diameter agar plate for analysis. Alternatively, 1:5 serial dilutions can be used for analysis where appropriate.
27. To prevent spots from running into each other, warm the YPD, CSM-URA, and 5-FOA plates prior to spotting and, after pipetting, let spots dry before moving plates. Using CSM-based plates for analysis is preferable to top spreading supplements onto YM-based plates. YM plates will often have a surface that is too slick for the 3 μL drop to maintain surface tension, causing the drop to spread, and sometimes fuse with adjacent drops.
28. 5-FOA is toxic to yeast expressing *URA3*. 5-FOA is an analog of orotic acid, a pyrimidine that is a precursor in nucleotide biosynthesis. Exposure to 5-FOA results in inhibition of nucleotide biosynthesis and DNA synthesis and is lethal. Significantly, a recent study has demonstrated that the sensitivity to 5-FOA observed while using the *URA3-TEL VII-L* reporter can reflect defects in silencing or, alternatively, be related to processes not associated with general telomeric silencing defects. For some mutants, previous results interpreted as silencing defects, may instead be related to the site of integration of the *URA3* reporter (*ADH4*) or 5-FOA-dependent transcriptional activation of ribonucleotide reductase and upregulation of transcription at *URA3-TEL VII-L* [49] (*see* also ref. 50). Thus, results obtained with the *URA3-TEL VII-L* reporter should be confirmed by an independent method such as the Quantitative Real-Time PCR Assay (*see* **Note 23** also).
29. Adapted from [51].

30. Cell pellets may be stored at −80 °C prior to use. If frozen, thaw the pellets in the presence of AE buffer plus SDS (*see* Subheading 3.5, RNA isolation from cells, **step 2**) to avoid degradation of RNA.
31. Pre-equilibrate phenol with AE buffer.
32. The samples will thaw during centrifugation.
33. The primers listed in Table 3 have been previously optimized for Real-Time PCR for monitoring transcription of ***a1*** from *HMR*, *α1* from *HML*, and *YFR057w* from *Tel VR* as well as from the transcriptionally active control loci *SCR1* and *ACT1*.

Acknowledgements

This work was supported by the National Science Foundation (A.L.K.) and the Purdue University Center for Cancer Research (A.L.K.).

References

1. Rusché LN, Kirchmaier AL, Rine J (2003) The establishment, inheritance, and function of silenced chromatin in *Saccharomyces cerevisiae*. Annu Rev Biochem 72:481–516
2. Palladino F, Laroche T, Gilson E et al (1993) SIR3 and SIR4 proteins are required for the positioning and integrity of yeast telomeres. Cell 75:543–555
3. Smith JS, Boeke JD (1997) An unusual form of transcriptional silencing in yeast ribosomal DNA. Genes Dev 11:241–254
4. Bi X, Broach JR (1997) DNA in transcriptionally silent chromatin assumes a distinct topology that is sensitive to cell cycle progression. Mol Cell Biol 17:7077–7087
5. Cheng T-H, Li Y-C, Gartenberg MR (1998) Persistence of an alternate chromatin structure at silenced loci in the absence of silencers. Proc Natl Acad Sci U S A 95:5521–5526
6. Weiss K, Simpson RT (1998) High-resolution structural analysis of chromatin at specific loci: *Saccharomyces cerevisiae* silent mating type locus *HMLalpha*. Mol Cell Biol 18:5392–5403
7. Ravindra A, Weiss K, Simpson RT (1999) High-resolution structural analysis of chromatin at specific loci: *Saccharomyces cerevisiae* silent mating-type locus *HMRa*. Mol Cell Biol 19:7944–7950
8. Loo S, Rine J (1994) Silencers and domains of generalized repression. Science 264:1768–1771
9. Gottschling DE (1992) Telomere-proximal DNA in *Saccharomyces cerevisiae* is refractory to methyltransferase activity *in vivo*. Proc Natl Acad Sci U S A 89:4062–4065
10. Singh J, Klar AJ (1992) Active genes in budding yeast display enhanced *in vivo* accessibility to foreign DNA methylases: a novel *in vivo* probe for chromatin structure of yeast. Genes Dev 6:186–196
11. Rusché LN, Kirchmaier AL, Rine J (2002) Ordered nucleation and spreading of silenced chromatin in *Saccharomyces cerevisiae*. Mol Biol Cell 13:2207–2222
12. Hoppe G, Tanny J, Rudner A et al (2002) Steps in assembly of silent chromatin in yeast: Sir3-independent binding of a Sir2/Sir4 complex to silencers and role for Sir2-dependent deacetylation. Mol Cell Biol 22:4167–4180
13. Luo K, Vega-Palas MA, Grunstein M (2002) Rap1-Sir4 binding independent of other Sir, yKu or histone interactions initiates the assembly of telomeric heterochromatin in yeast. Genes Dev 16:1528–1539
14. Buchberger JR, Onishi M, Li G et al (2008) Sir3-nucleosome interactions in spreading of silent chromatin in *Saccharomyces cerevisiae*. Mol Cell Biol 28:6903–6918
15. Yang B, Kirchmaier AL (2006) Bypassing the catalytic activity of SIR2 for SIR protein spreading in *S. cerevisiae*. Mol Biol Cell 17: 5287–5297
16. Young TJ, Kirchmaier AL (2012) Cell cycle regulation of silent chromatin formation. Biochim Biophys Acta 1819:303–312

17. Katan-Khaykovich Y, Struhl K (2005) Heterochromatin formation involves changes in histone modifications over multiple cell generations. EMBO J 24:2138–2149
18. Tham WH, Zakian VA (2002) Transcriptional silencing at *Saccharomyces* telomeres: implications for other organisms. Oncogene 21: 512–521
19. Jacobi JL, Kirchmaier AL (2011) Propagation of epigenetic states during DNA replication. In: Kusic-Tisma J (ed) Fundamental aspects of DNA replication. In Tech Publishing, Vienna, pp 245–270
20. Gottschling DE, Aparicio OM, Billington BL et al (1990) Position effect at *S. cerevisiae* telomeres: reversible repression of Pol II transcription. Cell 63:751–762
21. Aparicio OM, Billington BL, Gottschling DE (1991) Modifiers of position effect are shared between telomeric and silent mating-type loci in *S. cerevisiae.* Cell 66:1279–1287
22. Vega-Palas MA, Martin-Figueroa E, Florencio FJ (2000) Telomeric silencing of a natural subtelomeric gene. Mol Gen Genet 263:287–291
23. Li C, Mueller JE, Bryk M (2006) Sir2 represses endogenous polymerase II transcription units in the ribosomal DNA nontranscribed spacer. Mol Biol Cell 17:3848–3859
24. Yang B, Britton J, Kirchmaier AL (2008) Insights into the impact of histone acetylation and methylation on Sir protein spreading and silencing in *Saccharomyces cerevisiae.* J Mol Biol 381:826–844
25. Le S, Davis C, Konopka JB et al (1997) Two new S-phase-specific genes from *Saccharomyces cerevisiae.* Yeast 13:1029–1042
26. Axelrod A, Rine J (1991) A role for CDC7 in repression of transcription at the silent mating-type locus HMR in *Saccharomyces cerevisiae.* Mol Cell Biol 11:1080–1091
27. Xu EY, Kim S, Replogle K et al (1999) Identification of *SAS4* and *SAS5*, two genes that regulate silencing in *Saccharomyces cerevisiae.* Genetics 153:13–23
28. Singer MS, Gottschling DE (1994) *TLC1*: template RNA component of *Saccharomyces cerevisiae* telomerase. Science 266:404–409
29. Smith JS, Caputo E, Boeke JD (1999) A genetic screen for ribosomal DNA silencing defects identifies multiple DNA replication and chromatin-modulating factors. Mol Cell Biol 19:3184–3197
30. Herskowitz I, Rine J, Strathern J (1992) Mating-type determination and mating-type interconversion in *Saccharomyces cerevisiae.* In: Jones EW, Pringle JR, Broach JR (eds) The molecular and cellular biology of the yeast *Saccharomyces*: gene expression. Cold Spring Harbor Laboratory Press, Plainview, NY, pp 583–656
31. Herskowitz I (1988) Life cycle of the budding yeast *Saccharomyces cerevisiae.* Microbiol Rev 52:536–553
32. Sprague GF Jr, Blair LC, Thorner J (1983) Cell interactions and regulation of cell type in the yeast *Saccharomyces cerevisiae.* Annu Rev Microbiol 37:623–660
33. Kimmerly W, Buchman A, Kornberg R et al (1988) Roles of two DNA-binding factors in replication, segregation and transcriptional repression mediated by a yeast silencer. EMBO J 7:2241–2253
34. Miller A, Yang B, Foster T et al (2008) Proliferating cell nuclear antigen and ASF1 modulate silent chromatin in *Saccharomyces cerevisiae* via lysine 56 on histone H3. Genetics 179:793–809
35. Ehrenhofer-Murray AE, Kamakaka RT, Rine J (1999) A role for the replication proteins PCNA, RF-C, polymerase ε and Cdc45 in transcriptional silencing in *Saccharomyces cerevisiae.* Genetics 153:1171–1182
36. Miller A, Chen J, Takasuka TE et al (2010) Proliferating cell nuclear antigen (PCNA) is required for cell cycle-regulated silent chromatin on replicated and nonreplicated genes. J Biol Chem 285:35142–35154
37. Sprague GF Jr (1991) Assay of yeast mating reaction. Methods Enzymol 194:77–93
38. Jones EW, Fink GR (1982) Regulation of amino acid and nucleotide biosynthesis in yeast. In: Strathern J, Jones E, Broach JR (eds) The molecular biology of the yeast *Saccharomyces*: metabolism and gene expression. Cold Spring Harbor Laboratory, Cold Spring Harbor, pp 181–299
39. Rivier DH, Ekena JL, Rine J (1999) *HMR-I* is an origin of replication and a silencer in *Saccharomyces cerevisiae.* Genetics 151:521–529
40. Sussel L, Vannier D, Shore D (1993) Epigenetic switching of transcriptional states: cis- and trans-acting factors affecting establishment of silencing at the *HMR* locus in *Saccharomyces cerevisiae.* Mol Cell Biol 13:3919–3928
41. Yang B, Miller A, Kirchmaier AL (2008) HST3/HST4-dependent deacetylation of lysine 56 of histone H3 in silent chromatin. Mol Biol Cell 19:4993–5005
42. Bourns BD, Alexander MK, Smith AM et al (1998) Sir proteins, Rif proteins, and Cdc13p bind *Saccharomyces* telomeres *in vivo.* Mol Cell Biol 18:5600–5608
43. Miller AM (1984) The yeast *MATa1* gene contains two introns. EMBO J 3:1061–1065

44. Guthrie C, Fink GR (eds) (2002) Guide to yeast genetics and molecular and cell biology, part B, vol 350, Methods in enzymology. Academic, San Diego, CA, p 623
45. Sprague GF Jr, Herskowitz I (1981) Control of yeast cell type by the mating type locus. I. Identification and control of expression of the a-specific gene *BAR1*. J Mol Biol 153:305–321
46. Adams A, Gottschling DE, Kaiser CA et al (1997) Methods in yeast genetics. In: Dickerson MM (ed) A Cold Spring Harbor laboratory course manual. Cold Spring Harbor Laboratory Press, Plainview, NY, p 177
47. Hartwell LH (1980) Mutants of *Saccharomyces cerevisiae* unresponsive to cell division control by polypeptide mating hormone. J Cell Biol 85:811–822
48. Brachmann CB, Davies A, Cost GJ et al (1998) Designer deletion strains derived from *Saccharomyces cerevisiae* S288C: a useful set of strains and plasmids for PCR-mediated gene disruption and other applications. Yeast 14: 115–132
49. Rossmann MP, Luo W, Tsaponina O et al (2011) A common telomeric gene silencing assay is affected by nucleotide metabolism. Mol Cell 42:127–136
50. Takahashi YH, Schulze JM, Jackson J et al (2011) Dot1 and histone H3K79 methylation in natural telomeric and *HM* silencing. Mol Cell 42:118–126
51. Schmitt ME, Brown TA, Trumpower BL (1990) A rapid and simple method for preparation of RNA from *Saccharomyces cerevisiae*. Nucleic Acids Res 18:3091–3092
52. Cheng T-H, Gartenberg MR (2000) Yeast heterochromatin is a dynamic structure that requires silencers continuously. Genes Dev 14: 452–463
53. Hill JE, Myers AM, Koerner TJ et al (1986) Yeast/*E. coli* shuttle vectors with multiple unique restriction sites. Yeast 2:163–167
54. Sherman JM, Stone EM, Freeman-Cook LL et al (1999) The conserved core of a human *SIR2* homologue functions in yeast silencing. Mol Biol Cell 10:3045–3059

Chapter 18

A User's Guide to the Ribosomal DNA in *Saccharomyces cerevisiae*

Joseph M. Johnson, Jeffrey S. Smith, and David A. Schneider

Abstract

Messenger RNA synthesis (mRNA) accounts for a small fraction of total RNA synthesis in growing eukaryotic cells. The bulk of cellular transcription is devoted to ribosomal RNA (rRNA) synthesis (Warner, Trends Biochem Sci 24:437–440, 1999). Several unique characteristics of the rDNA and RNA polymerase I must be considered in order to accurately quantify the synthesis rate of rRNA or to characterize its processing. Indeed, an entirely different set of techniques must be applied to the study of rRNA synthesis than is routinely to study mRNA synthesis. Five of the most useful strategies for genetic and molecular analysis of rRNA synthesis and regulation are outlined in this chapter. The techniques described were developed for characterization of the model eukaryote *Saccharomyces cerevisiae*; however, many of these strategies can be adapted for studies in other eukaryotic cells.

Key words Ribosome, Ribosomal RNA, Transcription, Epigenetics, rRNA processing, RNA polymerase I

1 Introduction

Techniques for the analysis of gene expression in prokaryotic and eukaryotic cells have evolved and improved substantially over the past few decades. Sophisticated reporter-based assays permit detailed quantification of transcription rates of individual genes. Recently, methods for monitoring transcription of single genes within individual cells have emerged [1]. These techniques can be widely applied to mRNA encoding genes. Three unique properties of the ribosomal DNA and its metabolism limit the utility of these techniques for characterization of rRNA synthesis: the tandemly repeated structure of the rDNA, processing of Pol I-derived transcripts, and epigenetic silencing of a substantial fraction of the rDNA repeats.

The ribosomal DNA exists as a tandemly repeated array in one or more loci in all eukaryotic cells. In *S. cerevisiae* (referred to as "yeast" for simplicity), the rDNA occupies a single locus in chromosome XII.

Jeffrey S. Smith and Daniel J. Burke (eds.), *Yeast Genetics: Methods and Protocols*, Methods in Molecular Biology, vol. 1205, DOI 10.1007/978-1-4939-1363-3_18,

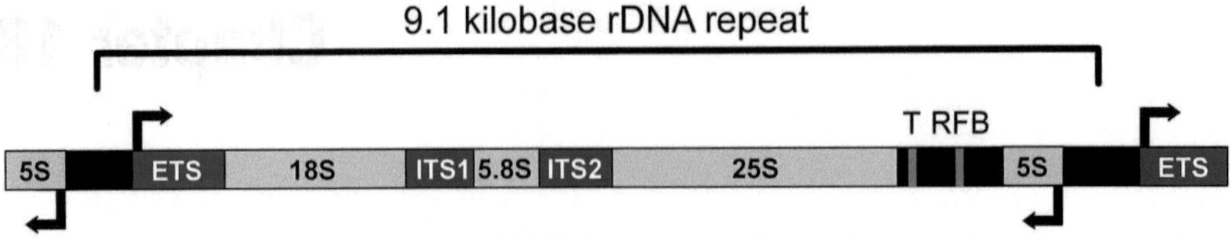

Fig. 1 Schematic diagram showing the yeast rDNA repeat structure. Each repeat is 9.1 kb in length. RNA polymerase I transcribes a 35S rRNA precursor, that is processed into the mature 18S, 5.8S, and 25S rRNAs. The intergenic spacers also include a Pol III-transcribed 5S rRNA gene. *T* terminator for Pol I transcription; *RFB* DNA replication fork block sequence that prevents DNA replication from moving into transcribed region of the rDNA genes, in the opposite direction of transcription; *ETS* external transcribed spacer; *ITS* internal transcribed spacers

In an average wild type (WT) strain there are approximately 200 tandemly repeated copies of the 9.1 kb rDNA transcription unit (Fig. 1). Within each repeat is the coding sequence for the 35S pre-rRNA (transcribed by RNA polymerase I (Pol I)) and the 5S rRNA (transcribed by RNA polymerase III (Pol III) in the opposite direction). In between the 35S and 5S genes is a replication fork block. Fob1p and other proteins associate with this region of the DNA to prevent progress of the replication machinery in the opposite direction of Pol I transcription. The presence of this replication fork block leads to hyper-recombination between rDNA repeats [2]. As a consequence, the number of rDNA repeats can vary substantially between strains and between individual cells. Thus, unlike most RNA polymerase II (Pol II) studies of mRNA synthesis, careful characterization of rRNA synthesis must include controls for changes in the rDNA copy number. Multiple assays have been developed to measure the rDNA repeat number. In this chapter, we will describe the use of contour-clamped homogenous field electrophoresis (CHEF).

Reporter-based assays such as luciferase gene expression do not work well with the rDNA promoter because Pol I-derived transcripts are neither capped nor polyadenylated, and thus they are not translated efficiently. Several studies have used reverse transcription of rRNA precursors, followed by quantitative real time PCR or primer extension to measure the abundance of short-lived pre-rRNA species (for example the ITS1 and ITS2 elements; [3]). An important caveat to these approaches is that changes in rRNA processing efficiency or decay can lead to altered abundance of these species without altered transcription rates by Pol I. In our experience, these approaches are useful for detecting trends in rRNA expression, but the values measured do not always accurately reflect rRNA synthesis rates. We routinely use isotopic labeling of the rRNA to measure its synthesis rate. Pre-rRNA is methylated cotranscriptionally in yeast, and thus we use [^{3}H-methyl] methionine in pulse-chase experiments to quantify the rRNA synthesis rate and processing of the rRNA [4, 5]. This method of labeling and simple methods to validate the results are described in this chapter. Metabolites other than methionine can be used in similar pulse-chase approaches to accurately measure rRNA synthesis rates; however, great care should be taken to control for potential differences in cellular uptake of the metabolite used, such as with ^{3}H-uridine labeling [6].

Although each yeast cell carries ~200 copies of the rDNA repeat, only approximately half of these repeats are actively transcribed during exponential growth; the remaining repeats are maintained in an inactive state. In yeast and mammals, the fraction of actively transcribed repeats can be regulated in response to growth or developmental stimuli [7–9]. Thus, methods to measure the fraction of inactive versus active repeats have been developed. In this chapter we describe a commonly used technique that exploits the differential sensitivity of the active versus inactive repeats to psoralen cross-linking.

In all cells, efficient control of ribosome biosynthesis is critical to survival. Thus, genetic studies that target the regulation of rRNA expression present unique challenges. Although several creative, successful screens have been devised over the years, this chapter will highlight two strategies that have led to substantial progress in the identification of genes whose products influence rRNA synthesis directly or indirectly. In 1991, Dr. Masayasu Nomura and colleagues developed a series of plasmids that could be used to complement mutations in the endogenous rDNA [10]. Importantly, they also constructed plasmids in which synthesis of rRNA could be achieved with an inducible Pol II promoter [11]. These vectors enabled the field to rapidly and thoroughly characterize the factors required for Pol I transcription of the rDNA. More recently, the Smith lab developed an effective genetic reporter system that can detect changes in Pol I transcription and can be used to screen for inhibitors or activators of rDNA transcription using the endogenous rDNA and no additional plasmids [12]. Both of these genetic tools are described herein. While not described in this chapter, a highly complementary approach to studying Pol I transcription is through the direct visualization of rDNA genes through electron microscopy of Miller chromatin spreads [13].

Although characterization of the mechanisms that control and optimize ribosome expression is difficult, the results of these studies are important for understanding metabolism of eukaryotic species [14]. Furthermore, the intimate connection between ribosome expression and cell proliferation rates demonstrates the direct impact of these studies on human health. There is much to learn regarding the molecular regulation of ribosome synthesis and the techniques described here will be useful in these ongoing studies.

2 Materials

2.1 CHEF Plug Preparation and Analysis

1. For all experiments, cells are grown with aeration at 30 °C, unless the strain requires alternative growth conditions. We use either water bath shakers or air shakers, depending on the application.
2. We use two general types of growth media for yeast. Complex medium is YEPD (1 % yeast extract, 2 % peptone and 2 % dextrose). Other carbon sources can be used. For defined medium we use: 0.17 % yeast nitrogen base (without amino acids or ammonium; Difco, Detroit, MI), 0.5 % ammonium sulfate, 2 % dextrose, and 0.087 % nutrient mixture. The complete nutrient mixture contains 0.8 g of adenine, 0.8 g of arginine, 4 g of aspartic acid, 0.8 g of histidine, 2.4 g of leucine, 1.2 g of lysine, 0.8 g of methionine, 2 g of phenylalanine, 8 g of threonine, 0.8 g of tryptophan, 1.2 g of tyrosine, and 0.8 g of uracil. Dropout media are prepared by omission of the candidate compound from the nutrient mix.
3. 0.05 M EDTA, pH 8.0.
4. SE buffer: 0.9 M sorbitol; 0.1 M EDTA; pH 8.0.
5. Light microscope, preferably with phase contrast, 40× objective. We use a Zeiss Primo Star.
6. Zymolase 20T suspended in water (10 mg/mL) and stored in frozen aliquots (specific manufacturer is not important).
7. Agarose: Lonza SeaKem LE and SeaPlaque low melt agarose.
8. 10× TBE (per liter): 108 g Tris base; 55 g boric acid; 40 mL 0.5 M EDTA, pH 8.0.
9. CHEF mapper: We use the Bio-Rad CHEF DrII, but other models can be used.
10. CHEF plug molds are small plastic molds made by Bio-Rad Laboratories specifically for CHEF gel plug preparation.
11. Sarkosyl buffer: 1 % sarkosyl, 0.1 M EDTA, pH 9.0.
12. Proteinase K, suspended in water (20 mg/mL), aliquoted and stored at −80 °C.
13. SYBR safe gel stain (Invitrogen, Carlsbad, CA). Diluted to 2× in water for gel staining solution.
14. Glass baking dishes: we use 9×9 Pyrex dishes for soaking the gels and 13×9 dishes for the transfer buffer reservoir.
15. Gel fixer: 0.25 M HCl.
16. Transfer buffer: 0.4 M NaOH, 1.5 M NaCl.
17. Whatman 3MM paper.
18. GeneScreen Plus membrane (Perkin Elmer, Waltham, MA).

19. Glass plate larger than CHEF gel; we use plates used for casting vertical acrylamide gels.
20. Plastic pipette tip box or other item for use as a capillary transfer base.
21. Stack of paper towels at least as wide as the CHEF gel (usually brown, bifold towels).
22. UV light source, we use a UV Stratalinker (Stratagene), 120 mJ total energy.
23. 20× SSC: 3 M NaCl, 0.3 M Sodium Citrate, pH 7.0.
24. Rediprime II kit (GE Healthcare Life Sciences, Uppsala Sweden).
25. ^{32}P-dCTP (Perkin Elmer, Waltham, MA).
26. Hybridization solution: 6× SSPE, 5× Denhardt's solution, 0.5 % SDS, 25 % formamide, 100 μg/mL salmon sperm DNA.
27. 20× SSPE: 3 M NaCl; 0.2 M $NaH_2PO_4.H_2O$; 0.02 M EDTA, pH 8.0.
28. 100× Denhardt's solution: 2 % Ficoll 400, 2 % polyvinylpyrrolidone 360, 2 % BSA.
29. 10 % SDS stock solution (made in water).
30. X-ray film and developer; or phosphorimager storage cassette and imager. We use GE Healthcare Lifesciences Storm 820 or Typhoon Trio scanners with associated cassettes.

2.2 rRNA Synthesis and Processing Measurements

1. Materials for cell growth as described in Subheading 2.1.
2. [^{3}H-methyl] methionine (Perkin Elmer, Waltham, MA).
3. Dry ice, crushed to powder.
4. Spectrophotometer for measuring cell density.
5. L-Methionine, 50 mg/mL in water.
6. TES: 10 mM Tris–HCl, pH 7.5; 10 mM EDTA; 0.5 % SDS.
7. Acidic Phenol, 100 %.
8. Chloroform 100 %.
9. Heat block (capable of 65 °C) with block fit to 1.5 mL tubes.
10. Vortex mixer.
11. Salty ethanol: 1 M ammonium acetate, prepared in 95 % ethanol; all RNase free.
12. 70 % ethanol, RNase free.
13. Horizontal gel apparatus (standard, but clean as possible).
14. 10× MOPS buffer: 0.4 M Morpholinopropanesulfonic acid (MOPS); 0.1 M sodium acetate; 0.01 M EDTA; pH to 7.2 with NaOH and store in the dark at 4 °C.
15. Agarose (any manufacturer, we use Fisherbrand).

16. 5× RNA loading dye, per 1.5 mL: 720 μL deionized formamide (RNase free); 260 μL 37 % formaldehyde; 80 μL saturated bromophenol blue solution; 200 μL of 40 % glycerol; 160 μL 10× MOPS; 80 μL deionized water (RNase free). One can add an RNA stain to this reagent if the gel will be visualized. We have used ethidium bromide (20 μg/mL final concentration) or SYBR RNA stain (Invitrogen, Carlsbad, CA).
17. 1 % agarose gel: 1 g agarose, boiled in 85 mL water; cooled to 65 °C; add 10 mL 10× MOPS and 5.4 mL 37 % formaldehyde; mix and pour into gel mold.
18. Whatman paper and GeneScreen membrane is the same as described in Subheading 2.1.
19. Semidry transfer apparatus: we use a GE TE-77 PWR (Uppsala Sweden).
20. UV source, as described in Subheading 2.1.
21. Fluoro-Enhance (Research Products International; Mount Prospect, IL).
22. X-ray film and exposure cassette.
23. Film developer.
24. Scintillation counter with appropriate vials and scintillation cocktail; we use EcoLume (MP Biomedicals; Solon, OH).
25. Flatbed scanner to make image of the film.
26. Transparent copy paper and copy machine.

2.3 Psoralen Cross-Linking

1. Materials for cell growth, as described in Subheading 2.1.
2. Spectrophotometer, we use (Shimadzu, UVmini-1240).
3. Ice-cold sterile water.
4. TE: 10 mM Tris–HCl, pH 8.0, 10 mM EDTA.
5. 4,5′,8-trimethylpsoralen (200 μg/mL in 100 % ethanol), Sigma T-6137.
6. UV lamp (UVP Black-Ray model B-100AP).
7. DNA Extraction Buffer: 0.5 M NaCl, 0.2 M Tris–HCl, pH 7.6, 0.01 M EDTA, 1 % SDS.
8. Acid-washed glass beads, 425–600 μm (Sigma, Cat. # G8772-1KG).
9. Proteinase K (20 mg/mL in water).
10. Phenol: Chloroform: Isoamyl Alcohol (PCI) Solution (25:24:1).
11. 3 M Sodium Acetate.
12. Ethanol (100 and 70 %).
13. RNase A (10 mg/mL).
14. 4 M Ammonium Acetate.

15. Agarose (standard grade, any manufacturer).
16. Stratalinker (Stratagene UV Stratalinker 2400).
17. Positively charged nylon blotting membrane. We use Immobilon-NY+ (Millipore, or GeneScreen Plus).
18. 10× SSC (autoclaved, 87.5 g/L NaCl, 44 g/L Na Citrate dihydrate).
19. Southern denaturation buffer: 0.2 M NaOH, 0.6 M NaCl.
20. Southern neutralization buffer: 0.5 M Tris–HCl, pH 7.5, 1.5 M NaCl.
21. ^{32}P-dCTP, we use EasyTidesRTM [α-^{32}P]-Deoxycytidine 5′-triphosphate, specific activity: 3,000 Ci/mmol. Cat. #: BLU513H250UC from PerkinElmer.
22. Salmon sperm DNA (Agilent Technologies, Cat. # 201190).
23. QuikHyb (Stratagene, Cat. # 201221-21).
24. Nylon Mesh (Bellco Glass, Inc. Cat. # 7910-02026).
25. Hybridization Bottles (Boekel Scientific).
26. Blotting Paper (Whatman 3MM).
27. Parafilm.
28. X-ray film and developer; or phosphorimager storage cassette and imager. We use GE Healthcare Lifesciences Storm 820 or Typhoon Trio scanners with associated cassettes.

2.4 Plasmids Used for Mutational Analysis of rDNA

1. Yeast grown as described in Subheading 2.1.
2. Manipulation of plasmid DNA should be performed using standard methods appropriate for your lab. An example of the plasmid structure is shown in Fig. 2.
3. Plasmids should be maintained in standard *E. coli* cloning strains, we use DH5α.
4. Growth medium for *E. coli* work is Luria-Bertani (LB) broth (or agar, for solid medium, with agar added to 2 % prior to autoclaving). LB: 1 % Tryptone, 0.5 % yeast extract, 1 % NaCl.

2.5 Genetic Screening for Activators and Repressors of Pol I Transcription

1. Yeast are grown as described in Subheading 2.1.
2. SC, SC – uracil, and SC + FOA plates.
3. Yeast strain YRH269 (Table 1).
4. Oligonucleotide primers for cDNA synthesis and real-time PCR.

   JS765 (5′-ATACGCAAACCGCCTCTCC-3′).

   JS766 (5′-TGTCGTGCCAGCTGCATTA-3′).

   JS769 (5′-AGTTGTAAAGCCGGTTTTGCC-3′).

   JS770 (5′-TGGGAAGACAGCACGAGGAG-3′).

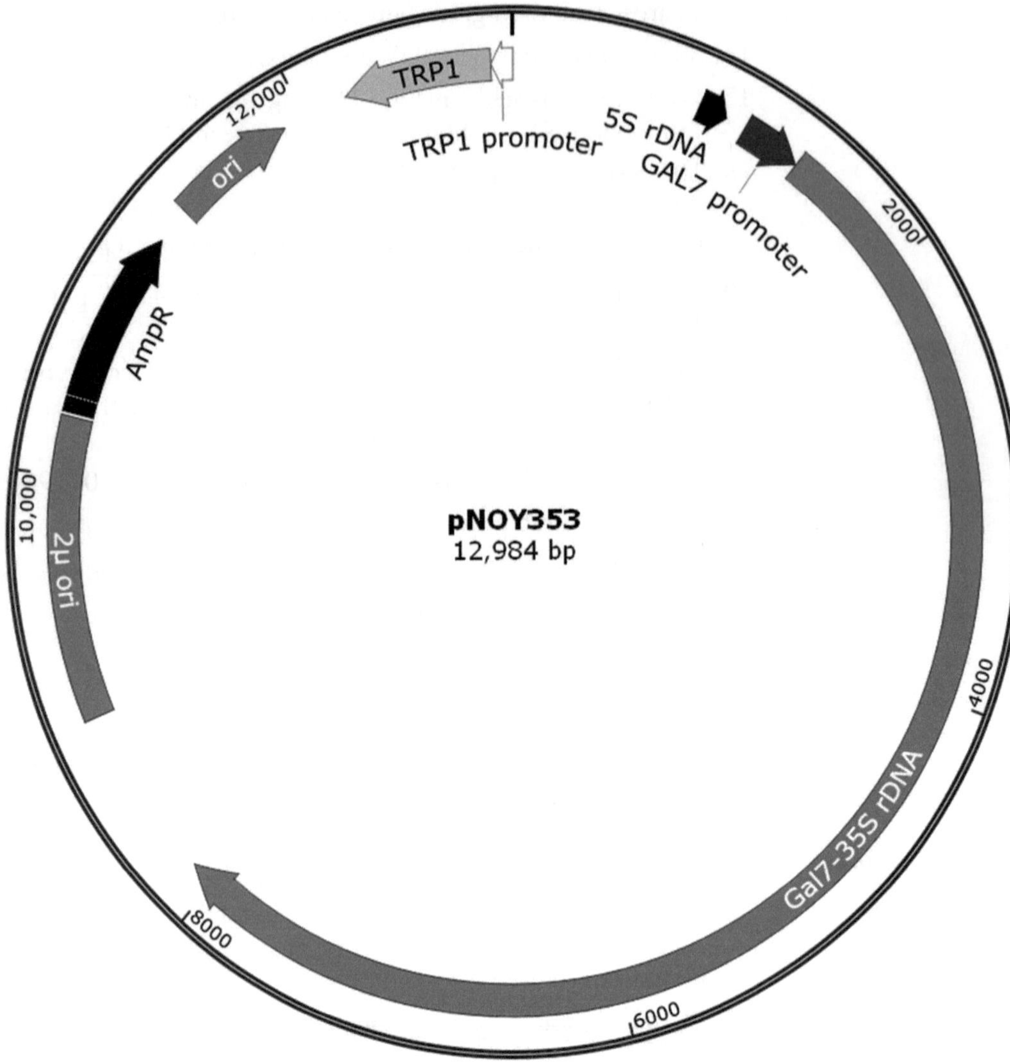

Fig. 2 Example of a pNOY vector that expresses the large 35S rRNA precursor from inducible *GAL7* promoter via RNA Pol II. This version (pNOY353) also encodes the 5S rRNA gene, and can be used to maintain cell viability when the entire rDNA tandem array has been deleted [18]

5. Superscript II reverse transcriptase (Invitrogen).
6. RNaseOUT (Invitrogen).
7. SYBR Green PCR Master Mix (Applied Biosystems).
8. EDTA (25 mM).
9. dNTP mix (10 mM stock).
10. Real-time PCR machine. We typically use an ABI 7300.

Table 1
Plasmids and strains used for genetic analysis of rDNA transcription

Plasmids	*Description*
pNOY102	High-copy-number plasmid carrying *GAL7*-35S rDNA, *URA3*, 2 μ, *amp*
pNOY199	High-copy-number plasmid carrying *GAL7*-35S rDNA, *TRP1*, 2 μ, *amp*
pNOY130	High-copy-number plasmid carrying *GAL7-35S rDNA 5S rDNA URA3* 2 μ, *amp*
pNOY353	High-copy-number plasmid carrying *GAL7-35S rDNA 5S rDNA TRP1* 2 μ, *amp*
pNOY373	YEp351 derivative carrying rDNA with promoter starting from -206 and with *Xho*I–*Not*I flanked enhancer, *LEU2*, *2μ*, *bla*
Strains	*Description*
NOY408-1a	*MAT*a *ade2-1 ura3-1 trp1-1 leu2-3,112 his3-11 can1-100 rpa135Δ::LEU2*+pNOY102
NOY773	*MAT*a *ade2-1 ura3-1 trp1-1 leu2-3,112 his3-11 can1-100 rdnΔΔ::HIS3*, +pNOY353
NOY984	*MAT*α *ade2-1 ura3-1 trp1-1 leu2-3,112 his3-11,15 can1-100 rdnΔΔ::hisG*, +pNOY130
YRH269	*MAT*a *his3Δ200 leu2Δ1 trp1Δ63 ura3-167 nts1Δ::mURA3-HIS3*

3 Methods

3.1 rDNA Copy Number

1. Grow cells in YEPD with aeration at 30 °C.
2. Harvest 20 mL of cell culture (A_{600} = ~0.5) by centrifugation, using 50 mL disposable tubes and a clinical centrifuge (5 min at 5,000 × *g*).
3. Remove supernatant and wash cell pellet in 10 mL 0.05 M EDTA, pH 8.0.
4. Spin again.
5. Remove supernatant and resuspend pellet in 1 mL 0.05 M EDTA pH 8.0 Transfer the cell suspension to a sterile 1.5 mL microfuge tube.
6. Spin 3 min at 10,000 × *g*.
7. Remove supernatant and resuspend in 1 mL SE buffer.
8. Spin 5 min at 10,000 × *g*.
9. Remove supernatant and resuspend pellet in 240 μL SE buffer.
10. Prep slide and observe cells (2–3 μL) using 40× objective and phase contrast.
11. Add 50 μL zymolase 200T (10 mg/mL) to start spheroplast preparation.

12. Incubate cells at 30 °C on a roller until yeast cells enlarge and become "dark" when viewed using phase contrast microscope. Typically this takes 30–60 min.
13. Melt agarose and incubate at 50 °C.
14. Once spheroplasting is complete, add 300 μL of low melt agarose in SE and keep warm (~45 °C).
15. Pipet 70 μL into CHEF mold.
16. Allow to solidify by incubation at room temperature.
17. Push plugs into 1.5 mL Eppendorf tube containing 1 mL sarkosyl buffer.
18. Incubate 15 min at room temperature.
19. Carefully remove supernatant, and wash two more times, with 15 min incubation between.
20. Remove supernatant and add 950 μL sarkosyl buffer plus 50 μL proteinase K (20 mg/mL).
21. Incubate at 50 °C overnight (>12 h).
22. Remove supernatant and wash plugs twice with 1 mL 0.05 M EDTA pH 8.0.
23. Resuspend in 0.05 M EDTA and store at 4 °C.
24. Prepare 100 mL of 0.8 % SeaKem LE agarose in 0.5× TBE (warmed to fully dissolve).
25. Cool to ~50 °C and pour into CHEF gel mold, setup according to manufacturer's directions.
26. Store at room temperature for ~2 h to solidify.
27. Carefully place plugs into wells (*see* **Note 1**) and seal with sealing melted sealing agarose (1 % SeaKem LE in 0.5× TBE). May include commercially available premade *S. cerevisiae* plugs as standards (Bio-Rad Laboratories; Hercules, CA).
28. Prepare 2.5 L of 0.5× TBE, pour into CHEF mapper and start pump to move buffer through all tubing. More than 2.5 L may be needed if long lengths of tubing are used with the apparatus.
29. Remove bubbles in tubing.
30. Turn on buffer cooler, set to 14 °C. Allow buffer to cool completely.
31. Place gel (with metal tray) into the gel tank of the CHEF mapper, oriented as directed by the manufacturer's directions. Be sure that gel is fully submersed. Add more buffer if needed.
32. Run gel for 68 h with the following settings: Angle 120°; linear ramp; initial switch time 300 s; final switch time 900 s; voltage = 3 V/cm.

33. Occasionally check on the instrument over the 3-day run to verify that bubbles are not blocking buffer flow. If flow is blocked, chiller may ice up. The manufacturer suggests changing the running buffer every 24 h, but we have not observed any benefit from doing this.
34. When run is finished, carefully remove gel and incubate in 2× SYBR safe at room temp with gentle agitation for 1 h.
35. Briefly wash gel with dH_2O and image with a gel documentation station.
36. We find that high quality plugs permit easy detection of chromosome XII (the uppermost chromosome) with simple SYBR staining.
37. rDNA repeat number determination. We include plugs from four control strains in each gel: NOY1064, 190 copies; NOY1071, 25 copies; NOY886, 42 copies; and NOY1051, 143 copies. These strains were previously constructed and characterized by the Nomura lab [15]. They carry the indicated number of rDNA repeats and the repeat number is locked by deletion of *FOB1*. We measure the migration distance of chromosome XII from these control samples and make a linear regression of migration versus copy number. We then measure the migration distance for the unknown strains and calculate the repeat number from the regression. If the cells have accumulated large numbers of repeats (greater than ~300) the resolution in the gel becomes poor and this method is not accurate. However, for most strains, this approach is easy and useful for determination of the rDNA repeat number.
38. If chromosome XII is not clearly visible (due to low quality plugs or excessive heterogeneity in rDNA repeat number within the culture) one can perform a Southern blot, probing for the rDNA. The following classical method in Subheading 3.2 can be used.

3.2 Southern Blotting

1. After visualizing the CHEF gel, the gel is depurinated in 0.25 M HCl for 10 min at room temperature with gentle agitation. We perform these operations in 9 in. × 9 in. glass baking dishes, with sufficient volume to completely cover the gel.
2. The HCl is removed and the gel is soaked in transfer buffer for 15 min at room temperature with gentle agitation. This transfer buffer contains NaOH and will denature the DNA.
3. Meanwhile, we cut two long sheets of Whatman 3MM paper (the width of the CHEF gel and at least three times the length of the gel) and two sheets of Whatman 3MM paper to the exact size of the gel. All paper is pre-soaked in transfer buffer.

4. We also cut a piece of GeneScreen Plus membrane to the exact size of the gel. The membrane is pre-soaked once in deionized water for 5 min and once in transfer buffer for 5 min.
5. Once the gel is ready, we assemble the capillary transfer stack in a large glass baking dish. Then stack the following in order from the bottom up: a base (we use an inverted plastic pipette tip box); a glass plate larger than the gel; the two long sheets of Whatman paper with both ends draped into the bottom dish; the gel, upside down; the membrane; two sheets of Whatman paper cut to the size of the gel; and a stack of paper towels at least 15 cm tall (the base of the stack must be at least as wide as the gel). Do not allow the towels to contact the lower Whatman paper directly. Fill the bottom dish with buffer, but not up to the glass plate.
6. The capillary transfer will take 2 days. Make sure the buffer does not run dry. One can use plastic wrap to seal the edges if desired.
7. Disassemble the stack and wash the membrane in 2× SSC for 2 min with gentile agitation.
8. Cross-link the DNA to the membrane with UV light; 120 mJ total energy.
9. To prep the rDNA probe we use standard PCR to generate a probe ~600 bp in length, specific to the rDNA (DAS314 forward 5′-GGGGCACCTGTCACTTTGGAAA-3′, DAS23 reverse 5′-GCTGATTTGAGAGGAGGTTAC-3′).
10. Label the probe using the Rediprime II kit, using α-^{32}P dCTP (*see* **Note 2**).
11. Rinse the blot in 6× SSC for 5 min at room temperature.
12. Transfer the blot to a vessel large enough to lay the blot flat, and add sufficient hybridization solution to cover the blot (but avoid excess, as this will be radioactive waste).
13. Incubate the blot at 41 °C with gentle agitation for 4 h.
14. Add the labeled probe (we prefer to boil the probe and snap cool on ice before adding).
15. Incubate overnight at 41 °C with gentle agitation.
16. The next day, wash the blot: in 2× SSC + 0.5 % SDS at room temperature for 5 min; then in 2× SSC + 0.1 % SDS at room temperature for 15 min, and finally in 0.1× SSC + 0.5 % SDS at 50 °C for 30 min (repeat last wash twice).
17. Dry the blot and visualize by exposure to film or phosphorimager cassette.

3.3 rRNA Synthesis and Processing

1. Grow cells in SD –met (*see* **Note 3**). The volume of cell culture depends on the number of time points and the volume of each sample. We routinely use 2.5 mL of culture per labeling, thus we grow 5–10 mL of culture.
2. Grow cells to mid-exponential phase (A_{600} = ~0.3).
3. While cells are growing, aliquot 125 μCi of [^{3}H-methyl] methionine into sterile 50 mL plastic conical tubes (50 μCi/mL final concentration). Prepare enough tubes for the entire day, and store at room temperature.
4. Label sterile 1.5 mL microfuge tubes appropriately, and set them in crushed dry ice to prechill the tubes.
5. 5 min before sampling the culture, place a tube containing the [^{3}H-methyl] methionine in an appropriate rack in the same shaker as the cell culture to pre-equilibrate the temperature.
6. At t = 0, carefully and rapidly transfer 2.5 mL of cells from the culture flask into the 50 mL conical tube and place back into the incubator. This initiates the pulse labeling.
7. Immediately determine the A_{600} of the starting (nonradioactive) culture and record. The rest of the culture can be discarded.
8. At t = 4 min, carefully and rapidly transfer 1 mL of the ^{3}H-labeled culture into one of the chilled microfuge tubes on dry ice (the sample rapidly freezes), and then immediately return the remaining culture to the incubator.
9. At t = 5 min, add cold methionine to the culture at a final concentration of 500 μg/mL. This initiates the chase.
10. At t = 10 min, transfer 1 mL of the culture to another prechilled microfuge tube. For most genetic backgrounds, this 5 min chase will permit completion of rRNA processing.
11. Carefully dispose of all waste.
12. The frozen radioactive samples can be stored at –80 °C for later processing.
13. To prepare RNA, samples should be thawed on ice. This will take several hours.
14. While thawing, label a set of four sterile, RNase-free microfuge tubes for each sample to be processed. To obtain clean RNA, we perform three acidic phenol extractions, one chloroform extraction, and an ethanol precipitation. The first phenol extraction is performed in the frozen tubes. Thus, two of the fresh tubes should be set up to carry 400 μL of phenol, one should be set up to carry 400 μL of chloroform, and the fourth tube should be set up to carry 1,000 μL of salty ethanol.

15. Once thawed, pellet the aliquot of cells at 17,000 × *g* for 5 min at 4 °C (*see* **Note 4**).
16. Carefully remove the supernatant and dispose in radioactive waste container (all liquids and solids are radioactive and must be handled carefully (*see* **Note 5**)). The pellets must be kept cold at all times.
17. Resuspended the pellets in 400 μL ice cold TES and immediately treat with 400 μL of phenol. Each tube is vigorously mixed using a vortex mixer at the maximum setting for 30 s and placed in a 65 °C heat block.
18. Every 15 min for 1 h, vigorously mix the samples for 10 s each.
19. After 1 h, incubated the samples on ice for 10 min.
20. Centrifuge the samples at 17,000 × *g* for 5 min to separate the aqueous phase from the phenol.
21. Carefully transfer the upper aqueous phase to a clean tube carrying 400 μL of phenol and mix by pipetting up and down repeatedly (*see* **Note 6**), and then centrifuge as in **step 20**.
22. Transfer the aqueous phase to another tube of 400 μL phenol for the third phenol treatment. Mix by pipetting up and down repeatedly, and then centrifuge as in **step 20**.
23. Transfer the aqueous phase to the tube of chloroform. Mix by pipetting up and down, and then centrifuge as in **step 20**.
24. Carefully transfer the aqueous phase to microfuge tubes containing the salty ethanol. Be careful to avoid transferring any chloroform. We generally set the pipet to a lower volume than present in the tube (typically ~200 μL recovered per sample). Equal volumes must be transferred for later data to remain quantitative.
25. Incubate samples overnight at −20 °C to precipitate RNA.
26. The next day, centrifuge the samples at 17,000 × *g* for 15 min at 4 °C.
27. Carefully remove the supernatant and dispose in a radioactive waste container.
28. Wash the RNA pellet one time with 1,000 μL of ice cold 70 % ethanol.
29. Incubate on ice for 5 min.
30. Spin at 17,000 × *g* for 5 min to preserve the pellet.
31. Remove all ethanol and dispose in radioactive waste.
32. Resuspend pellets in a volume proportional to A_{600} of the harvested culture. For example, if one culture was harvested at $A_{600} = 0.35$ and another was harvested at 0.4, those pellets should be resuspended in 35 and 40 μL of RNA loading dye respectively. In this way, the amount of labeled RNA per A_{600} can be measured.

33. Incubate the RNA for >30 min on ice to resuspend the pellet. Mix by pipetting up and down to ensure resuspension.
34. Cast a 1 % formaldehyde–agarose horizontal gel using carefully cleaned apparatus. We use 7 cm × 10 cm gels, but any gel should work if sufficiently large to separate the RNAs (*see* **Note 7**).
35. Equilibrate the gel in the electrophoresis tank in 1× MOPS running buffer for ~30 min.
36. Incubate the RNA samples in a 95 °C heat block ≥5 min.
37. Transfer immediately to ice.
38. Load equal volumes to each lane. We load less than the total amount heated in order to maintain equal volume between samples.
39. Run the gel at 80 V until the bromophenol blue band is 2/3 to ¾ of the distance through the gel.
40. Visualize the RNA using a standard gel documentation station and UV light. You should readily detect the 25S (upper) and 18S (lower) bands if SYBR stain or ethidium bromide is included in the loading dye.
41. Pre-soak one piece of GeneScreen Plus membrane and the RNA gel in 1× MOPS buffer for 20 min. Pre-soak ten pieces of Whatman 3MM paper in 1× MOPS for 1–2 min before using (membrane and Whatman paper all cut exactly to size of the gel).
42. Using a semidry transfer apparatus, stack pre-soaked five sheets of Whatman paper, the membrane, the gel, and an additional pre-soaked five sheets of Whatman. Assemble the "sandwich" according to the manufacturer's directions.
43. Ensure that the stack is moist/wet and transfer according to the manufacturer's directions. We transfer at 56 mA for 4 h (gel size 7 cm × 10 cm).
44. Remove the blot and cross-link the RNA to the blot with UV light and dry. We use a Stratalinker UV lightbox on the "auto" setting (120 mJ total energy). Dispose of the remaining material in radioactive waste.
45. Soak the blot in a low-energy radioactivity enhancement solution for 1 h (we use Fluoro-Enhance).
46. Thoroughly dry the blot.
47. Expose to film at −80 °C for 1–7 days, depending on the intensity of the signal.
48. Develop the film, scan with a flatbed scanner and analyze.
49. Notes on data analysis: From the film, one can *qualitatively* assess whether there is variation in rRNA synthesis rates between samples. The samples that were chased for 5 min with cold methionine are best for this analysis or rRNA synthesis rate.

To analyze rRNA processing, the pulse (without chase) samples should be analyzed. Variation in the intensity of precursors versus mature products can be analyzed. In the 4 min pulse, one usually observes roughly equal amounts of 27S versus 25S and 20S versus 18S (unless processing is impaired substantially). *Do not quantify* the data from the scanned image. To quantify the data, copy the film onto a transparency using a standard copy machine without any "zoom". Overlay the image with the blot and pin the two together to Styrofoam block. Cut out individual, well-isolated bands (mature or precursors) and place each blot fragment in individual scintillation vials. Keep track of the order. Add the appropriate amount of scintillation cocktail (we use EcoLume) and count the samples on the ^3H setting of scintillation counter. Duplicate strains and samples should be included for error analysis. These data are remarkably reproducible.

50. To validate the synthesis rate data, we generally grow separate, nonradioactive cultures in the same medium and carefully measure growth rate. We then extract total RNA at the same stage of growth and quantify the amount of RNA recovered per A_{600} with a spectrophotometer (A_{260}). Since total RNA is typically reflective of the amount of rRNA, one can approximate the rRNA synthesis rate as follows (as a function of the control strain): (fraction of WT growth rate) × (fraction of WT total RNA) = predicted rRNA synthesis rate. For example, if a cell grows at ½ the WT rate and carries ½ as much total RNA, the predicted rRNA synthesis rate would be 0.25 that of WT. These data disregard rRNA degradation rate, but the contribution of degradation is typically low. The latter method is not subject to isotopic uptake or specific activity of the "pool" of methionine, so it is a good, independent control for the described technique.

51. To validate rRNA processing measurements, one should prepare total RNA from growing cells and perform a series of Northern blots probing for precursor rRNA species as well as mature products. Several good examples of this approach are published [16].

3.4 Psoralen Cross-Linking

1. Inoculate an overnight YPD culture with a strain of interest. We typically use 10 mL cultures in glass test tubes (metal caps), rotating at 30 °C in a roller drum. This is your starter culture.
2. Using the spectrophotometer, measure absorbance of the starter culture at 600 nm. Re-inoculate a fresh 50 mL YPD culture in a 250 mL flask to an A_{600} of 0.05.
3. Incubate this culture at 30 °C in a shaker until it reaches mid-log phase (A_{600} = ~0.3–0.5). For time course experiments, start with a larger culture volume and larger flask, to allow for multiple aliquot removal.

4. Transfer the culture to a 50 mL disposable conical tube on ice and then spin in a refrigerated (4 °C) tabletop centrifuge (swinging bucket rotor) at 2,500 rpm.
5. Wash the cell pellets two times with 20 mL of ice-cold water.
6. Resuspend the pellets in 1 mL of ice-cold TE and transfer to microfuge tubes.
7. Pellet cells in microfuge tube 30 s at full speed and discard the TE supernatant.
8. Resuspend cell pellets in 1.4 mL of ice-cold TE.
9. Pipette 0.7 mL cell solution into individual wells of a 24-well tissue culture plate on ice. The extra cells can be pelleted, flash-frozen, and stored at −80 °C for later use.
10. Add 70 μL psoralen solution (200 μg/mL) to each well and mix gently by pipetting up and down or using an orbital plate shaker.
11. UV-irradiate cells at a distance of 6 cm with a UVP B-100-A lamp for 5 min, with the plate sitting on ice. It is important to remove the purple filter from the lamp before cross-linking, otherwise the intensity will not be strong enough. The lamp gets extremely bright, so we typically do the crosslinking in a darkroom. Wear protective UV face shield or goggles.
12. Repeat **steps 10** and **11** four more times, for a total of five cross-linking treatments.
13. Transfer the cross-linked cell solution to a 2.0 mL screw cap microcentrifuge tube and pellet (full speed for 30 s at 4 °C). The screw cap will prevent leakage during cell disruption in **step 17**.
14. Remove and discard the supernatant and wash the cell pellet once with 1 mL ice-cold TE.
15. Spin down cells and discard supernatant. (Optional flash freeze and storage at −80 °C here).
16. Resuspend cell pellets in 200 μL DNA extraction buffer and 200 μL 1× TE. Add acid-washed glass beads to the meniscus.
17. Disrupt the yeast cells by vortexing vigorously. We use a Mini-beadbeater (Biospec Products) and shake three times for 30 s at 4 °C, with a 30 s pause in between to prevent overheating.
18. Recover the supernatants by carefully punching a hole in the bottom of each tube with a 25G×5/8″ (0.5 mm×16 mm) needle, and placing the punctured tube into a 13×100 mm glass collection tube. The screw cap tube should sit at the top of the glass tube. Spin in a swinging bucket rotor in the tabletop centrifuge at ~1,000×*g* for 2 min at 4 °C. The cell lysate will drain into the glass tube.

19. Transfer the collected material into a new standard 1.5 mL microfuge tube.
20. Add 25 μL of Proteinase K solution and incubate at 50 °C for 3 h.
21. Add 400 μL of PCI and vortex for 30 s. Then spin at full speed in a microcentrifuge for 5 min. Transfer the aqueous phase to a fresh microfuge tube. Repeat the PCI extraction two more times.
22. After the last extraction, add 1/10th volume (~30 μL) 3 M sodium acetate, mix gently, and add then add 1 mL 100 % ethanol. Mix and place inside −20 °C freezer for at least 1 h to precipitate DNA.
23. Spin in a microfuge at 17,000 × *g* (full speed) for 5 min at 4 °C to pellet the precipitated DNA.
24. Remove the ethanol and resuspend the pellet in 400 μL TE.
25. Add 3 μL of 5 mg/mL RNase A and incubate at 37 °C for 15 min.
26. Add 10 μL of 4 M ammonium acetate and 1 mL of 100 % ethanol. Mix and allow to DNA to precipitate in a −20 °C freezer for at least 1 h.
27. Pellet the DNA in a microfuge (full speed) for 5 min at 4 °C.
28. Wash the DNA pellets with 0.5 mL cold 70 % ethanol and then spin the tubes again at full speed for 5 min. Carefully siphon off the ethanol and allow the DNA pellet to dry either in a SpeedVac or sitting at room temperature. Resuspend the dried DNA pellet in 40 μL of TE.
29. Quantify the DNA concentration by diluting 2 μL of DNA prep into 998 μL water and reading A_{260} in spectrophotometer (in a quartz microcuvette).
30. Digest 4 μg of the recovered genomic DNA with *Eco*RI:

    3 μL 10× EcoRI buffer.

    0.5 μL RNase A (5 mg/mL).

    x μL Water.

    1 μL *Eco*RI (20 U).

    x μL genomic DNA.

    30 μL total volume.

    Incubate at 37 °C for 5 h.
31. Separate the digested DNA on a 1.3 % agarose gel (ethidium bromide is not added to the gel or running buffer). We typically use a large 25 × 15 cm recirculating gel box apparatus from Owl Scientific, though any system will work. Run the gel slowly: For this large-sized gel, we use 80 V for 22 h (*see* **Note 8**).

32. After electrophoresis, stain the gel by washing with ethidium bromide solution (0.8 μg/mL in water) for 10 min, Slowly rocking or rotating on a platform at room temperature.
33. Destain the gel in distilled water for 30 min, slowly rocking or rotating on a platform.
34. Place the gel on plastic Saran wrap and reverse the psoralen cross-links in a Stratagene Stratalinker at an energy setting of 6000. This is equal to 0.12 J/cm^2.
35. Submerse the gel in 200 mM HCl for 10 min at room temperature, gently rocking or shaking (*see* **Note 9**).
36. Pour off the HCl solution, then submerse the gel in Southern *denaturation* buffer for 45 min, slowly rocking or shaking.
37. Pour off Southern denaturation buffer and replace with Southern *neutralization* buffer, and slowly rock/shake for 45 min.
38. Transfer to positively charged nylon membrane by Southern blotting overnight in 10× SSC transfer buffer. The Smith lab uses Immobilon-NY+ (Millipore) and the Schneider lab uses Genescreen Plus (PerkinElmer). Stack transfer apparatus from bottom to top: Whatman paper across top of gel tray with edges immersed in 10× SSC, agarose gel, membrane, two sheets of wet Whatman paper, single-fold paper towels, a plastic tray with small weight. Place Parafilm or Saran wrap strips around edges of the gel between the gel to ensure capillary action does not bypass the gel–membrane interface (*see* **Note 10**).
39. Following the overnight transfer, remove the membrane and rinse it with 2× SSC to remove any agarose debris sticking to it when removed from the transfer. Then allow the membrane to air-dry on a piece of Whatman 3MM or other blotting paper.
40. Place the dried membrane face up, being mindful of orientation, into the Stratalinker, and run the autocrosslink cycle to fix DNA to the membrane.
41. Preheat 20 mL of QuickHyb in a hybridization tube that is slowly rotating in a hybridization oven at 68 °C.
42. After re-wetting the membrane with 2× SSC, sandwich it between two mesh sheets, roll it up, and insert into the hybridization tube containing the preheated QuikHyb. Rotate in the oven for 20 min at 68 °C. This is the blocking step.
43. While the membrane is blocking, mix 100 μL of sheared salmon sperm DNA (10 mg/mL stock) with 25 μL of a ^{32}P-labeled rDNA probe in a microfuge tube. The Smith lab typically uses a 3.6 kb *Xba*I fragment of the rDNA that spans the 35S transcribed region. Probes to the 35S region can also be easily generated using PCR. We also use Invitrogen's Random Primer DNA Labeling System (Invitrogen Cat# 18187-013) with Perkin

Elmer EasyTidesRTM [α-^{32}P]-Deoxycytidine 5′-triphosphate, 3,000 Ci/mmol (Cat # BLU513H250UC).

44. Boil this DNA-probe mix for 5 min. We use a lid lock on the microfuge to prevent the cap from popping open during boiling and potentially spraying radioactivity across the workspace.
45. Remove 1 mL of the pre-warmed QuikHyb from the prehybridization and mix it with the boiled probe mixture. Then add this combined probe mixture back to the hybridization tube/membrane and rotate for 1 h at 68 °C. This is the hybridization step.
46. Rinse the bottle with 50 mL of room temperature 2× SSC/ 0.1 % SDS two times to remove excess radioactive probe. Then transfer the membrane to a plastic tray.
47. Wash the membrane twice (5 min each) with enough 2× SSC, 0.1 % SDS to cover the membrane while slowing shaking on a platform.
48. Wash the membrane two more times (5 min each) with 0.1× SSC, 0.1 % SDS at 60 °C), while rotating on the shaker platform.
49. Seal the washed membrane in plastic and expose to X-ray film in an autoradiography cassette. Place the cassette in −80 °C freezer during the exposure period.
50. Develop the film (*see* **Note 11**). Examples of this assay can be found in the following references [9, 17].

3.5 Nomura Vectors

1. Dr. Masayasu Nomura and colleagues have developed a complete set of vectors required for genetic manipulation of Pol I transcription or manipulation of the rDNA. The minimal set of vectors that one can use is shown in Table 1.
2. If the goal of the experiment is to analyze the consequences of mutations in the Pol I transcription apparatus, one can use the plasmids that lack the 5S rDNA. However, if the endogenous rDNA will be deleted or manipulated, it is critical that the 5S gene be included in the helper plasmid.
3. There are countless experiments that one can devise with these vectors. Many screens have been published, including the Nomura lab's seminal work that genetically defined the Pol I transcription machinery [10]. Subsequent studies have used these vectors to mutate the rDNA and characterize rRNA processing effects. If mutating the rDNA in chromosome XII, one can complement the mutation with the Pol I helper (pNOY373) or the Pol II helper (pNOY102 or pNOY199). The Pol II helper must be grown with galactose as the sole carbon source to fully induce the *GAL7* promoter. Other labs have subsequently

devised similar plasmids with alternative features. Ultimately, the handling tips described will be applicable to all related plasmids.

4. When manipulating or storing the plasmids in *E. coli*, care should be taken to avoid long-term incubation of the strain. Plasmids should be analyzed on a gel (uncut as well digested with enzymes to generate a known pattern). Once the integrity of the plasmid sample is confirmed, transform the plasmid into an *E. coli* cloning strain (e.g., DH5α) using standard techniques, and select for growth on LB + amp. We routinely pool all transformants and immediately freeze at −80 °C (in 20 % glycerol). Cells can be grown from this stock to produce fresh plasmid (which would be confirmed again).
5. Very few unique restriction sites remain in these plasmids, thus cloning is difficult. In principal, recombineering strategies should work, but great care should be taken to ensure integrity of the resulting plasmids.
6. When transforming into yeast, if any rDNA exists in the recipient strain, recombination between the plasmid and the host will be efficient. Strains like NOY773 or NOY984 can be used as they are confirmed to be free of endogenous rDNA, however there still can be recombination between the host helper plasmid and the newly transformed plasmid (even if the host helper is quickly shuffled out).
7. We recommend the following approach. Transform NOY984 with a derivative of pNOY353 selecting for growth on SD-trp. The organization of pNOY353 is shown in Fig. 2. Plate several transformants immediately on SD-trp +5-FOA. This will cure the cells of pNOY130. Isolate several strains and characterize their growth. Typically, the slow growing cells are the right ones, but this depends on the experiment. Once a set of putatively correct clones are identified, re-isolate the plasmid from these strains, retransform *E. coli*, and confirm the plasmid integrity.
8. Although laborious, these steps are critical to insure that the anticipated plasmid is indeed the plasmid that is intended to be characterized. Exceptional care *must* be taken in working with these vectors and strains.

3.6 Screening for Modifiers of RNA Pol I Transcription

1. The YRH269 strain harbors a modified *URA3* reporter gene (*mURA3*) positioned adjacent to the centromere-proximal rDNA gene on chromosome XII. The Pol I terminator region of this rDNA gene was deleted and replaced with a stuffer DNA fragment of the same length (Fig. 3a). RNA polymerase I reads through this stuffer fragment and interferes with RNA Pol II-driven transcription of *mURA3*. The result is that YRH269

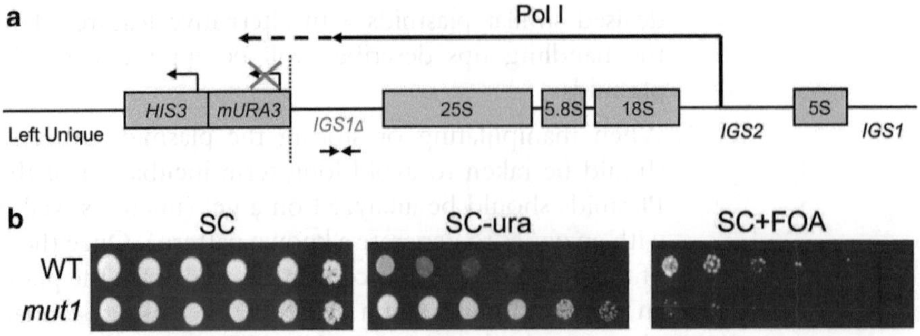

Fig. 3 Promoter interference system for detecting changes in rDNA transcription. (**a**) The transcriptional terminator sequences in IGS1 were deleted from the centromere-proximal rDNA gene, which allows Pol I transcription to read-through and interfere with Pol II transcription of an adjacent modified *URA3* reporter gene (*mURA3*). (**b**) The read-through transcription interferes with expression of *mURA3*, resulting in partial growth on SC – ura and SC + FOA plates. An example is shown for a mutant (*mut1*) that causes a decrease in Pol I transcription, thus resulting in more *mURA3* expression and a more Ura$^+$ phenotype compared to the WT parent (YRH269)

grows poorly on SC plates lacking uracil (SC – ura), but the amount of repression is not sufficient enough to make the cells strongly resistant to 5-FOA (Fig. 3b). Mutations, compounds, or growth conditions that reduce Pol I transcription of the left-most rDNA gene result in less interference with *mURA3* and a more Ura$^+$/FOA-sensitive phenotype. Conversely, conditions that improve Pol I transcription result in a more Ura$^-$/FOA-resistant phenotype (*see* **Note 12**).

2. This strain can be easily mutated with a variety of mutagens and the colonies then screened for reciprocal changes in Ura$^+$ and FOA-resistant growth phenotypes. We have previously used a transposon insertion library to screen for nonessential modifiers of Pol I transcription [12]. Following mutagenesis, the cells are spread onto either YPD, in the case of chemical or UV, or selective SC plates in the case of transposon mutagenesis. The colonies are then replica-plated to SC, SC – ura, and SC + FOA plates and incubated at 30 °C. Some mutants will become very Ura$^+$ or FOA-resistant within 1 day, so it is best to identify candidates every 24 h over 3 or 4 days. Weaker mutants will be noticeable later. Regardless of the day the candidate mutants are identified, it is important to only pick those mutants that show reciprocal changes in growth on SC – ura and SC + FOA. This will help reduce the number of false positive mutants that are simply altered in their resistance to 5-FOA, and not changing *mURA3* expression.

3. One major advantage of this reporter system is that the *mURA3-HIS3* cassette is stable and not subjected to the high rate of recombination that occurs between rDNA genes within the rDNA tandem array. This allows one to carry out backcrosses, tetrad dissections, and other standard genetic manipulations common in the follow-up to genetic screens.

4. Pick the candidate colonies and restreak for single colonies on the appropriate plates, either YPD or selective. After the colonies have grown for 2 or 3 days, replica plate to SC, SC – ura, and SC + FOA again to confirm the original phenotypes.
5. A common false positive in this type of screen is a colony in which the size of the rDNA array has become significantly longer or shorter. If the array is shorter, then the frequency of Pol I transcription from the remaining rDNA genes, including the left-most repeat, is increased, resulting is more efficient interference with *mURA3* expression and a Ura^-/FOA^R phenotype. A longer tandem array results in the opposite, a more Ura^+/FOA^S phenotype. The changes in array size could be directly due to the induced mutation, but the change in Pol I transcription is likely indirect. Therefore, it is important to determine the rDNA copy number using the CHEF gel method in Subheading 3.1.
6. As with any genetic screen, the researcher should also determine whether the mutation is dominant or recessive, and confirm the mutant phenotype segregates 2:2 following standard backcrossing and tetrad dissections.
7. Quantitative RT-PCR can be used to confirm an appropriate change in read-through transcription from the leftmost rDNA gene, in the mutant compared to the parental strain (*see* **Note 13**).
8. Isolate total RNA using the hot-acid phenol method as in Subheading 3.2, although in this case there is no radioactivity involved.
9. Synthesize cDNA specific to the leftmost rDNA gene from 5 μg RNA with 1 μM primer JS766 using SuperScript reverse transcriptase. For a control, synthesize cDNA from *ACT1* RNA using primer JS770. DNase I is added to the reactions to ensure the absence of contaminating DNA, which could be a potential problem with repetitive rDNA. Set up the following reaction in an RNase-free microfuge tube:

   1 μL 10× DNase I buffer.

   1 μL DNase I (New England Biolabs).

   x μL (5 μg) total RNA.

   x μL H_2O (RNase-free).

   10 μL total volume.

   Incubate at room temperature for 15 min.
10. Add 1 μL of 25 mM EDTA and incubate at 65 °C for 10 min.
11. Add 1 μL dNTPs (10 mM stock), and 1 μL of cDNA primer (2 μM stock), then heat to 65 °C for 5 min, followed by chilling on ice. This anneals the primer to RNA.

12. Add 4 μL 5× single-strand buffer that comes with Superscript II, 2 μL of 0.1 M DTT, and 1 μL of RNaseOUT, an RNase inhibitor. Then incubate at 42 °C for 2 min.
13. Add 1 μL (200 U) of Superscript II reverse transcriptase and continue incubating at 42 °C for another 50 min.
14. Heat-inactivate for 15 min at 70 °C.
15. Add 1 μL of RNase H (2 U) and incubate at 37 °C for 20 min. The final volume at the end of this procedure is 22 μL.
16. Dilute the PCR primers to 5 μM. Primers for PCR amplification of the rRNA read-through product are JS765 and JS766. Primers for *ACT1* are JS769 and JS770. The cDNA preparations should be diluted 1:625 in H_2O (*see* **Note 14**).
17. Each 20 μL PCR reaction contains the following components:

    0.8 μL forward primer

    0.8 μL reverse primer.

    10 μL SYBR Green PCR Master Mix (Applied Biosystems).

    7.4 μL H_2O.

    1 μL cDNA (of a 1:625 dilution).
18. Run the PCR reactions in a real-time PCR system. We use an Applied Biosystems 7300. The PCR reactions utilize a four stage profile: stage 1: 50 °C 2 min; stage 2: 95 °C 10 min; stage 3: 40 cycles of 95 °C 15 s, 60 °C 1 min; stage 4: (dissociation): 95 °C 15 s, 60 °C 30 s, 95 °C 15 s.
19. Run the samples in triplicate. The change in rRNA is expressed relative to the change in *ACT1* (rRNA signal/*ACT1* signal).

4 Notes

1. The CHEF plugs fit perfectly into the wells. It may be difficult to get the plugs inserted at first. We typically use two small spatulas, one in each hand. We guide the plug with one spatula and push with the other. It is important to push the plugs to remove bubbles in the well. Once the bubbles are gone, liberally cover the wells with sealing agar.
2. As with any experiment using radioactivity, use all possible precautions to limit exposure and to minimize the amount of radioactivity used. All personal protective measures should be employed.
3. For radioactive labeling of growing cultures, we grow cells in an air shaker. We have modified the air shaker to permit continuous shaking, even with the lid opened. With this modification the cells are not exposed to excessively long drops in agitation during sampling.

4. Avoid getting tritiated material on or near the lid of the eppendorf tube. If radioactivity is near the lid it will coat the centrifuge during the spin.
5. Tritium is a low-energy beta emitter. Geiger counters cannot detect tritium contamination. Thus, all potentially contaminated surfaces must be swiped and measured in a scintillation counter to determine if there is contamination.
6. In first phenol extraction, attempt to remove as much of the aqueous phase as possible. Later, remove only enough to limit carryover of the interface. On final transfer to ethanol, set pipet to volume lower than expected volume of the aqueous phase and transfer equal volume of each sample.
7. We pour the formaldehyde–agarose gels as thin as possible. Thin gels transfer much better to blots using the semidry transfer described herein.
8. When transferring DNA from gel to membrane, the backside of the gel is often smoother than the top due to the pouring conditions. This makes for a smoother transfer and ultimately a cleaner film image. Thus, place the gel upside down when assembling the transfer apparatus. Additionally, after placing the membrane and the two sheets of wet Whatman paper on top of the gel, carefully roll a test tube across the gel–membrane–paper stack to press air bubbles out the side. You may choose to roll the tube both ways to ensure bubbles are entirely removed.
9. This diluted acid treatment partially depurinates the DNA fragments, breaking them into small pieces that are more efficiently transferred to the membrane.
10. The Smith lab typically uses a wicking procedure for the Southern transfer, though other setups are available, including stacks of blotting paper, dense sponges, and even vacuum blotting if you have that kind of apparatus. Though the genomic DNA was psoralen-treated, the Southern blot is run like any other typical experiment.
11. Autoradiography is preferred for hard copies of the blot, but for quantifying the percentage of active genes, it will be necessary to expose the membrane to a phosphorimager cassette.
12. When performing a genetic screen with this system, it is sometimes useful to increase the concentration of 5-FOA to 0.2 %, which decreases the background level of FOA-resistance, making it easier to identify positive colonies following replica-plating.
13. This assay is used for measuring changes in Pol I transcription from the left-most rDNA gene, but it does not necessarily mean that the same changes are occurring at all rDNA genes

within the tandem array. It is therefore important to follow up with other methods of measuring Pol I transcription, such as the in vivo labeling with ^{3}H-methionine in Subheading 3.2.

14. While we have settled on a 1:625 cDNA dilution as being centered in the linear range with our experiments, it is still suggested that you run a dilution series with the cDNA you have generated for the first time.

References

1. Larson DR, Zenklusen D, Wu B et al (2011) Real-time observation of transcription initiation and elongation on an endogenous yeast gene. Science 332:475–478
2. Kobayashi T, Heck DJ, Nomura M et al (1998) Expansion and contraction of ribosomal DNA repeats in *Saccharomyces cerevisiae*: requirement of replication fork blocking (Fob1) protein and the role of RNA polymerase I. Genes Dev 12:3821–3830
3. Clemente-Blanco A, Mayan-Santos M, Schneider DA et al (2009) Cdc14 inhibits transcription by RNA polymerase I during anaphase. Nature 458:219–222
4. Anderson SJ, Sikes ML, Zhang Y et al (2011) The transcription elongation factor Spt5 influences transcription by RNA polymerase I positively and negatively. J Biol Chem 286: 18816–18824
5. Zhang Y, Smith AD IV, Renfrow MB et al (2010) The RNA polymerase-associated factor 1 complex (Paf1C) directly increases the elongation rate of RNA polymerase I and is required for efficient regulation of rRNA synthesis. J Biol Chem 285:14152–14159
6. Schneider DA, French SL, Osheim YN et al (2006) RNA polymerase II elongation factors Spt4p and Spt5p play roles in transcription elongation by RNA polymerase I and rRNA processing. Proc Natl Acad Sci U S A 103: 12707–12712
7. Conconi A, Widmer RM, Koller T et al (1989) Two different chromatin structures coexist in ribosomal RNA genes throughout the cell cycle. Cell 57:753–761
8. Dammann R, Lucchini R, Koller T et al (1993) Chromatin structures and transcription of rDNA in yeast *Saccharomyces cerevisiae*. Nucleic Acids Res 21:2331–2338
9. Sandmeier JJ, French SL, Osheim YN et al (2002) *RPD3* is required for the inactivation of yeast ribosomal DNA genes in stationary phase. EMBO J 21:4959–4968
10. Nogi Y, Vu L, Nomura M (1991) An approach for isolation of mutants defective in 35S ribosomal RNA synthesis in *Saccharomyces cerevisiae*. Proc Natl Acad Sci U S A 88:7026–7030
11. Nogi Y, Yano R, Nomura M (1991) Synthesis of large rRNAs by RNA polymerase II in mutants of *Saccharomyces cerevisiae* defective in RNA polymerase I. Proc Natl Acad Sci U S A 88:3962–3966
12. Hontz RD, Niederer RO, Johnson JM et al (2009) Genetic identification of factors that modulate ribosomal DNA transcription in *Saccharomyces cerevisiae*. Genetics 182:105–119
13. Osheim YN, French SL, Sikes ML et al (2009) Electron microscope visualization of RNA transcription and processing in *Saccharomyces cerevisiae* by Miller chromatin spreading. Methods Mol Biol 464:55–69
14. Warner JR (1999) The economics of ribosome biosynthesis in yeast. Trends Biochem Sci 24: 437–440
15. Cioci F, Vu L, Eliason K et al (2003) Silencing in yeast rDNA chromatin: reciprocal relationship in gene expression between RNA polymerase I and II. Mol Cell 12:135–145
16. Schneider DA, Michel A, Sikes ML et al (2007) Transcription elongation by RNA polymerase I is linked to efficient rRNA processing and ribosome assembly. Mol Cell 26:217–229
17. Hontz RD, French SL, Oakes ML et al (2008) Transcription of multiple yeast ribosomal DNA genes requires targeting of UAF to the promoter by Uaf30. Mol Cell Biol 28: 6709–6719
18. Oakes M, Aris JP, Brockenbrough JS et al (1998) Mutational analysis of the structure and localization of the nucleolus in the yeast *Saccharomyces cerevisiae*. J Cell Biol 143:23–34

Chapter 19

Two-Dimensional Agarose Gel Electrophoresis for Analysis of DNA Replication

Sandra K. Villwock and Oscar M. Aparicio

Abstract

The initiation, elongation, and termination of DNA replication are each associated with distinct, nonlinear DNA structures that can be resolved and identified by two-dimensional (2D) agarose gel electrophoresis. This method involves: isolation of genomic DNA while preserving fragile replication structures, digestion of the DNA with a restriction enzyme, separation of DNA by size and shape through two distinct stages of agarose gel electrophoresis, and Southern blotting to probe for the specific sequence(s) of interest. The method has been most commonly used to determine the activity level of putative replication origin-containing sequences, and has also been used to analyze replication timing, fork progression, fork pausing, fork stalling and collapse, termination, and recombinational repair.

Key words DNA replication, Replication fork, Replication origin, Replication initiation, Replication elongation, Replication termination, Restriction fragment, Agarose gel electrophoresis

1 Introduction

Bell and Byers first demonstrated the utility of 2D agarose gel electrophoresis for the resolution of branched recombination intermediates (Holliday Junctions) from linear molecules based on the differential migration of linear versus branched DNA molecules under appropriate electrophoresis conditions [1]. Brewer and colleagues and Huberman and colleagues derived independent 2D gel approaches for analysis of replication intermediates [2, 3]. Replication initiation involves unwinding of the DNA duplex at each replication origin and assembly of a pair of diverging replication forks, which synthesize the complementary strands and form an expanding replication "bubble" at each origin. Thus, the presence of a bubble within a restriction fragment is indicative of an origin within the fragment. Flanking restriction fragments are typically replicated by a single replication fork and thus characterized by a "Y"-shaped, partially replicated DNA molecule during their replication. When replication forks converge and terminate replication

Jeffrey S. Smith and Daniel J. Burke (eds.), *Yeast Genetics: Methods and Protocols*, Methods in Molecular Biology, vol. 1205, DOI 10.1007/978-1-4939-1363-3_19, © Springer Science+Business Media New York 2014

within a particular restriction fragment, the fragment is characterized by "double-Y" or "X"-shaped structures. Each of these structures can be distinguished by 2D gel analysis.

The most commonly used procedures for 2D gel analysis of replication structures, including the one described here, are based on the neutral–neutral method pioneered by Brewer and Fangman, in which both gel electrophoresis steps are run under neutral (non-denaturing) conditions [2]. Briefly, the method involves "fixation" of replication structures by treatment with sodium azide combined with rapid chilling of cells. Cells are broken by bead-beating, the nuclear fraction is isolated and deproteinized, and the DNA is isolated on a CsCl density gradient. Following purification, the DNA is digested with an appropriate restriction enzyme(s) prior to gel electrophoresis. The protocol we present incorporates an additional step at this point, absent from some protocols, that enriches for ssDNA-containing restriction fragments (such as replication intermediates) and allows more replication intermediates to be loaded per gel lane. The digested DNA is loaded into a single lane of a low concentration agarose gel and electrophoresed at low voltage (first dimension). The gel-lane containing the size-fractionated DNA is excised, rotated 90°, placed atop a freshly poured high concentration agarose gel containing ethidium bromide, and electrophoresed at high voltage (second dimension). These second dimension conditions favor separation based on shape, causing slower migration of more extensively branched DNA molecules. The DNA is analyzed by (Southern) blotting to a nylon membrane, which is probed for the specific restriction fragment of interest. The blot may be stripped and reprobed for analysis of additional loci.

The most significant variable to consider is the choice of restriction enzyme(s), as it determines the regions that can be effectively probed, and affects the detection of bubble and fork pause sites. Most importantly, clear detection of bubble structures at origins requires that the origin be located within approximately the middle third of the restriction fragment; otherwise, one of the two replication forks migrates outside of the restriction fragment much more rapidly than the other and mainly a single fork structure will be detected. Conversely, pause sites are best located asymmetrically within the restriction fragment as they are best visualized on the ascending and/or descending Y-arcs, rather than at the apex of the Y-arc where signal tends to accumulate with or without pausing. Fork pauses are confirmed by analysis with a second restriction enzyme that moves the pause within the restriction fragment and hence, changes its position within the Y-arc. More in-depth discussion of these and related considerations have been presented by Friedman and Brewer [4].

The 2D gel method has limitations. It is cumbersome and lengthy, and involves the use of radioactive probes for maximum sensitivity. Perhaps more significantly, in the era of genomics, the method does not appear to be scalable for genome-wide analysis. Nevertheless, 2D gels for analysis of replication structures remain a powerful tool for physical analysis of DNA replication and recombination structures.

2 Materials

Solutions

1. EDTA–glycerol solution: 0.2 M EDTA, 17 % glycerol.
2. 10 % sodium azide.
3. Nuclear Isolation Buffer (NIB): 17 % (v/v) glycerol, 50 mM MOPS, 150 mM KOAc, 2 mM $MgCl_2$, 500 μM spermidine, 150 μM spermine, pH 7.2.
4. TEN Buffer: 50 mM Tris–HCl pH 8.0, 50 mM EDTA, 100 mM NaCl.
5. 10 % Sarkosyl.
6. NET Buffer: 1 M NaCl, 1 mM EDTA, 10 mM Tris–HCl, pH 8.0.
7. Depurination Solution: 0.25 N HCl.
8. Denaturation Solution: 0.5 N NaOH, 1.5 M NaCl.
9. Neutralization Solution: 1 M Tris–HCl, 1.5 M NaCl, pH 7.4.
10. 20× SSC: 3 M NaCl, 0.3 M Na Citrate, pH 7.0.
11. 1 M Phosphate Buffer: 0.28 M NaH_2PO_4, 0.72 M Na_2HPO_4.
12. Hybridization Solution: 7 % SDS, 0.5 M Phosphate Buffer, 1 mM EDTA (SDS will precipitate out, so heat before use)
13. Blot Wash Solution: 1 % SDS, 40 mM Phosphate Buffer, 1 mM EDTA.
14. Blot Stripping Solution: 0.5 % SDS.

3 Methods

3.1 Yeast Cell Harvest

1. Freeze 80 mL EDTA–glycerol solution in 500 mL centrifuge bottles, preparing three bottles per liter of culture to be harvested. Place bottles at −80 °C lying nearly horizontally to increase surface area of solution.
2. Inoculate an overnight culture of yeast from a single colony.
3. Use the overnight culture to inoculate 1 L medium (e.g., YPD). Grow until $OD_{600} \approx 1$ (~2×10^7 cells/mL).
4. Add 10 mL of 10 % sodium azide stock solution to 1 L culture (0.1 % final) to kill cells.
5. Immediately pour culture into bottles with frozen EDTA–glycerol pellets. Divide 1 L culture evenly between three

500 mL centrifuge bottles. Place the centrifuge bottles on ice and shake gently until frozen EDTA–glycerol pellet thaws completely.

6. Pellet cells 8,000 rpm (~10,000 × *g*) in GS3 rotor or equivalent for 7 min at 4 °C. Carefully discard supernatant promptly as pellet loosens quickly.
7. Resuspend cells in each bottle in 10 mL ice-cold H_2O and pool the three fractions of the same culture in a 50 mL conical, screw-cap (Falcon) tube. Pellet cells at 2,500 rpm (~1,500 × *g*) for 5 min at 4 °C in tabletop centrifuge with swinging-bucket rotor. Discard the supernatant. Pellet may be frozen in dry ice–ethanol bath and stored at −80 °C.

3.2 DNA Isolation

1. Resuspend the cells in ice-cold NIB. For 1 L culture, resuspend in 12 mL NIB and split into two 50 mL conical tubes.
2. Add an equal volume (~6 mL) of acid-washed glass beads (450–600 μm diameter) to each tube. Keep on ice.
3. Vortex samples at maximum power until >90 % of cells have been broken. Keep samples on ice between vortexing cycles or vortex continuously at 4 °C (*see* **Note 1**).
4. Remove lysate from beads with a 10 mL pipet by drawing from the bottom of the tube, and transfer to an ~35 mL Oak Ridge centrifuge tube (do not worry if a few beads transfer also). Rinse beads with 6.5 mL of NIB and combine rinse with lysate; repeat rinse. Pool all lysates from the same sample, balance tubes with NIB, and centrifuge at 8,000 rpm (~10,000 × *g*) in SA-600 rotor or equivalent for 20 min at 4 °C.
5. Discard supernatant. Resuspend pellet of nuclei, unbroken cells, and debris in 8.0 mL TEN buffer.
6. Add 1.5 mL 10 % Sarkosyl stock (final 1.5 %). Mix gently.
7. Add 300 μL 20 mg/mL Proteinase K stock (final 600 μg/mL); incubate at 37 °C ≥2 h.
8. Centrifuge at 5,000 rpm (~3,500 × *g*) in SA-600 rotor or equivalent for 5 min to pellet cells and debris.
9. Transfer the supernatant to a graduated 15 mL conical, screw-cap tube and measure volume (expect ~10.5 mL). Add 1.05 g CsCl per mL of supernatant. Dissolve by inverting or rotating tubes gently; may incubate at 37 °C to hasten solvation.
10. After CsCl dissolves, measure volume again. Add 0.025× volumes 5 mg/mL Hoechst 33258 stock. Mix gently.
11. Using Pasteur pipet, transfer sample to a 13.5 mL Quick-Seal polyallomer tube for ultracentrifugation. Tube must be filled to prevent collapse; prepare a blank solution (TEN/CsCl/Sarkosyl/Hoechst) if necessary to top-off and balance tubes.

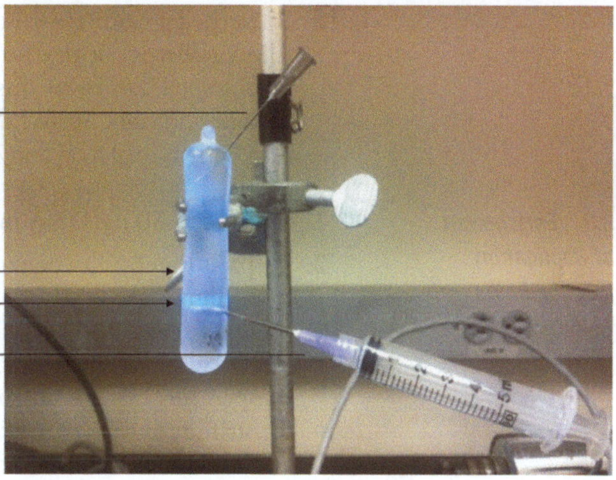

Fig. 1 Extraction of chromosomal DNA from cesium chloride gradient. mDNA is mitochondrial DNA; gDNA is genomic DNA, which includes a narrow band of ribosomal DNA at the bottom of the band, which is not apparent in this image. Note upward-facing bevel of 16G needle

12. Centrifuge at 55,000 rpm (~240 k × *g*) in NVT65 rotor or equivalent for 12–18 h at 18 °C to form density gradient; remove from centrifuge and avoid disturbing the gradient.
13. Immobilize tube in clamp with long-wave UV light directed at it (remember to use eye protection) (*see* **Note 2**).
14. Insert a 22G needle into top of tube to prevent vacuum when pulling band. Insert a 16G needle attached to a 5 mL syringe about 1 cm below the chromosomal and rDNA bands (Fig. 1). Angle the needle up to contact the bottom of the band and pull slowly on the syringe plunger to avoid shearing the DNA while removing from gradient. Remove the syringe from the needle and expel the DNA from the syringe into a 14 mL polypropylene snap-cap tube (*see* **Note 3**). From here on, continue to treat DNA very gently to avoid shearing. Use wide bore pipets, do not vortex, and do not use excess heat.
15. Add an equal volume of 5:1 isopropanol–H_2O solution. Mix gently by rolling the tubes on the benchtop or by gentle inversion; do not vortex. After the phases separate, aspirate to remove the upper (alcohol) layer. Repeat the extraction two times.
16. Add three volumes ice-cold 70 % ethanol slowly to the side of the tube. Gently swirl or invert capped tube until the phases are fully mixed. The DNA should precipitate out of solution as a fibrous clump.
17. Centrifuge at 8,000 rpm (~10,000 × *g*) in SA-600 rotor or equivalent for 20 min. Discard supernatant.

18. Rinse pellet with ice-cold 70 % ethanol and discard ethanol. Allow pellet to air-dry briefly. Add 500 μL TE. The pellet may take several hours to dissolve; dissolve by gently tapping the tube periodically. Store DNA at 4 °C. Measure DNA concentration with NanoDrop and/or by running 1 μL on agarose gel.

3.3 Enrichment of Replication Intermediates

1. Digest 25–50 μg DNA in 500 μL volume ~2 h at 37 °C.
2. Prepare the BND (benzoylated naphthoylated DEAE) cellulose (Sigma):
   Weigh out 4 g BND cellulose in a 50 mL conical tube. Boil the BND cellulose in 20 mL H_2O for 5 min. Let cool to room temperature. Break up the particles with a rubber policeman. Centrifuge at ~500 × *g* for 2 min. Discard supernatant. Suspend and wash the BND cellulose once with 20 mL H_2O, and pellet by centrifugation. Wash two times with 20 mL NET buffer. Store at 4 °C in 20 mL NET buffer.
3. Prepare a disposable column for each sample (e.g., Bio-Rad Poly-Prep Chromatography Columns). With large-bore pipet, add BND cellulose suspension to the column to make 0.5–1 mL packed column volume (Fig. 2). Let the column sit for a few minutes and allow the liquid to drain by gravity. Wash the column with 10 bed volumes NET buffer. Minimize disturbance to the column bed (*see* **Note 4**).
4. Add 0.25 volumes 5 M NaCl stock to the DNA digest (final 1 M NaCl). Load the DNA solution onto the column and allow

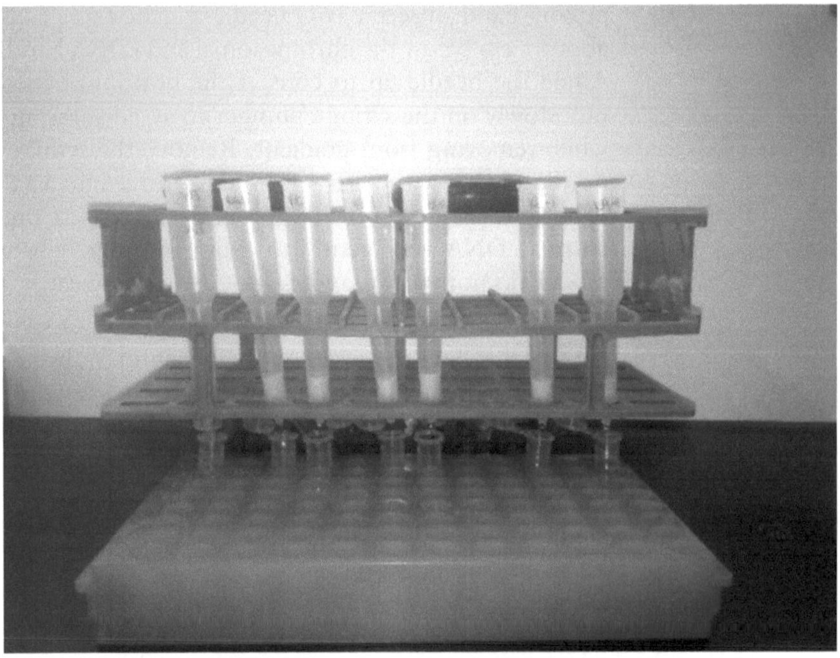

Fig. 2 BND cellulose column setup. Note the weights (*red* doughnuts) to keep the rack stable (Color figure online)

it to enter the resin by gravity. Collect the flow-through in a microcentrifuge tube and apply to the column two more times.

5. Wash the column with 3–5 volumes NET buffer; discard flow-through.
6. Elute bound DNA with 2–3 volumes pre-warmed (50 °C) 1.8 % caffeine in NET buffer. Collect the eluate and divide into two or more 1.5 mL microcentrifuge tubes (750 μL maximum volume per tube).
7. Add 1 volume isopropanol to the eluate(s) and invert slowly to mix.
8. Centrifuge for 10 min at full speed in microcentrifuge (≥10,000 × *g*).
9. Wash pellet with ice-cold 70 % ethanol, centrifuge for 2 min at full speed in microcentrifuge. Remove ethanol and air-dry pellet briefly.
10. Dissolve in 20 μL TE; do not vortex. Add loading dye to the DNA.

3.4 2D Gel Electrophoresis

1. Melt a 0.4 % agarose, 1× TBE solution. Allow solution to cool to ~55 °C before pouring gel. Make wells using a 1 mm × 7 mm comb or smaller. Do not add ethidium bromide to either the gel solution or running buffer. Add 1 μL of dye to each well to test integrity.
2. If possible, load alternate lanes of gel to avoid contamination of samples. Carry out the electrophoresis at ~1.0 V/cm (distance between electrodes) for >15 h at room temperature. Run at about 470 V h for bands ~5 kb, about 425 V h for bands ~3 kb. The dye should have run roughly 5 cm.
3. Stain gel for 20 min in 1× TBE, 0.3 mg/mL ethidium bromide.
4. Photograph the gel (Fig. 3).
5. Cut the gel with sharp razor to separate lanes. Then cut gel 1 cm below MW band of interest and at the well.
6. Melt a 1.0 % agarose, 1× TBE, 0.3 μg/mL ethidium bromide gel solution (*see* **Note 5**). Allow gel solution to cool to ~55 °C before pouring gel. Place excised gel slice in gel casting tray, rotated 90° from first gel so that the largest MW band of each slice is on the left (Fig. 4). Slowly pour agarose around the gel slice. Allow the agarose to harden around the gel fragment.
7. Prepare running buffer: 1× TBE, 0.3 μg/mL ethidium bromide. Run gel 5 V/cm (based on distance between electrodes) at 4 °C recirculating the buffer at ~20 mL/min until the low MW DNA has run 6–7 cm. Run gel for ~5–8 h; check progress using long-wave UV.
8. Photograph the gel (Fig. 5).

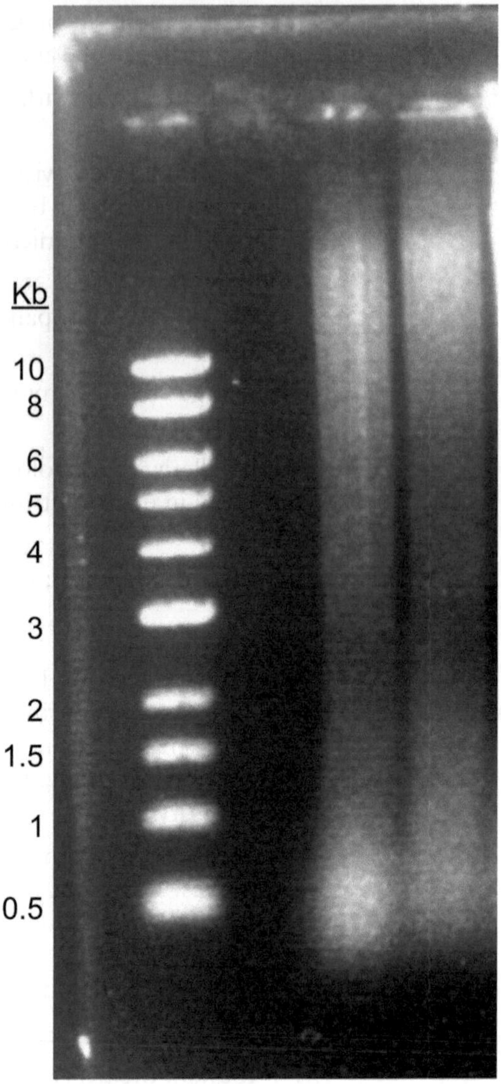

Fig. 3 Image of first dimension gel. Gel was stained with EtBr after DNA separation. Two sample lanes are shown

3.5 Southern Transfer

1. Soak the gel in Depurination Solution 8 min, gently rocking.
2. Pour off liquid. Soak the gel in Denaturation Solution 20 min, gently rocking.
3. Pour off liquid. Soak the gel in Neutralization Solution 20 min, gently rocking.
4. Cut nylon (e.g., Hybond N+or equivalent) membrane to the size of the gel. Wet the membrane by floating on the surface of dH_2O, then submerge and leave for 5 min. Then soak in 20× SSC for a minute before assembling blot.
5. Set up the blot on an elevated platform in a large glass dish (*see* **Note 6** and Fig. 6). Pour 20× SSC into the glass dish; the

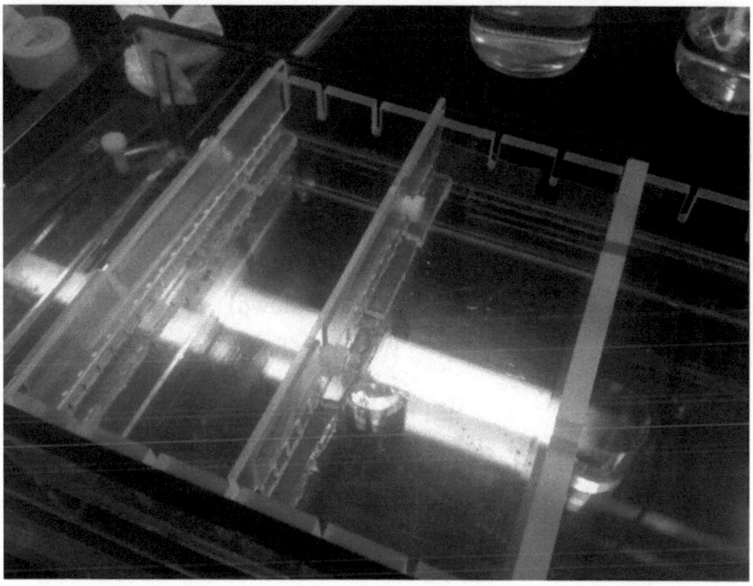

Fig. 4 Second dimension gel setup. Excised bands from first dimension gel were rotated 90° and placed in rows against the combs (shown). Next, gel is poured from the bottom of the gel tray to flow up against the supported gel slices. After the gel is poured, the combs are removed

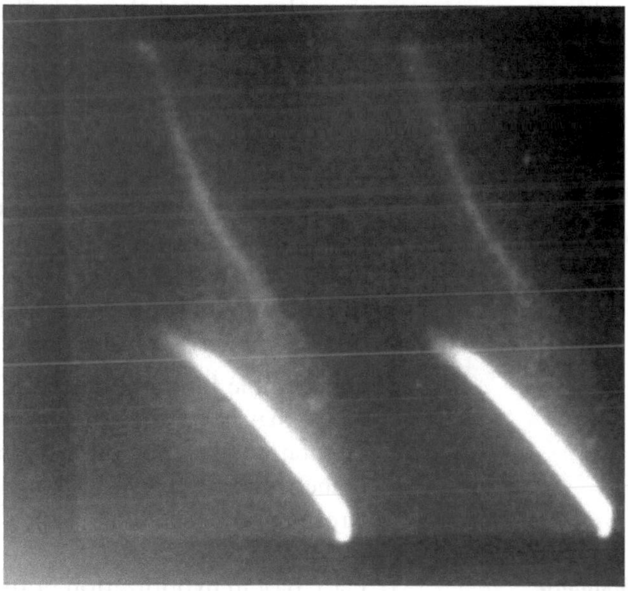

Fig. 5 Image of second dimension gel. The bright arcs contains linear DNA species (large to small from *right* to *left*) and the lighter, slower-migrating species (above the main arc) are structured DNA molecules. Two samples were run side by side

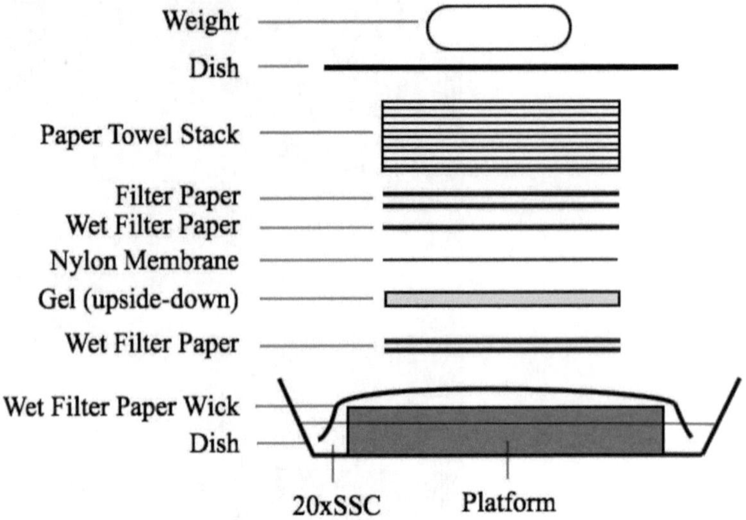

Fig. 6 Southern transfer setup. See text for instructions

level of the buffer should be lower than that of the platform. Cut a piece of 3MM Whatman paper to act as wick for transfer solution; it should be longer than the solid support so that two ends of the paper are submerged in the buffer and wide enough to support the gel. Wet the paper entirely with 20× SSC. Place two pieces of 3MM Whatman paper, cut to the dimensions of the gel, on top of the wick filter paper. Place the gel upside-down on top of filter paper (avoid trapping bubbles) so that the smooth side will contact the nylon membrane. Place nylon membrane on top of gel. Remove any air bubbles by rolling a glass pipet across surface. Place another filter paper, wetted with 20× SSC, on top of the membrane. Place a few more dry filter papers on top. Place a stack of paper towels at least 10 cm high, cut to dimensions of gel, on top of filter papers. Place a dish or flat tray (e.g., gel tray) on top of the stack of paper towels and add some weight on top of the flat tray (~2 kg for 20×25 cm gel). Transfer overnight.

6. Disassemble blot and discard all except for one piece of filter paper that is on top of the membrane. Mark membrane for orientation. Do not rinse membrane. Place the damp membrane with filter paper backing with the DNA side up in a UV Cross-linker. Cross-link using 254 nm bulbs 1200 μJ/cm^2.

3.6 Southern Hybridization

1. Place blot in hybridization bottle with DNA side facing inward and incubate with ~25 mL Hybridization Solution at 65 °C ≥1 h (*see* **Note 7**).
2. Add labeled, denatured probe to Hybridization Solution (*see* **Note 8**).

3. Incubate at 65 °C ≥12 h in hybridization oven with rotation of hybridization bottle.
4. Pour out Hybridization Solution with labeled probe into appropriate waste container. Wash the blot in the hybridization bottle, rotating in the oven with ~50 mL pre-warmed (65 °C) Wash Solution 2 × 15 min, then 3 × 10 min.
5. Wrap the blot with one layer of saran wrap and expose it to phosphor screen overnight or longer.
6. Scan the phosphor screen using a phophorimager.
7. For reprobing, strip the blot by washing blot twice in ~250 mL boiling 0.5 % SDS solution, and repeat **steps 1–6** of this section.

4 Notes

1. Use a powerful vortexer. The FastPrep Instrument (MP Biomedicals) is very effective and will disrupt cells with two 45 s cycles at high power. Multivortexers may also be used, but may require up to 30 min. Monitor percent cell breakage by checking for the fraction of ghosts in a phase contrast microscope; keep a sample from before breakage for comparison.
2. Three bands should be visible in the gradient. The top band is mitochondrial DNA; the middle, prominent band is chromosomal DNA; the faint band just below the chromosomal DNA is rDNA; it is possible, though unnecessary to separate the rDNA band from chromosomal. Pull both bands together. A band of particulate matter may be visible between the nuclear and mitochondrial bands; try to avoid this, but not at the expense of losing DNA.
3. While the DNA band is being pulled, the band will travel down; adjust the needle accordingly until the band is entirely extracted. To minimize leaking of the remaining gradient, remove the 22G needle and place a piece of tape over the hole before removing the 16G syringe needle from the tube. Remove the tube together with the 16G needle and syringe from the clamp setup, then remove the needle with syringe while disposing of the tube as hazardous waste. Remove needle from syringe and transfer sample to a 14 mL polypropylene snap-cap tube.
4. Set up the columns on an elevated rack to facilitate collection of the flow-through and eluates. Estimate that replication intermediates (RI) will be 1–5 % of the digested DNA and a 0.5 mL packed column volume will bind about 5 mg of RI. About 1 mL BND cellulose suspension compacts to form a 0.5 mL column.

5. A gel ~10 cm in length is sufficient, however, multiple rows of samples may be run in a single large gel, spacing apart the rows of gel slices by at least 10 cm (Fig. 4). It is helpful to use a gel comb to align your gel slices. Keep the comb in the gel while pouring the agarose so the slices do not shift; remove comb once all the agarose is poured. This gel should be of the same thickness as the first dimension gel (~7–8 mm).
6. The platform needs to be a flat surface large enough for your gel (e.g., two microtube racks placed next to each other).
7. Roll up the blot around a clean pipet to facilitate placing inside hybridization bottle.
8. Prepare labeled probe using Multiprime DNA Labeling System (GE Healthcare) or equivalent. DNA fragment used for probe should be located fully within the restriction fragment to be analyzed.

Acknowledgement

Work in the Aparicio lab is supported by NIH grant GM065494.

References

1. Bell L, Byers B (1983) Separation of branched from linear DNA by two-dimensional gel electrophoresis. Anal Biochem 130: 527–535
2. Brewer BJ, Fangman WL (1987) The localization of replication origins on ARS plasmids in *S. cerevisiae*. Cell 51:463–471
3. Nawotka KA, Huberman JA (1988) Two-dimensional gel electrophoretic method for mapping DNA replicons. Mol Cell Biol 8:1408–1413
4. Friedman KL, Brewer BJ (1995) Analysis of replication intermediates by two-dimensional agarose gel electrophoresis. Methods Enzymol 262:613–627

Chapter 20

Replicative Life Span Analysis in Budding Yeast

George L. Sutphin, Joe R. Delaney, and Matt Kaeberlein

Abstract

Identifying and characterizing the factors that modulate longevity is central to understanding the basic mechanisms of aging. Among model organisms used for research related to aging, the budding yeast has proven to be an important system for defining pathways that influence life span. Replicative life span is defined by the number of daughter cells a mother cell can produce before senescing. Over the past 10 years, we have performed replicative life span analysis on several thousand yeast strains, identifying several hundred genes that influence replicative longevity. In this chapter we describe our method for determining replicative life span. Individual cells are grown on solid media and monitored from their initial undivided state until they undergo senescence. Daughter cells are manually removed using a fiber optic needle and quantified to determine the total number of times each mother cell divides.

Key words Yeast, Aging, Replicative life span, Micromanipulation, Dissection, Mother cell, Daughter cell

1 Introduction

Aging in the budding yeast *Saccharomyces cerevisiae* was described more than 50 years ago with the observation that mother cells produce a finite number of daughter cells and then cease replication [1, 2]. This finite mitotic capacity is referred to as replicative life span [3]. A second model of aging has also been described in yeast, referred to as chronological aging. Chronological aging is a post-mitotic paradigm where yeast cells are maintained in a non-dividing, nutrient-limited state. Chronological life span is defined by the length of time that a cell retains the ability to reenter the cell cycle upon return to growth-promoting conditions [4, 5].

Life span is a key phenotype of interest in aging-related studies of yeast and other organisms. In general, genetic or environmental interventions that extend life span are considered to be more relevant than those that reduce life span, because decreased longevity can result from a variety of nonspecific factors while increased longevity requires that the intervention ameliorate the processes limiting life span in the wild type organism. Life span shortening

Jeffrey S. Smith and Daniel J. Burke (eds.), *Yeast Genetics: Methods and Protocols*, Methods in Molecular Biology, vol. 1205, DOI 10.1007/978-1-4939-1363-3_20, © Springer Science+Business Media New York 2014

interventions can be of interest in cases where a strong argument can be made that the reduction results from accelerating the normal aging process. The ability to measure life span in a reproducible and robust manner on a relatively short time-scale is one of the features that make budding yeast a particularly powerful model organism for aging-related studies. Studies from many different labs have contributed to a growing list of genes and interventions that modulate longevity, some of which play a conserved role in modulating the life span of multicellular eukaryotes.

Replicative life span is typically measured by the time and labor consuming micro dissection of daughter cells away from a dividing mother cell [6]. Methods have been developed for enriching populations of yeast for cells in their mid to late replicative life span [7–10]. One method, termed the Mother Enrichment Program, or MEP, has been modified to allow high-throughput measurement of replicative life span using genetic switches that are lethal to new daughter cells [10]. While this method holds promise, it has yet to become widely used or validated and requires several genetic modifications to the background being tested, making it less than ideal for genetic studies. Manual micromanipulation of daughter cells away from dividing mother remains the most robust and commonly used method for measuring replicative life span. To offset the labor-intensive nature inherent to this approach to measuring replicative life span, we previously developed an iterative strategy that allows moderate-throughput screening of a large numbers of interventions starting with only a small number of cells for each intervention [11], and have since demonstrated the robustness of this approach by identifying more than 50 replicatively long-lived single gene deletion mutants [12–17]. In this chapter we describe the basic underlying protocol for measuring replicative life span by manual micromanipulation of daughter cells away from dividing mothers, including our system for managing numerous such experiments simultaneously.

2 Materials

1. YPD plates: (1 % yeast extract, 2 % Bacto-Peptone, 2 % glucose, 2 % agar).
2. Synthetic Complete (SC) plates: (2 % glucose, 0.17 % yeast nitrogen base, 0.5 % ammonium sulfate, 2 % agar, amino acids). *See* Table 1 for complete list of components, including individual amino acids. Exclude appropriate amino acid for auxotrophic marker selection plates.
3. AxioScop.A1 HAL 50 or equivalent microscope configured with the following features designed for yeast dissection.

Table 1
Synthetic complete (SC) media composition

Component	Concentration (g/L)
D-glucose	20
Yeast nitrogen base (w/o ammonium sulfate or amino acids)	1.7
Ammonium sulfate, $(NH_4)_2SO_4$	5
Agar	20
Adenine	0.04
L-arginine	0.02
L-aspartic acid	0.1
L-glutamic acid	0.1
L-histidine	0.1
L-leucine	0.3
L-lysine	0.03
L-methionine	0.02
L-phenylalanine	0.05
L-serine	0.375
L-threonine	0.2
L-tryptophan	0.04
L-tyrosine	0.03
L-valine	0.15
Uracil	0.1

This table provides the mass of each component of SC solid media per unit liter

(a) Stage capable of two-dimensional (*x* and *y*) translation with horizontal and vertical demarcations every 1 mm and click stops every 5 mm.

(b) 10 cm petri plate mount on the stage.

(c) 50 μm fiber optic needle (from Cora Styles) mounted with two-dimensional (*y* and *z*) controls for micromanipulation of cells on the plate surface.

3 Methods

For the methods detailed in this chapter, we refer to yeast cultured at 30 °C on YPD medium. In principle, these methods can be used to measure replicative life span for yeast on virtually any desired solid medium and temperature. For dietary restriction, medium containing a reduced concentration of glucose (typically 0.5 % or 0.05 % instead of the 2 % found in YPD) or an alternative carbon source (e.g., 3 % glycerol instead of 2 % glucose) can be used. For pharmacological experiments, compounds of interest can be added to the YPD medium at the desired concentration. In these cases, the experiments are performed as described, except cells should be patched onto the alternative medium beginning the night prior to positioning cells on life span plates (Subheading 3.3).

3.1 Design Replicative Life Span Experiment

This section outlines considerations to take when designing a replicative life span experiment, including the number of cells to include, the number of plates needed for a given number of cells, and how to divide cells among life span plates.

1. Determine the strains and number of mother cells to be assayed. A typical replicative life span experiment in the Kaeberlein Lab contains 30 strains with 20 mother cell life spans determined for each strain, for a total of 600 mother cell life spans per experiment. Multiple biological replicates are often assayed for the same strain in order to obtain 40, 60, 80, or 100 mother cell life spans in the same experiment.
2. Code the strain so that the dissectors are blind to the identity of each strain. We use a simple numerical code (strain 1, strain 2, … strain 30) for each experiment.
3. Determine the number of life span plates needed for the experiment, and how strains will be divided among the plates. The following points should be taken into consideration:
   (a) Each life span plate will typically contain 60 cells encompassing three strains with 20 cells each. However, a plate can hold upwards of 100 vertically oriented cells and smaller or larger cell sets per strain if needed for a particular experiment design.
   (b) If one experiment contains haploid strains of both mating types, make sure that each plate contains only *MATa* or *MATα* cells. Secreted mating factor from cells of one mating type can impact cell cycle control in cells of the opposite mating type and alter replicative life span.
   (c) Each experiment should be designed such that only one media condition is tested per plate.

3.2 Pour Plates, Prepare Strains, and Perform Initial Quality Control

Prior to beginning the quantification of replicative life span, prepare all media needed for the experiment, streak all strains from frozen stocks, and perform initial quality control to ensure that the strains used for analysis have the expected genotypes. Strains are always freshly streaked from frozen stocks prior to each experiment to ensure consistency between strains and between experiments, and to minimize the accumulation of spontaneous mutations that can result from maintaining live strains.

1. Prepare initial YPD plates, quality control plates, pre-patch plates, and life span plates (*see* Subheading 2 for base YPD and SC plate recipes).
   (a) Initial YPD plates: Prepare enough YPD plates to streak all strains from the frozen stock. Up to 8 strains can be streaked onto each YPD plate.
   (b) Quality control plates: Prepare enough YPD plates to patch all strains for quality control. Up to 10 strains can be patched onto each plate. For each YPD plate, prepare one selection plate corresponding to each genetic marker used in the experiment. Common markers include *KanMX*, which is selected using YPD + G418, and auxotrophic markers (*HIS3*, *LEU2*, *URA3*, etc.), which are selected using SC media lacking the appropriate amino acid. Plates containing 3 % glycerol in place of glucose can also be included to determine whether a strain is capable of undergoing respiration. To test mating type, patched cells are replica plated onto lawns of a specific "mating test" strain which contains an auxotrophy complementary to those in the strain being tested (e.g., a strain auxotrophic for arginine is used for the BY background, which is auxotrophic for histidine, uracil, leucine, methionine, and lysine). The mating test strain should be of the opposite mating type as that expected for the strain to be tested. After strains are allowed to mate for 12–24 h at 30 °C, they are then replica plated onto basic synthetic media (SC media with no amino acids added). Only cells which successfully mate will be capable of growing on basic media.
   (c) Pre-patch plates: For each media type used in the experiment, prepare enough additional plates so that all strains can be pre-grown on the appropriate media prior to being transferred to the plates used in the experiment. This allows strains to acclimate to the media for several cell cycles prior to starting life span measurement (*see* **Note 1**).
   (d) Life span plates: Pour life span plates a minimum of 72 h prior to selection of virgin cells (Subheading 3.4) to allow sufficient drying time (*see* **Note 2**).
2. Streak all strains for single colonies onto initial YPD plates from frozen stocks 5 days prior to selection of virgin cells (Subheading 3.4). Strains with severe growth defects can be streaked from frozen stocks on preceding days as needed to ensure that colonies from all strains are ready on the same day. From this point forward, strain designators assigned in Subheading 3.1, **step 2** (strain 1, strain 2, etc.) should be used to keep the life span blinded.

3. Incubate freshly streaked plates at 30 °C for 2 days to allow cells to form colonies.
4. For each strain, pick a single colony from the initial YPD plate and patch onto the quality control YPD plate. Take care to select colonies of the appropriate size based on the phenotype of the strain. For example, select small colonies for slow-growing strains that have a tendency to acquire mutations that suppress the slow growth phenotype. Select large colonies for strains that tend to lose their mitochondrial DNA, resulting in the formation of petite colonies.
5. Incubate patched quality control plate at 30 °C overnight to allow patched cells to grow.
6. Replica-plate each quality control plate to each type of selection plate.
7. Incubate selection plates at 30 °C overnight.
8. Assess growth on selection plates to verify that each strain contains expected markers. Remove strains that do not display expected markers and redesign or restart experiment if needed.
9. Two days prior to selection of virgin cells (Subheading 3.4), pick a single colony from the initial YPD plate and patch onto the appropriate pre-patch plate.
10. Incubate pre-patch plates at 30 °C overnight to allow patches to grow.
11. The evening before selection of virgin cells, pick a small quantity of cells from the edge of each patch on the pre-patch plate and patch to one side of the appropriate life span plate as shown in Fig. 1a. Be sure to patch lightly so that cells do not become overgrown prior to initiating the life span experiment. Up to six patches may be included on a single life span plate.
12. Incubate life span plate at 30 °C overnight to allow patches to grow.

3.3 Position Cells on Life Span Plates

In this section individual cells will be arrayed in a single vertical line for rapid manipulation and quantification of number of divisions. To mark which patch each set of cells came from, the fiber optic needle is used to physically punch holes in the agar of the plate, with a single hole marking cells from the first patch, two holes marking cells from the second patch, and so on. To making counting easier, each set of 20 cells is divided into 2 sets of 10 cells separated by a horizontal line of needle holes. Figure 1 provides an illustration of the life span plate setup for one 20-cell set.

1. Place life span plate on the microscope plate mount with the patches to the left.
2. Mark the edge of the plate relative to the microscope stage so that the plate can be identically positioned in future steps.

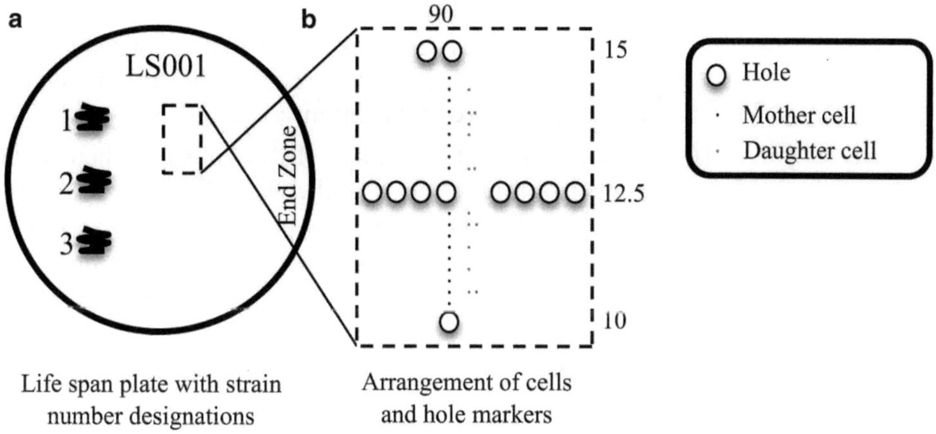

Fig. 1 Life span plate layout. (**a**) Each life span plate has three main areas: (1) patched cells for growth prior to analysis, (2) the life span analysis area, which contains cells arrayed for division quantification and marker holes, and (3) the "End Zone," where excess cells are discarded well away from the analysis area to prevent colonies from obscuring cells included in experiment. (**b**) Expanded diagram of cell analysis area

3. Position the stage so that the fiber optic needle is near the edge of the plate away from cells (see the "End Zone" in Fig. 1a). Punch several deep holes through the surface of the plate to ensure that the needle is clean and free of cells.
4. Using the fiber optic needle, pick up ~40 cells from the first patch and drop near a predetermined position in the top center of the life span plate. For our microscope, all life spans are set up with the first set of cells beginning at the 90 mm horizontal and 10 mm vertical position (Fig. 1b).
5. Ensure that the needle is clear of cells by touching the tip to the surface several times.
6. Punch a single hole to mark the starting position for cells from the first patch on this plate (Fig. 1b). Be careful not to allow any cells to fall into the hole. If they do, they will form a colony that will overtake and obscure the cells being examined for life span.
7. Pick 10 individual cells and position them in a vertical line toward the center of the plate, leaving ~100 μm (2–3 needle widths) between cells (Fig. 1b). Cell clusters can be separated by lightly touching the needle to the plate surface over the cells and gently tapping the side of the microscope to cause the needle to vibrate.
8. Ensure that the needle is clear of cells by touching the tip to the surface several times.
9. Punch a horizontal line of eight holes several needle widths beyond the line of 10 cells. These holes are used to indicate

the halfway point for the first strain. Leave a gap between the center-most holes (Fig. 1b) to allow a continuous needle sweep while cleaning up excess cells at the end of each replicative life span score point (Subheading 3.4, **step 6**).

10. Pick 10 additional individual cells and position them in a vertical line toward the center of the plate (Fig. 1b).
11. Separate 5–10 additional cells and leave them off to the side. These cells will be used to replace cells that fail to divide during the first round of division. More cells can be added for strains with defects in cell division.
12. Ensure that the needle is clear of cells by touching the tip to the surface several times.
13. Punch two horizontal holes several needle widths beyond the second line of 10 cells. These holes will mark the starting position for cells from the second patch on this plate (Fig. 1b).
14. Discard excess cells in the End Zone.
15. Repeat the Subheading 3.3, **steps** 7 through **14** for all remaining strains in the experiment.
16. Once all cells have been positioned, wrap life span plates in Parafilm and incubate at 30 °C for 90–120 min.

3.4 Select Virgin Cells for Life Span Analysis

The cells arrayed in Subheading 3.3 were picked directly from a growing patch. As such, there is no way to know how many times each cell divided prior to selection. Following the 90–120 min incubation in Subheading 3.3, **step 16**, each arrayed cell will have divided once on average. To obtain a population with a defined replicative life span, newly budded (virgin) daughter cells are picked for analysis, and the original cells arrayed in Subheading 3.3 discarded.

1. Prepare replicative life span score sheets. Score sheets typically consist of a grid with each cell represented by a single row, and each time point represented by a single column. In our current practice, scores are tracked using a spreadsheet program on a netbook equipped with a number pad. Each plate is assigned a separate file. This allows scores to be immediately digitized and uploaded into a replicative life span database for storage and analysis. In the past, graph paper was used to record replicative life span data, with one plate represented per sheet, and all data manually entered at the end of the experiment. Figure 2 provides a representative score sheet for one strain.
2. Place the first plate in the life span analysis on the microscope stage and locate the first strain (e.g., the single hole punched at location 90, 10 in Subheading 3.3, **step 6**).

	Mother		RLS
	1	n 0 0 0 n 0 0 1 1 1 0 0 0 2 1 2 1 2 2 2 2 1 2 1 2 1 2 1 0 0 0 0 0 0 0 0 0 0 x U	27
	2	n 0 0 0 2 1 0 1 0 1 0 1 2 1 1 2 1 2 1 2 1 1 0 2 2 1 2 1 2 2 1 1 2 2 1 0 0 0 x S	39
	3	n 0 0 0 1 1 0 1 1 1 1 1 2 1 2 1 1 2 1 0 1 0 0 0 0 0 0 0 0 0 0 0 0 0 0 0 0 0 x U	18
	4	n 0 0 0 1 1 0 2 0 1 1 1 0 2 0 1 1 2 1 2 2 1 0 2 2 2 2 2 2 2 1 1 2 2 1 0 0 0 x U	40
	5	n 0 0 0 1 1 1 1 1 1 0 2 1 2 0 2 1 2 1 2 2 1 1 2 2 0 1 0 0 0 0 0 0 0 0 0 0 0 x U	28
	6	n 0 0 0 1 2 0 1 1 1 1 1 1 2 1 1 2 1 1 1 1 0 0 0 0 0 0 0 0 0 0 0 0 0 0 0 0 0 x U	19
	7	n 0 0 0 1 1 1 1 0 0 1 1 2 0 0 0 0 0 0 0 0 0 0 0 0 0 0 0 0 0 0 0 0 0 0 0 0 0 x L	8
	8	n 0 0 0 2 1 0 1 0 0 1 0 1 2 1 1 1 2 1 0 0 0 0 0 0 0 0 0 0 0 0 0 0 0 0 0 0 0 x U	14
	9	n 0 1 0 2 1 1 1 2 1 0 2 1 2 1 2 1 2 1 2 1 1 1 1 1 2 1 1 1 1 1 1 1 0 0 0 0 0 x U	37
Strain	10	n 0 0 0 1 1 0 1 1 1 1 1 1 2 1 1 1 2 1 2 0 0 0 0 0 0 0 0 0 0 0 0 0 0 0 0 0 0 x U	18
1	11	n 0 0 0 2 1 0 2 0 0 1 0 1 2 1 1 0 2 1 2 2 1 1 1 2 1 0 2 1 0 0 0 0 0 0 0 0 0 x U	27
	12	n 0 0 0 1 0 0 2 0 1 1 2 1 2 1 2 1 2 2 2 2 1 1 1 2 2 2 2 1 2 2 1 0 0 0 0 0 0 x C	39
	13	A n 0 0 0 1 1 1 0 0 0 0 0 0 0 0 0 0 0 0 0 0 0 0 0 0 0 0 0 0 0 0 0 0 0 0 0 0 x U	3
	14	A n 0 0 0 1 0 1 0 0 0 0 0 0 0 0 0 0 0 0 0 0 0 0 0 0 0 0 0 0 0 0 0 0 0 0 0 0 x U	2
	15	A n 0 0 0 n 0 1 0 0 0 1 0 0 0 0 0 0 0 0 0 0 0 0 0 0 0 0 0 0 0 0 0 0 0 0 0 0 x U	2
	16	A n 0 0 1 1 1 0 1 0 0 2 1 2 1 2 1 2 1 2 1 1 0 1 0 1 1 0 0 0 0 0 0 0 0 0 0 0 x U	23
	17	A n 0 0 1 1 1 0 1 0 0 2 1 2 0 2 0 2 1 2 1 1 1 1 2 1 2 2 1 0 0 0 0 0 0 0 0 0 x U	28
	18	A n 0 0 0 n 0 0 0 2 1 2 2 2 1 0 1 0 0 0 0 0 0 0 0 0 0 0 0 0 0 0 0 0 0 0 0 0 x U	11
	19	A n 0 0 0 n 0 2 1 L O S T	
	20	A n 0 0 2 1 0 1 2 0 0 2 2 2 0 2 1 1 2 2 1 1 1 0 1 0 0 0 0 0 0 0 0 0 0 0 0 0 x U	24

Fig. 2 Representative score sheet. The replicative life span score sheet consists of rows representing individual cells within the experiment and columns representing time points. Each point in the grid represents the number of divisions that were recorded for the corresponding cell on the corresponding round of quantification

3. Examine each cell individually and perform the following actions:

   (a) If the cell has divided, separate the newly formed daughter cell from the original cell using the needle. Leave the daughter cell and move the original cell a few needle widths to the right to be discarded. On the score sheet, record that a virgin daughter cell was selected by entering "n" in the box for the first time point of the appropriate cell.

   (b) If the original cell has not yet divided, or if the new daughter cannot be removed, first examine the extra set of cells that were set off to one side in Subheading 3.3, **step 11**. If a virgin daughter can be separated from one of these cells, use it to replace the undivided original cell. Move the undivided original to the right to be discarded and record that a virgin daughter cell was selected by entering "n" on the score sheet.

(c) If a virgin daughter cell cannot be obtained from either the original cell or one of the extra cells, move the original cell a few needle widths to the left and record that a virgin cell was not selected by entering "A" on the score sheet. The cells for a given strain are often rearranged so that all "A" cells are grouped consecutively in the array (e.g., positions 8–10). This ensures that all "A" cells are dealt with appropriately during the next round of scoring, and not treated as newly dividing virgin cells.

4. Separate any remaining extra cells for the current strain. Leave five to ten individual cells and move the rest to the side to be discarded. These extra cells will be used later to replace selected cells that fail to divide.
5. Continue until all cells on the current plate have been examined.
6. Clean up excess cells by sweeping the needle across the surface of the plate. Collect and discard all cells except selected virgin daughter cells, undivided original cells, and the few remaining extra cells. All discarded cells on a plate can be simultaneously collected in this fashion and moved to the End Zone (*see* **Note 3**).
7. Repeat the Subheading 3.4, **steps 2** and **6** for all remaining plates in the experiment.

3.5 Quantify Replicative Life Span

Once virgin cells have been selected and arrayed, life span quantification can begin. The virgin cells selected in Subheading 3.4 are the individuals that will be followed for replicative life span, and will be referred to from this point forward as the "mother cells." This process consists of repeated rounds of incubating the life span plates and quantifying the number of divisions completed by each cell during incubation. The incubation time is targeted to allow each cell to divide 1 to 2 times. As the life span progresses and cell division slows, the incubation time is increased. An incubation time of 2 h is typically used for the first ~16 divisions and then increased in 1 h increments to 4 h (*see* **Note 4**).

During these rounds, the plates are kept at 30 °C during the incubation period and at room temperature while being scored. Plates can be stored at 4 °C overnight without affecting replicative life span to prevent cell division and allow researchers to rest. Each experiment should be advanced by a minimum of one round every day, with at least three rounds of dissection per plate optimal, including on weekends. Under no circumstances should an experimental plate to be stored at 4 °C for more than one day without incubation and dissection.

1. Incubate life span plates at 30 °C for a number of hours appropriate for the current stage of the life span and rate of cell division. If this is the first round of life span quantification, incubate for 2 h.

2. Place the first plate in the life span on the microscope stage and locate the first strain (i.e., the single hole punched at location 90, 10 in Subheading 3.3, **step 6**).
3. Examine each cell individually and perform the following actions:
   (a) If this is one of the first few rounds of the life span and a virgin cell has not yet been selected for the current position (i.e., the original cell is still present set off to the left, and an "A" was recorded for the previous time point on the score sheet), perform the actions outlined in Subheading 3.4, **step 3** for this cell.
   (b) If no bud is visible for the cell, record "0" on the score sheet. If the cell has not yet divided, and this is the third "0" recorded, replace the cell with a virgin cell and record "n" on the score sheet to indicate that a new virgin cell was selected at this time point. The virgin cell can be taken either from a neighboring cell within the life span, or from one of the extra cells. Refer to cells 1, 15, 18, and 19 in Fig. 2 for examples.
   (c) Remove all daughter and granddaughter cells from the mother cell by lightly touching the needle to the cells and gently tapping the side of the microscope. The mother cell will almost always be the largest cell present.
   (d) Determine the number of divisions that the mother cell has undergone since the last time point. Daughter and granddaughter cells will begin to divide as soon as they have reached sufficient size, so the number of cells present is not equivalent to the number of divisions. The pattern of cells present can be used to determine the number of mother cell divisions (*see* Fig. 3). Record the appropriate score on the score sheet.
   (e) Occasionally, older mother cells will lose the ability to complete cytokinesis and daughter cells cannot be removed (Fig. 3b). In this case, it is difficult to determine whether a new cell was produced by the mother cell or the attached daughter cell. Unless a definitive determination can be made that the removed cell was produced by the mother cell, remove any loose cells and record "0" on the score sheet.
   (f) Move all non-mother cells off to the right to be discarded.
   (g) If no cell is present at the current position, indicate that the cell has been lost by entering "LOST" on the score sheet for the current position and highlighting the row associated with that cell. Punch a single hole in the plate to indicate that a cell has been lost in that location. This cell will be excluded from analysis (e.g., cell 19 in Fig. 2).

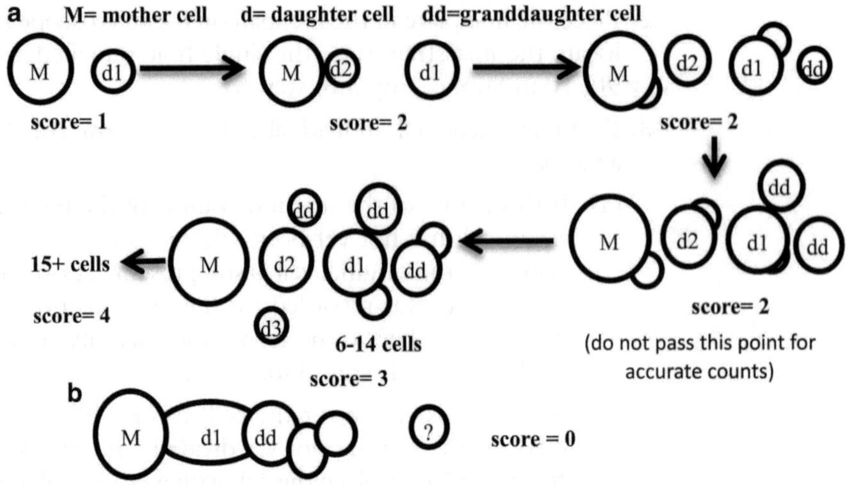

Fig. 3 Scoring number of mother cell divisions. Daughter cells produced by the mother cell will often begin dividing during the incubation time, complicating the quantification of mother cell divisions. (**a**) Reference chart for determining number of mother cell divisions based on cell pattern. (**b**) Mother cells that fail to complete cytokinesis are scored as having undergone 0 divisions for all future time points, because any cells which detach are unlikely to arise from the original mother cell

4. Once every cell has divided at least once for a given strain, all extra cells can be discarded.
5. Once all cells have been dissected and scored on a plate, clean up all excess cells as described in Subheading 3.4, **step 6**.
6. Review the scores given for each strain. For a given strain, if three rounds of quantification have been completed without a single cell division (i.e., all scores are recorded as "0" for all cells for three concurrent rounds), the strain is considered dead. Proceed to Subheading 3.6.
7. Wrap the plate in Parafilm and return to 30 °C.
8. Repeat the Subheading 3.5, **steps 1–7** for all remaining plates in the current experiment.
9. Review the score sheet for the completed time point. If the majority of strains were given a score of "0" or "1," increase the incubation time by 1 h to a maximum of 4 h.
10. Repeat the steps outlined in Subheading 3.5 until all strains have completed 8 concurrent hours of incubation without a single cell division.

3.6 Complete Replicative Life Span Measurement

For a given cell set, each cell is continually monitored until all cells in that set have ceased dividing. Once a set of cells is judged "dead" in Subheading 3.5, **step 6**, perform the steps outlined in this section to finalize the data.

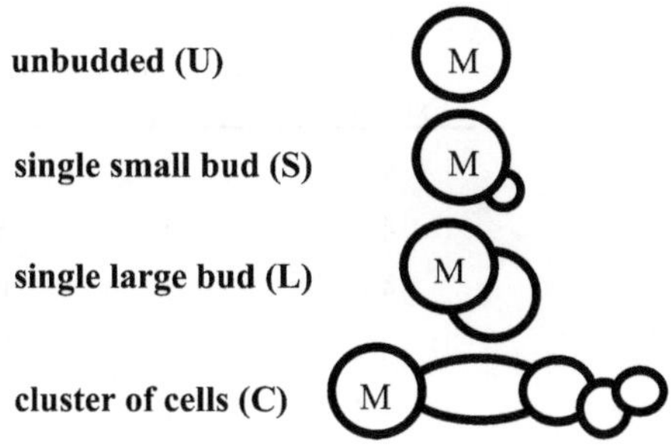

Fig. 4 Terminal cell states. Yeast cells can cease dividing at various points in the cell cycle. Provided are codes used to indicate common terminal cell states

1. Record "x" in the next available time point for all cells in the dead cell set except cells labeled as "LOST."
2. Cross off the patch number on the life span plate to indicate that quantification for that cell set has been completed.
3. Cells can cease division at various points in the cell cycle. Terminal phenotypes of mother cells can be recorded if desired. Our lab uses the following designations in the score sheets: "U" for unbudded, "S" for single small bud, "L" for single large bud, or "C" for cluster of cells (as in the case when a mother cannot perform cytokinesis and forms a string of connected cells). *See* Fig. 4 or a representation of each terminal cell state.
4. Once all cell sets in an experiment have been judged dead, the experiment is complete.

3.7 Manage Multiple Experiments Simultaneously

The previous sections describe the protocol for determining the replicative life span for cells included in one experiment. We have developed a system to simultaneously monitor replicative life span for multiple experiments. Our current setup consists of four dedicated microscopes with a 5th microscope for overflow, and a team of approximately 20 individuals, each contributing to replicative life span dissection part-time. At present, the team typically completes 1,200–1,400 mother cell life spans per week.

Analysis for a single experiment typically requires 3–4 weeks. The workload decreases over time as cell division slows and dissection is ceased on strains that undergo senescence. By staggering start dates, we are able to pair new experiments that require a large amount of dissection time with older experiments that require less time. On a typical week, two 600–800 cell experiments are started, with timing planned such that selection of virgin cells (Subheading 3.4) occurs on Monday for the first experiment and Tuesday for the second, allowing the maximum number of rounds of quantification to be completed prior to the first weekend.

Experiment	Incubation Time	Incubation Start Time	Dissection Start Time	Rounds Done Today	Note
485	4	9:00	1:00	I	
486	4	9:30	1:30	I	
487	3	10:00	1:00	I	
488	2	10:30	12:30	I	
489	2	12:00	2:00	II	
490	1.5	10:30	12:00		Pick daughters

Fig. 5 Example chart for managing multiple experiments simultaneously. When multiple experiments at different stages of analysis are progressing in parallel, a common area to record the current status of each is useful. For each experiment, we track the current incubation time each round, the start of the current incubation step, the time at which the next round of division scoring should begin, the number of rounds completed on the current day so far, and any notes relevant to the current or next scoring round. In the Kaeberlein lab, this information is recorded on a white board and updated as experiments are progressed through rounds of replicative life span quantification

The following section describes our process for managing multiple experiments at different stages of analysis simultaneously.

1. To keep track of the current status of each experiment, prepare a chart or table with columns for experiment name, current length of incubation time between rounds, incubation start time, targeted incubation end time, rounds of quantification completed on the current day, and current notes (*see* Fig. 5). We use a whiteboard in the microscope room, so that all team members can quickly refer to the current work status and so that the current status can be quickly updated as needed.
2. Designate three sections of the 30 °C incubator as "In progress," "Done," and "Incubating."
3. At the beginning of a day of replicative life span analysis, move life span plates for all experiments to the "Incubating" section of the 30 °C incubator. On setup days (Monday and Tuesday) some experiments may be left at 4 °C until setup is done, as time and personnel permit.
4. Update the status board as follows:
   (a) Change the incubation start time to the current time.
   (b) Change the target time to begin the next round of quantification based on both the current length of incubation required for each experiment, and the amount of incubation time that each experiment received at the end of the previous day.

(c) Reset the number of rounds completed today to 0.

(d) Add any experiments that are starting today to the board.

5. When the dissection start time for the first experiment arrives, move all plates for that experiment to the "In progress" section of the 30 °C incubator.
6. Take the first plate in the experiment back to the microscope and quantify the number of daughter cells produced by each mother cell on that plate (Subheading 3.5). We use a spreadsheet program in a shared network folder and a netbook computer at each microscope so that all team members have access to all score sheets at all times. For paper score sheets, some other method of sharing will have to be employed.
7. Return completed plate to the "Done" section of the 30 °C incubator. If this was the first plate in the experiment, change the incubation start time to reflect the current time, and the target dissection time to reflect the beginning of the next dissection time point. Add a tally mark to the number of rounds completed today.
8. Repeat Subheading 3.7, **steps 5–7** for all remaining plates in the current experiment.
9. Once scoring is completed for an experiment, move all life span plates for that experiment to the "Incubating" section of the 30 °C incubator.
10. Continue scoring experiments and updating the board throughout the day.
11. At the end of the dissection time for the day, perform the following steps:

(a) Move life span plates for all experiments to 4 °C.

(b) Update the "Dissection Start Time" section of the status board to reflect the time that the plates were moved to 4 °C. This information will be used the following day to determine how much remaining incubation time is needed for each experiment.

(c) Fill in any notes that will be relevant to the person starting the experiments in the morning.

3.8 Analyze Data

This section describes using data collected in the replicative life span assay described above to generate basic life span statistics and survival curves for a given cell set.

1. Sum the divisions across each row of the replicative life span score sheet to obtain the total number of divisions undergone by each individual cell in the experiment (Fig. 2). For each cell set, this will provide a list of replicative ages at death (i.e., senescence) for each individual in that set.

2. The list of individual replicative life spans can be used directly to generate basic life span statistics, such as mean, median, or standard deviation. The Wilcoxon Rank Sum test can be used with the lists of individual replicative life spans from two cell sets to calculate the probability that the two cell sets came from the same population. The mean replicative life span for wild type yeast in the BY4742 background is about 26 divisions.
3. To generate survival curves, complete the following steps:
   (a) Tally the number of individuals in each set that lived to each replicative age.
   (b) Invert the tally to determine the proportion of individuals within the population alive at each progressively increasing number of divisions.
   (c) Plot the proportion of individuals alive at each replicative age (*y*-axis) against replicative age (*x*-axis).

4 Notes

1. Plan to patch MAT*a* and MATα strains on different plates to avoid exposure to of strains to opposite mating factors. Up to 10 can be patched to a single plate.
2. Plates should be left unwrapped, face up on the bench with lids on. Cells are difficult to manipulate with the fiber optic needle if the plates are too wet. This drying period may include the time during which cells are patched and grown on the life span plate prior to life span analysis. If the plates will not be used immediately following the 72 h drying period, wrap in Parafilm and store inverted to minimize desiccation.
3. When moving cells on the life span plates, be sure not do drag them long distances, as this will often result in a cell falling off the needle and creating a colony, which can be problematic later in the life span. Lift the cells fully off the plate when moving long distances.
4. Note that the incubation time is approximate, and variations of 30–60 min from the desired incubation time are often acceptable, as long as most mother cells have not produced more than two daughter cells since the previous time point.

Acknowledgements

This work was supported by NIH grant R01AG039390 to MK. GLS and JRD were supported by NIH Training Grant T32AG000057. MK is an Ellison Medical Foundation New Scholar in Aging.

References

1. Mortimer RK, Johnston JR (1959) Life span of individual yeast cells. Nature 183: 1751–1752
2. Kaeberlein M (2010) Lessons on longevity from budding yeast. Nature 464:513–519
3. Steinkraus KA, Kaeberlein M, Kennedy BK (2008) Replicative aging in yeast: the means to the end. Annu Rev Cell Dev Biol 24:29–54
4. Fabrizio P, Longo VD (2003) The chronological life span of *Saccharomyces cerevisiae*. Aging Cell 2:73–81
5. Fabrizio P, Longo VD (2007) The chronological life span of *Saccharomyces cerevisiae*. Meth Mol Biol 371:89–95
6. Steffen KK, Kennedy BK, Kaeberlein M (2009) Measuring replicative life span in the budding yeast. J Vis Exp (28). doi: 1209 [pii] 10.3791/1209
7. D'Mello NP, Childress AM, Franklin DS et al (1994) Cloning and characterization of *LAG1*, a longevity-assurance gene in yeast. J Biol Chem 269:15451–15459
8. Laun P, Pichova A, Madeo F et al (2001) Aged mother cells of *Saccharomyces cerevisiae* show markers of oxidative stress and apoptosis. Mol Microbiol 39:1166–1173
9. Lesur I, Campbell JL (2004) The transcriptome of prematurely aging yeast cells is similar to that of telomerase-deficient cells. Mol Biol Cell 15:1297–1312
10. Lindstrom DL, Gottschling DE (2009) The mother enrichment program: a genetic system for facile replicative life span analysis in *Saccharomyces cerevisiae*. Genetics 183:413–422
11. Kaeberlein M, Kennedy BK (2005) Large-scale identification in yeast of conserved aging genes. Mech Ageing Dev 126:17–21
12. Kaeberlein M, Powers RW 3rd, Steffen KK et al (2005) Regulation of yeast replicative life span by TOR and Sch9 in response to nutrients. Science 310:1193–1196
13. Smith ED, Tsuchiya M, Fox LA et al (2008) Quantitative evidence for conserved longevity pathways between divergent eukaryotic species. Genome Res 18:564–570
14. Kaeberlein M, Kirkland KT, Fields S et al (2004) Sir2-independent life span extension by calorie restriction in yeast. PLoS Biol 2:E296
15. Kaeberlein M, Kirkland KT, Fields S et al (2005) Genes determining replicative life span in a long-lived genetic background. Mech Ageing Dev 126:491–504
16. Steffen KK, MacKay VL, Kerr EO et al (2008) Yeast life span extension by depletion of 60s ribosomal subunits is mediated by Gcn4. Cell 133(2):292–302
17. Managbanag JR, Witten TM, Bonchev D et al (2008) Shortest-path network analysis is a useful approach toward identifying genetic determinants of longevity. PloS one 3:e3802

Chapter 21

Metabolomic and Lipidomic Analyses of Chronologically Aging Yeast

Vincent R. Richard, Simon D. Bourque, and Vladimir I. Titorenko

Abstract

Metabolomic and lipidomic analyses of yeast cells provide comprehensive empirical datasets for unveiling mechanisms underlying complex biological processes. In this chapter, we describe detailed protocols for using such analyses to study the age-related dynamics of changes in intracellular and extracellular levels of various metabolites and membrane lipids in chronologically aging yeast. The protocols for the following high-throughput analyses are described: (1) microanalytic biochemical assays for monitoring intracellular concentrations of trehalose and glycogen; (2) gas chromatographic quantitative assessment of extracellular concentrations of ethanol and acetic acid; and (3) mass spectrometric identification and quantitation of the entire complement of cellular lipids. These protocols are applicable to the exploration of the metabolic patterns associated not only with aging but also with many other vital processes in yeast. The described here methodology complements the powerful genetic approaches available for mechanistic studies of fundamental aspects of yeast biology.

Key words Yeast, Trehalose, Glycogen, Ethanol, Acetic acid, Lipids, Microanalytic biochemical assays, Gas chromatography, Mass spectrometry, Lipidomics

1 Introduction

The replicative and chronological age of a eukaryotic cell is defined by the spatiotemporal dynamics of a plethora of cellular processes [1–5]. To infer the relative contribution of each of these processes to cellular aging and to understand how certain genetic, dietary and pharmacological interventions delay cellular aging by modulating the chronology of critical cellular processes, the methodologies of metabolomic and lipidomic analyses can be advantageous [5–9]. The application of metabolomics and lipidomics to the acquisition and quantitative analysis of vast empirical data on the metabolic history of a cell progressing through consecutive stages of aging process enables the systems level study of cellular aging [5–9]. It is conceivable that with the wealth of data that metabolomics and lipidomics can provide us, some of the most fundamental

Jeffrey S. Smith and Daniel J. Burke (eds.), *Yeast Genetics: Methods and Protocols*, Methods in Molecular Biology, vol. 1205, DOI 10.1007/978-1-4939-1363-3_21, © Springer Science+Business Media New York 2014

questions in cellular aging—such as whether it is a programmed process [10–14] or merely a result of the lifelong accumulation of unrepaired cellular and molecular damage [15–17]—can be answered.

Because longevity signaling pathways and mechanisms of their modulation by genetic, dietary and pharmacological interventions are evolutionarily conserved [1–5], the budding yeast *Saccharomyces cerevisiae* is a valuable unicellular model organism for studying mechanisms underlying cellular aging in multicellular eukaryotes [1–5, 18]. Due to the relatively short and easily monitored replicative and chronological life spans of this genetically and biochemically manipulable unicellular eukaryote with annotated genome, it has been successfully used to (1) identify numerous novel longevity genes, many of which have been later implicated in regulating longevity of multicellular eukaryotic organisms; (2) establish the chemical nature of molecular damage that causes cellular and organismal aging and accelerates the onset of age-related diseases; and (3) identify a number of longevity-extending small molecules, many of which have been later shown to slow down aging, improve health, attenuate age-related pathologies and delay the onset of age-related diseases in multicellular eukaryotes [3, 5, 8, 9, 18–28].

Recent metabolomic and lipidomic analyses of the metabolic history of chronologically aging yeast cells strongly suggest that their longevity is defined by a pattern of metabolism and organelle dynamics established prior to cell entry into a non-proliferative state in a genotype-, diet-, and pharmacological intervention-dependent manner [8, 9, 20–23, 25–28]. This chapter describes detailed protocols for such high-throughput analyses that have been used to measure the levels of trehalose, glycogen, ethanol, and acetic acid as well as to quantitatively assess the entire complement of cellular lipids. These metabolites have been shown to play a pivotal role in defining the chronological life span of a yeast cell [8, 9, 21–23, 25–28]. It should be emphasized that the described here metabolic and lipidomic analyses provide powerful empirical tools for studying mechanisms that underly not only aging but also many other paradigms of yeast biology. As such, these high-throughput analyses are complementary to the great power of yeast genetics in exploring fundamental biological phenomena.

1.1 Trehalose Concentration Measurement

Trehalose is a nonreducing disaccharide that has been for a long time considered only as a reserve carbohydrate [29]. However, recent findings suggested a role for trehalose metabolism in regulating a variety of cellular processes, including redox homeostasis and protein folding [9, 30, 31]. Our studies on the effect of caloric restriction on the metabolic history of chronologically aging yeast provided evidence for the essential role of trehalose in defining yeast life span through its effects on protein homeostasis (proteostasis) [8, 23]. Therefore, the existence of a modulated by trehalose regulatory network of cellular proteostasis has been suggested; by maintaining proper synthesis, posttranslational modifications, folding, trafficking, degradation, and turnover of proteins within a cell, this network defines its chronological and replicative age [23]. In this chapter, we describe a robust microanalytic biochemical assay for monitoring trehalose concentration in chronologically aging yeast cells.

1.2 Glycogen Concentration Measurement

Glycogen, which is known to be the major storage form of yeast glucose [29], has recently been implicated in defining yeast longevity. Our metabolomic and proteomic analyses of chronologically aging yeast provided evidence that a proper balance between the biosynthesis and degradation of glycogen is obligatory for life span extension by caloric restriction [8]. In this chapter, we describe a robust microanalytic biochemical assay for monitoring glycogen concentration in chronologically aging yeast cells.

1.3 Ethanol and Acetic Acid Concentration Measurement

A reduction of initial glucose concentration in nutrient-rich growth medium from 2 to 0.5 % almost doubles the chronological life span of yeast [8]. Ethanol, a product of glucose fermentation by yeast cells, operates as a redox sink that regulates energy flux from deposited in lipid bodies neutral lipids to mitochondria [25–28]. By suppressing peroxisomal enzymes involved in fatty acid oxidation, high ethanol concentrations reduce the energy flux to mitochondria [8, 25–28]. The resulting decline of the electrochemical potential across the inner mitochondrial membrane promotes mitochondrial fragmentation, which in turn stimulates the release of pro-apoptotic proteins from the mitochondrial intermembrane space and ultimately causes apoptotic cell death [25–28]. Low ethanol concentrations promote peroxisomal fatty acid oxidation, thereby elevating the energy flux to mitochondria. The resulting increase of the electrochemical potential across the inner mitochondrial membrane promotes mitochondrial fusion and maintains mitochondrially produced reactive oxygen species at a level that is insufficient to damage cellular macromolecules but can activate several longevity-extending stress response pathways [8, 25–28].

Acetate is another product of glucose fermentation by yeast cells. Akin to ethanol amassed by yeast cultured in nutrient-rich growth medium, acetic acid accumulated by yeast cells grown in minimal medium accelerates chronological aging [3, 5, 21, 22].

Both ethanol and acetic acid pass freely across the cell membrane and are thus easily assayed in the growth media. Gas chromatography (GC) is an ideal way to measure these two volatile species due to its speed, high sensitivity, large linear dynamic range, and ease of sample preparation. Furthermore, ethanol and acetic acid can be monitored using a flame ionization detector (FID), with which virtually every GC system is equipped. Like other chromatographic techniques, GC relies on the differential solubility of molecules between the mobile (gas) phase and the stationary (liquid) phase on the inside surface of the column. Unlike other chromatographic techniques, however, the elution of molecules in GC is controlled by changing the temperature of the GC oven, not by changing the mobile phase composition. Thus, separation by GC is controlled by the vapor pressure of the separated molecules and the linear flow rate of the carrier gas. The method described in this chapter, quantitative assessment of extracellular concentrations of ethanol and acetic acid with the help of GC, takes advantage of this fact to separate them from non-volatile contaminants, such as amino acids and sugars.

1.4 Mass Spectrometric Quantitative Assessment of the Yeast Lipidome

Lipids are one of the major classes of biomolecules and play important roles in membrane dynamics, energy storage, and signaling [32]. A body of evidence supports the view that lipid metabolism plays an important role in longevity regulation across phyla [25–28]. The budding yeast *Saccharomyces cerevisiae* is a valuable model organism for studying molecular mechanisms linking cellular aging and metabolism of various lipid species in multicellular eukaryotes [25–28, 33]. In studying these mechanisms, it is crucial to have an analytical method for the robust and accurate quantitative assessment of the entire complement of yeast lipids (lipidome) [34, 35]. In this chapter we describe such a method, namely quantitative shotgun mass spectrometry (MS) using a high resolution Thermo Orbitrap Velos instrument. The method employs a modified version of the Bligh and Dyer lipid extraction [36] followed by Fourier transform tandem mass spectrometry (FT-MS/MS) to separate numerous lipid species. The raw data are then imported into the open source software LipidXplorer [37], which interprets extensive datasets of shotgun mass spectra to enable the quantitative characterization of all lipid species comprising the yeast lipidome.

2 Materials

2.1 Trehalose Concentration Measurement

2.1.1 Buffers, Solutions, and Growth Mediums

Prepare the following solutions in distilled water unless otherwise stated.

1. YP medium: 1 % (w/v) yeast extract and 2 % (w/v) bactopeptone. The media should be autoclaved at 15 psi/121 °C for 45 min prior to use. We typically use 0.2 %, 0.5 %, 1 % or 2 % glucose (w/v) as the carbon source.
2. PBS: 20 mM KH_2PO_4/KOH (pH 7.5) and 150 mM NaCl.
3. SHE solution: 50 mM NaOH and 1 mM EDTA.
4. THA solution: 100 mM Tris–HCl (pH 8.1) and 50 mM HCl.
5. Trehalose reagent: 25 mM KH_2PO_4/KOH (pH 7.5) and 0.02 % BSA; with or without 15 mU trehalase (Sigma-Aldrich, St. Louis MO).

6. Glucose reagent: 100 mM Tris/HCl (pH 8.1), 2 mM $MgCl_2$, 1 mM DTT, 1 mM ATP, 0.2 mM $NADP^+$, and mixture of hexokinase (7 U) and glucose-6-phosphate dehydrogenase (8 U) (Sigma-Aldrich, St. Louis MO).

2.1.2 Yeast Strains and Growth Conditions

Wild-type (WT) strain BY4742 (*MATα his3Δ1 leu2Δ0 lys2Δ0 ura3Δ0*) and single-gene-deletion mutant strains in the BY4742 genetic background (all from Open Biosystems) can be grown in YP medium containing 0.2 %, 0.5 %, 1 %, or 2 % glucose (w/v) as the carbon source. Cells should be cultured at 30 °C with rotational shaking at 200 rpm in Erlenmeyer flasks at a flask volume–medium volume ratio of 5:1.

2.2 Glycogen Concentration Measurement

2.2.1 Buffers, Solutions, and Growth Mediums

Prepare the following solutions in distilled water unless otherwise stated.

1. YP medium: 1 % (w/v) yeast extract and 2 % (w/v) bactopeptone. The media should be autoclaved at 15 psi/121 °C for 45 min prior to use. We typically use 0.2 %, 0.5 %, 1 % or 2 % glucose (w/v) as the carbon source.
2. PBS: 20 mM KH_2PO_4/KOH (pH 7.5) and 150 mM NaCl.
3. SHE solution: 50 mM NaOH and 1 mM EDTA.
4. THA solution: 100 mM Tris–HCl (pH 8.1) and 50 mM HCl.
5. Glycogen reagent: 50 mM sodium acetate (pH 4.6) and 0.02 % BSA; with and without 5 μg/mL amyloglucosidase 14 U/mg (Roche, Basel, Switzerland).
6. Glucose reagent: 100 mM Tris/HCl (pH 8.1), 2 mM $MgCl_2$, 1 mM DTT, 1 mM ATP, 0.2 mM $NADP^+$, and mixture of hexokinase (7 U) and glucose 6 phosphate dehydrogenase (8 U) (Sigma-Aldrich, St. Louis MO).

2.2.2 Yeast Strains and Growth Conditions

Wild-type (WT) strain BY4742 (*MATα his3Δ1 leu2Δ0 lys2Δ0 ura3Δ0*) and single-gene-deletion mutant strains in the BY4742 genetic background (all from Open Biosystems) can be grown in YP medium containing 0.2 %, 0.5 %, 1 % or 2 % glucose (w/v) as the carbon source. Cells should be cultured at 30 °C with rotational shaking at 200 rpm in Erlenmeyer flasks at a flask volume–medium volume ratio of 5:1.

2.3 Ethanol and Acetic Acid Concentration Measurement

1. Sample vials with cap/septum are available from many suppliers.
2. GC system equipped with an FID detector (*see* **Note 1**).
3. Column: The Equity-1 column from Supelco (0.32 mm I.D. × 30 m). This column has a simple stationary phase chemistry and provides good separation for ethanol and acetic acid. It is robust and widely available (*see* **Note 2**).

2.4 Mass Spectrometric Quantitative Assessment of the Yeast Lipidome

2.4.1 Yeast Strains and Growth Conditions

1. YP medium: 1 % (w/v) yeast extract and 2 % (w/v) bactopeptone. Media should be autoclaved at 15 psi/121 °C for 45 min prior to use. We typically use 0.2 %, 0.5 %, 1 % or 2 % glucose (w/v) as the carbon source.
2. Wild-type strain BY4742 (*MATα his3Δ1 leu2Δ0 lys2Δ0 ura3Δ0*) and single-gene-deletion mutant strains in the BY4742 genetic background (all from Open Biosystems) can be grown in YP medium containing 0.2 %, 0.5 %, 1 % or 2 % glucose (w/v) as the carbon source. Cells should be cultured at 30 °C with rotational shaking at 200 rpm in Erlenmeyer flasks at a flask volume/medium volume ratio of 5:1.

2.4.2 Lipid Extraction and Mass Spectrometry

1. CHROMASOLV HPLC (>99.9 %) chloroform and methanol or equivalent (Sigma-Aldrich, St. Louis MO).
2. 28 % ammonium hydroxide (Sigma-Aldrich, St. Louis MO).
3. Glass beads, acid-washed, 425–600 μM (Sigma-Aldrich, St. Louis MO).
4. Vortex with appropriate adapter.
5. 15 mL high-speed glass centrifuge tubes with Teflon lined caps (Fisher Scientific).
6. Chloroform–Methanol mixtures: 17:1, 2:1, and 1:2 mixtures of chloroform and methanol, as well as a chloroform–methanol (2:1) mixture with 0.1 % ammonium hydroxide (v/v).
7. ABC: 155 mM ammonium bicarbonate (pH 8.0).
8. Internal Standards: Triacylglycerol (13:0/13:0/13:0) (Larodan, Malmo, Sweden) as well as various species of phospholipids—including phosphatidylcholine (13:0/13:0), phosphatidylethanolamine (14:0/14:0), phosphatidylserine (14:0/14:0), phosphatidic acid (14:0/14:0), and cardiolipin (14:0/14:0/14:0/14:0) (all from Avanti Polar Lipid, Alabaster, AL, USA).
9. Glass 2 mL sample vials with Teflon lined caps.

2.4.3 Software

1. LipidXplorer: (https://wiki.mpi-cbg.de/wiki/lipidx/index.php/Main_Page).
2. MSConvert: http://proteowizard.sourceforge.net/.

3 Methods

3.1 Trehalose Concentration Measurement

3.1.1 Preparation of Alkali Cellular Extract

1. Measure the cell titer by taking 10 μL aliquots of cell culture, diluting appropriately, and counting cells with a hemacytometer.
2. Harvest 2×10^9 cells by centrifugation for 1 min at $16{,}000 \times g$ at 4 °C.
3. Wash the cell pellet three times in ice-cold PBS and then resuspend in 200 μL of ice-cold SHE solution. Add an additional

800 μL of ice-cold SHE solution to the cell suspension (~1 mL total volume).

4. Incubate the resulting alkali extract at 60 °C for 30 min to destroy endogenous enzyme activities and pyridine nucleotides.
5. Neutralize the extract by adding 500 μL of THA solution.
6. Divide the extract into 150 μL aliquots and quickly freeze them in liquid nitrogen. Store at −80 °C prior to use.

3.1.2 A Microanalytic Biochemical Assay for Measuring Trehalose Concentration

1. Add 50 μL of alkali extract to 150 μL of trehalose reagent with and without trehalase.
2. Incubate the mixture for 60 min at 37 °C. Add 800 μL of glucose reagent and incubate the mixture for 30 min at 25 °C.
3. Measure the NADPH generated from $NADP^+$ fluorimetrically (excitation at 365 nm, emission monitored at 460 nm).

3.2 Glycogen Concentration Measurement

3.2.1 Preparation of Alkali Cellular Extract

1. Measure the cell titer by taking 10 μL aliquots of cell culture, diluting appropriately, and counting cells with a hemacytometer.
2. Harvest 2×10^9 cells by centrifugation for 1 min at $16{,}000 \times g$ at 4 °C.
3. Wash the cell pellet three times in ice-cold PBS and then resuspend in 200 μL of ice-cold SHE solution. Add an additional 800 μL of ice-cold SHE solution to the cell suspension (~1 mL total volume).
4. Incubate the resulting alkali extract at 60 °C for 30 min to destroy endogenous enzyme activities and pyridine nucleotides.
5. Neutralize the extract by adding 500 μL of THA solution.
6. Divide the extract into 150 μL aliquots and quickly freeze them in liquid nitrogen. Store at −80 °C prior to use.

3.2.2 A Microanalytic Biochemical Assay for Measuring Glycogen Concentration

1. Add 50 μL of alkali extract to 500 μL of glycogen reagent and incubate at 25 °C for 30 min.
2. Add 500 μL of glucose reagent and incubate at 25 °C for 30 min.
3. Measure the NADPH generated from $NADP^+$ fluorimetrically (excitation at 365 nm, emission monitored at 460 nm).

3.3 Ethanol and Acetic Acid Concentration Measurement

3.3.1 Yeast Strains and Growth Conditions

The wild-type strain BY4742 (*MATα his3Δ1 leu2Δ0 lys2Δ0 ura3Δ0*) was grown in YP medium (1 % yeast extract, 2 % bactopeptone) containing 0.2 %, 0.5 %, 1 % or 2 % glucose as carbon source. Cells were cultured at 30 °C with rotational shaking at 200 rpm in Erlenmeyer flasks at a flask volume/medium volume ratio of 5:1.

3.3.2 Sample Preparation for Gas Chromatography

1. Collect aliquots of yeast cultures and harvest cells by centrifugation at 16,000 × *g* for 2 min at room temperature.
2. The supernatant collected can either be used immediately or stored at −80 °C prior to use.

3.3.3 Gas Chromatographic Measurement of Ethanol and Acetic Acid

1. Place 100 μL of supernatant in auto-sampler vials; to avoid growth in the vials, samples must be run at once for approximately 4 h. If you have a cooled auto sampler, you can load more samples.
2. Inject 1 μL of sample in a splitless injector (*see* **Note 3**), with constant back pressure, nitrogen as carrier gas and a temperature curve of:
   (a) 45 °C for 30 s.
   (b) 45–72 °C for 45 s.
   (c) 72–80 °C for 2 min.
   (d) Increase to maximal temperature at highest rate and hold for 3 min to elute higher boiling point contaminants.

Using this temperature profile, ethanol and acetic acid should elute after about 3 min, and the total run time will be about 20 min, including the cool-down time for the oven (*see* **Note 4**).

3.4 Mass Spectrometric Quantitative Assessment of the Yeast Lipidome

3.4.1 Yeast Culture Conditions and Glass Bead Lysate

1. Culture yeast in YPD medium (1 % yeast extract, 2 % bacto-peptone, 2 % glucose) at 30 °C with rotational shaking at 200 rpm in Erlenmeyer flasks at a flask volume/medium volume ratio of 5:1. Cultures should be grown to whichever growth phase is desired for the study.
2. Harvest 5×10^7 cells (~5 mL of the culture for cells recovered in logarithmic growth phase) by centrifugation at 3,000 × *g* for 5 min at room temperature.
3. Wash the cell pellet with ice-cold ABC, transfer to a 2 mL microfuge tube; centrifuge at 16,000 × *g* for 1 min at 4 °C.
4. Resuspend the pellet in 1 mL of ice-cold ABC.
5. Add 200 μL of glass beads and vortex for 5 min, then place the sample on ice for 1 min. Repeat this step three times.
6. Assess protein content of cell lysate using standard methods.
7. The cell lysate should be stored at −80 °C until lipid extraction.

3.4.2 Lipid Extraction

1. Calculate the amount of the lysate needed for approximately 50 μg of protein equivalent per replicate. Transfer to 15 mL glass centrifuge tubes.
2. Add 1 mL of nanopure water.
3. Add 20 μL of internal standard mix prepared in chloroform according to Table 1.

Table 1
Lipid standards

	Chain composition	MW	M/Z	Concentration (μg/μL)	Ion
PA	14:0/14:0	592.41	591.4	0.1	M–H
PE	14:0/14:0	635.45	634.44	0.1	M–H
PG	14:0/14:0	666.45	665.44	0.02	M–H
PS	14:0/14:0	679.43	678.43	0.1	M–H
CL	14:0/14:0/14:0/14:0	1241.83	619.91	0.1	M–H
CM	17:0/18:1	551.53	550.52	0.05	M–H
FFA	19:00	298.28	297.27	0.02	M–H
DAG	14:0/14:0	512.44	511.44	0.05	M–H
TAG	13:0/13:0/13:0	680.6	698.63	0.1	$M+NH_4$
PC	13:0/13:0	649.47	650.48	0.1	M+H

4. Add 3 mL of chloroform–methanol (17:1) mixture and vortex at 4 °C for 2 h (*see* **Notes 5–7**).
5. Separate phases by centrifugation at 3,000×*g* for 5 min at room temperature. Transfer the lower organic phase to a new 15 mL glass tube.
6. Add 1.5 mL of chloroform–methanol (2:1) mixture to the remaining aqueous phase and vortex at 4 °C for 2 h.
7. Separate phases by centrifugation at 3,000×*g* for 5 min at room temperature and combine with the organic phase from the chloroform–methanol (17:1) extract.
8. Evaporate off the solvent under nitrogen flow or in a vacuum evaporator.
9. Dissolve the lipid film in 100 μL of methanol–chloroform (2:1) mixture and transfer to a 2 mL glass sample vial with either Teflon or aluminum lined caps.
10. Samples should be stored at −20 °C until ready to be analyzed by MS (*see* **Note 8**).

3.4.3 Mass Spectrometric Analysis of Yeast Lipids

1. Prior to running samples for the first time, it is recommended to run a standard curve for your lipid standards (*see* **Notes 9** and **10**). This can be done by preparing dilutions spanning ~0.005–50 μM and analyzing them in both negative and positive ion modes.

Table 2
Instrument settings

Tune information		
Instrument polarity	**Positive**	**Negative**
Source voltage (kV)	3.9	4
Capillary temp (°C)	275	275
Sheath gas flow	5	5
Aux gas flow	1	1
FT-MS injection time (ms)	100	500
FTMS microscans	3	1

2. Dilute the lipid extract appropriately in chloroform–methanol (1:2) mixture with 0.1 % NH_4OH, so that your signal falls within the range of your calibration curve.
3. Resolve lipids by direct injection using a Thermo OrbitrapVelos mass spectrometer equipped with a HESI-II ion source (Thermo Scientific, Waltham, MA, USA) at a flow rate of 5 μL/min (*see* **Note 11**). Use an instrument method for data-dependent acquisition. The first segment is a full survey scan using the FT-MS; it is followed by 9 dependent MS/MS scans to sequentially perform MS/MS on all peaks above the set threshold from high to low intensity. Those peaks are then added to an exclusion list and the cycle will continue for 5 min (*see* **Note 12**). The tune settings and instrument methods that we routinely use are summarized in Tables 2 and 3.

3.4.4 Identification and Quantitation of Lipids Following Their Separation by Mass Spectrometry

1. Convert mass spectra to an open format (mzXML, mzML) using the ProteoWizard MSConvert software, which is freely available from http://proteowizard.sourceforge.net/.
2. Spectra can then be imported into LipidXplorer for the automated detection of lipid species. (https://wiki.mpi-cbg.de/wiki/lipidx/index.php/Main_Page; *see* Fig. 1).
3. Data are normalized by taking the ratio of signal intensity of precursor ions to that of their respective lipid class-specific internal standard (spiked standard), multiplied by the concentration of that standard to give a molar quantity. The ratio of an individual molar quantity by the sum of that contributed by all other detected lipid species will give the molar ratio, which should be expressed as a percentage.

Table 3
Instrument method for data-dependent acquisition

Acquisition time (min)	5 (+0.25 delay)	
Instrument polarity	**Positive**	**Negative**
MS (segment 1)		
Analyzer	FTMS	FTMS
Mass range	Normal	Normal
Resolution	100,000	100,000
Data type	Centroid	Centroid
Scan range	400–1,200	200–1,400
Data dependent MSMS (segments 2–10)		
Analyzer	FTMS	FTMS
Resolution	30,000	30,000
Data type	Centroid	Centroid
Activation	HCD	HCD
Activation time (ms)	0.1	0.1
Isolation width	1	1
Collision energy	35	65
Mass range	Normal	Normal
Data type	Centroid	Centroid
Scan range	–	–

4 Notes

1. Because the FID detects sample by burning it and reading the current across the flame, it is sensitive to how oxidized your analyte is—which is why it is more sensitive to ethanol than acetic acid.
2. The method and instrumentation are very robust and no further sample preparation is required beyond spinning down the cells. Unfortunately, this means the samples are quite dirty and your local GC owner may be hesitant to let you use their equipment. Offering to supply your own column, injection syringes and vaporization chambers may help.
3. Be sure to rinse the injection syringe several times after each run and, at the end of the day, remove it and clean it very well—otherwise sugars and other contaminants will render it unusable. It is good practice to run a standard curve at the beginning of each day. If many samples are analyzed, run a single point standard now and then to ensure the machine is operating properly; 0.1 % (v/v) for both ethanol and acetic acid work well. If you start seeing drift in retention times or differences in sensitivity, you should replace the vaporization

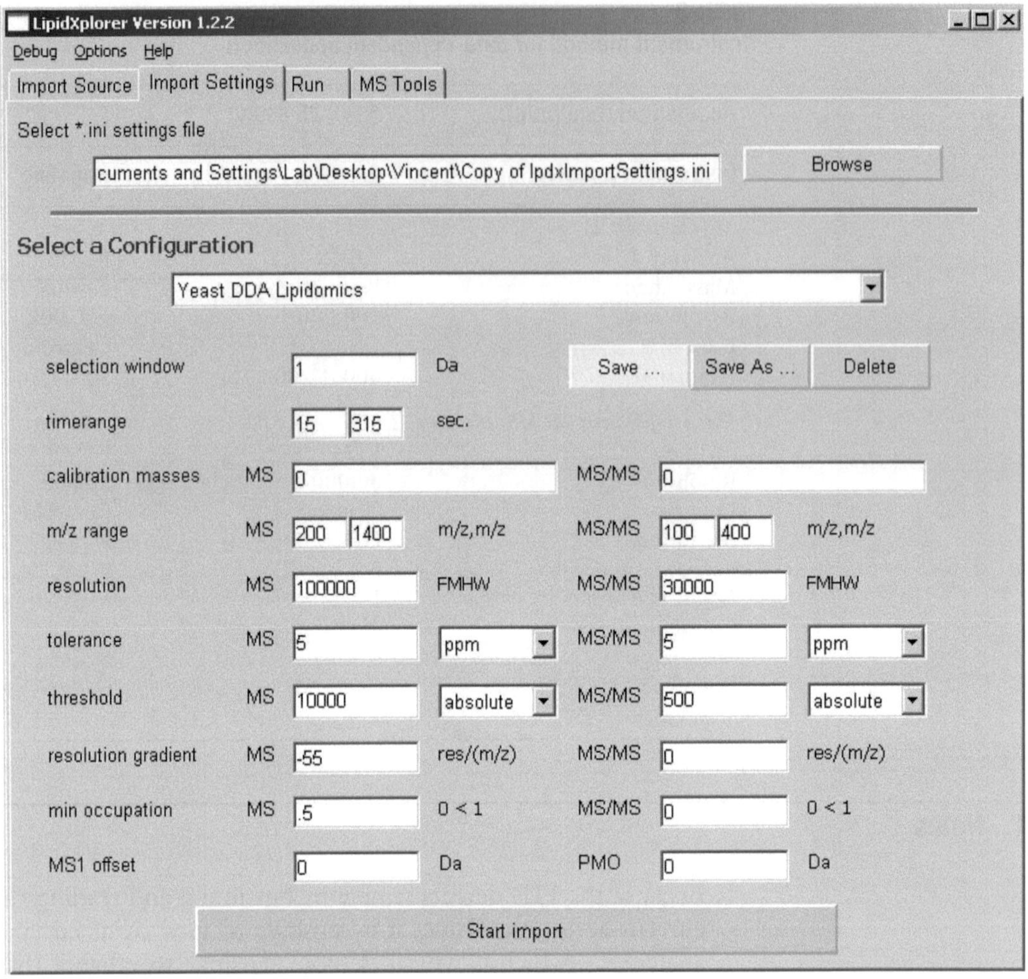

Fig. 1 Import setting for LipidXplorer. Following conversion to an open format (mzXML, mzML) using the ProteoWizard MSConvert software (http://proteowizard.sourceforge.net/), mass spectra can be imported into LipidXplorer (https://wiki.mpi-cbg.de/wiki/lipidx/index.php/Main_Page) for the automated detection of lipid species

chamber or the glass fiber inside it. At worst you will have to remove a few centimeters at the beginning of the column, since you have 30 m to work with; it will not seriously affect its separation abilities. Vaporization chambers can also be rejuvenated with an intensive acid wash. Over the course of several hundred injections, we replaced or washed the vaporization chamber perhaps twice and never had to cut the column.

4. In nutrient-rich cultural media initially containing 2 % glucose, we can see ethanol concentrations up to ~6 %, and acetic acid ~0.3 % in the cultures collected at logarithmic growth phase. Detection limits and linear dynamic range will vary from

instrument to instrument, but 0.001–10 % + for ethanol and 0.01–10 % + for acetate are reasonable.

5. Handle chloroform and methanol with caution as they are both fairly toxic and readily leech contaminants from a number of sources including plastics, your skin, the air etc. If you have these solvents in large volumes (more than half a liter), it is advised to carefully transfer a smaller portion to a meticulously cleaned smaller glass container.
6. Never access solvents directly! Pour the appropriate volume into smaller glassware first. This will reduce the likelihood of accidental contamination of your solvents. Any contaminants within the solvent will be concentrated into the lipid film following extraction, so for that reason it is important to not skimp on the grade of solvents which you use!
7. It is very important to avoid the use of plastics in steps that involve any kind of manipulation of organic solvents—since they will leech contaminants into your samples, greatly complicating your spectra and potentially causing ion suppression effects. Use borosilicate glass pipettes for these steps. It is recommended to rinse pipettes with chloroform and methanol prior to use. When handling lipid standards, it is a good idea to let them warm to room temperature and sonicate them prior to spiking your samples. This will help ensure the solubility of your standards.
8. Because the relative stabilities of different lipids are not very well known, it is important to be careful during all steps of the procedure to prevent sample degradation. Extraction and drying of lipid films should be done at 4 °C. When injecting samples, try to keep those that will not be immediately run cold. When finished running the sample, it is advisable to return any remainder to −20 °C for any further analysis.
9. Prior to running samples, it is a good idea to record baseline spectra of your solvents (in this case, it is a chloroform–methanol (1:2) mixture with 0.1 % NH_4OH).
10. For the same reason as point 5, it is not recommended to store lipid extracts for extended periods of time prior to running. In our workflow, we try to arrange extractions to take place during the same week that we will be running the samples.
11. Although we have access to an LTQ OrbitrapVelos, any reasonably high resolution mass spectrometer can be used with some modification to this protocol.
12. We typically use an instrument method for data-dependent acquisition, which automatically performs tandem MS on peaks in primary spectra. Depending on your specific needs, this may not be necessary.

References

1. Narasimhan SD, Yen K, Tissenbaum HA (2009) Converging pathways in lifespan regulation. Curr Biol 19:R657–R666
2. Fontana L, Partridge L, Longo VD (2010) Extending healthy life span – from yeast to humans. Science 328:321–326
3. Kaeberlein M (2010) Lessons on longevity from budding yeast. Nature 464:513–519
4. Kenyon C (2010) The genetics of ageing. Nature 464:504–512
5. Longo VD, Shadel GS, Kaeberlein M, Kennedy B (2012) Replicative and chronological aging in *Saccharomyces cerevisiae*. Cell Metab 16:18–31
6. Kirkwood TBL, Boys RJ, Gillespie CS et al (2003) Towards an e-biology of ageing: integrating theory and data. Nat Rev Mol Cell Biol 4:243–249
7. Murphy MP, Partridge L (2008) Toward a control theory analysis of aging. Annu Rev Biochem 77:777–798
8. Goldberg AA, Bourque SD, Kyryakov P et al (2009) Effect of calorie restriction on the metabolic history of chronologically aging yeast. Exp Gerontol 44:555–571
9. Ocampo A, Liu J, Schroeder EA et al (2012) Mitochondrial respiratory thresholds regulate yeast chronological life span and its extension by caloric restriction. Cell Metab 16:55–67
10. Antebi A (2005) Physiology. The tick-tock of aging? Science 310:1911–1913
11. Blagosklonny MV (2007) Paradoxes of aging. Cell Cycle 6:2997–3003
12. Blagosklonny MV (2007) Program-like aging and mitochondria: instead of random damage by free radicals. J Cell Biochem 102:1389–1399
13. Budovskaya YV, Wu K, Southworth LK et al (2008) An elt-3/elt-5/elt-6 GATA transcription circuit guides aging in *C. elegans*. Cell 134:291–303
14. Longo VD, Mitteldorf J, Skulachev VP (2005) Programmed and altruistic ageing. Nat Rev Genet 6:866–872
15. Kirkwood TBL (2005) Understanding the odd science of aging. Cell 120:437–447
16. Kirkwood TBL (2008) Understanding ageing from an evolutionary perspective. J Intern Med 263:117–127
17. Lithgow GJ (2006) Why aging isn't regulated: a lamentation on the use of language in aging literature. Exp Gerontol 41:890–893
18. Lin S-J, Sinclair D (2008) Molecular mechanisms of aging: insights from budding yeast. In: Guarente LP, Partridge L, Wallace DC (eds) Molecular biology of aging. Cold Spring Harbor Laboratory Press, Cold Spring Harbor, NY, pp 483–516
19. Mair W, Dillin A (2008) Aging and survival: the genetics of life span extension by dietary restriction. Annu Rev Biochem 77:727–754
20. Eisenberg T, Knauer H, Schauer A et al (2009) Induction of autophagy by spermidine promotes longevity. Nat Cell Biol 11:1305–1314
21. Fabrizio P, Gattazzo C, Battistella L et al (2005) Sir2 blocks extreme life-span extension. Cell 123:655–667
22. Burtner CR, Murakami CJ, Kennedy BK, Kaeberlein M (2009) A molecular mechanism of chronological aging in yeast. Cell Cycle 8:1256–1270
23. Kyryakov P, Beach A, Richard VR et al (2012) Caloric restriction extends yeast chronological lifespan by altering a pattern of age-related changes in trehalose concentration. Front Physiol 3:256
24. Goldberg AA, Richard VR, Kyryakov P et al (2010) Chemical genetic screen identifies lithocholic acid as an anti-aging compound that extends yeast chronological life span in a TOR-independent manner, by modulating house-keeping longevity assurance processes. Aging 2:393–414
25. Goldberg AA, Bourque SD, Kyryakov P et al (2009) A novel function of lipid droplets in regulating longevity. Biochem Soc Trans 37: 1050–1055
26. Titorenko VI, Terlecky SR (2011) Peroxisome metabolism and cellular aging. Traffic 12: 252–259
27. Beach A, Titorenko VI (2011) In search of housekeeping pathways that regulate longevity. Cell Cycle 10:3042–3044
28. Beach A, Burstein MT, Richard VR et al (2012) Integration of peroxisomes into an endomembrane system that governs cellular aging. Front Physiol 3:283
29. François J, Parrou JL (2001) Reserve carbohydrates metabolism in the yeast *Saccharomyces cerevisiae*. FEMS Microbiol Rev 25: 125–145
30. Trevisol ET, Panek AD, Mannarino SC, Eleutherio EC (2011) The effect of trehalose on the fermentation performance of aged cells of *Saccharomyces cerevisiae*. Appl Microbiol Biotechnol 90:697–704
31. Jain NK, Roy I (2010) Trehalose and protein stability. Curr Protoc Protein Sci 59: 4.9.1–4.9.12
32. Simons K (2011) The biology of lipids: trafficking, regulation, and function. Cold Spring

Harbor Laboratory Press, Cold Spring Harbor

33. Henry SA, Kohlwein SD, Carman GM (2012) Metabolism and regulation of glycerolipids in the yeast *Saccharomyces cerevisiae.* Genetics 190:317–349

34. Ejsing CS, Sampaio JL, Surendranath V et al (2009) Global analysis of the yeast lipidome by quantitative shotgun mass spectrometry. Proc Natl Acad Sci U S A 106:2136–2141

35. Klose C, Surma MA, Gerl MJ et al (2012) Flexibility of a eukaryotic lipidome – insights from yeast lipidomics. PLoS One 7: e35063

36. Bligh EG, Dyer WJ (1959) A rapid method of total lipid extraction and purification. Can J Biochem Physiol 37:911–917

37. Herzog R, Schuhmann K, Schwudke D et al (2012) LipidXplorer: a software for consensual cross-platform lipidomics. PLoS One 7:e29851

Index

Jeffrey S. Smith and Daniel J. Burke (eds.), *Yeast Genetics: Methods and Protocols*, Methods in Molecular Biology, vol. 1205, DOI 10.1007/978-1-4939-1363-3, © Springer Science+Business Media New York 2014

V

W

Y

Z

MIX
Papier aus verantwortungsvollen Quellen
Paper from responsible sources
FSC® C105338

If you have any concerns about our products,
you can contact us on
ProductSafety@springernature.com

In case Publisher is established outside the EU,
the EU authorized representative is:
Springer Nature Customer Service Center GmbH
Europaplatz 3, 69115 Heidelberg, Germany

Printed by Libri Plureos GmbH
in Hamburg, Germany